SCIENCE of WHITEWARES II

Related titles published by The American Ceramic Society:

Glazes and Glass Coatings
By Richard A. Eppler and Douglas R. Eppler
©2000, ISBN 1-57498-054-8

The Magic of Ceramics
By David W. Richerson
©2000, ISBN 1-57498-050-5

Science of Whitewares
Edited by Victoria E. Henkes, George Y. Onoda, and William M. Carty
©1996, ISBN 1-57498-011-4

Monolithic Refractories - A Comprehensive Handbook
By Subrata Banerjee
(Co-published with World Scientific Publishing, Inc.)
©1998, ISBN 981-02-3120-2

For information on ordering titles published by The American Ceramic Society, or to request a publications catalog, please contact our Customer Service Department at 614-794-5890 (phone), 614-794-5892 (fax),<customersrvc@acers.org> (e-mail), or write to Customer Service Department, 735 Ceramic Place, Westerville, OH 43081, USA.

Visit our on-line book catalog at <www.ceramics.org>.

SCIENCE *of* WHITEWARES II

Edited by

William M. Carty and Christopher W. Sinton

Published by
The American Ceramic Society
735 Ceramic Place
Westerville, Ohio 43081

Proceedings of the Science of Whitewares II Conference, Alfred University, Alfred, New York, May 31–June 2, 1998.

COVER PHOTO: Cover image is courtesy of William M. Carty.

Library of Congress Cataloging-in-Publication Data
A CIP record for this book is available from the Library of Congress.

ISBN 1-57498-067-X

For information on ordering titles published by The American Ceramic Society, or to request a publications catalog, please call 614-794-5890.

Printed in the United States of America.

4 3 2 1 – 03 02 01 00

Contents

Processing

Forming

Banquet Guest Lecture

Preface

Science of Whitewares I, held July 16–20, 1995 at Alfred University, Alfred, New York, was the first U.S. scientific conference on porcelains in more than 40 years. The conference was enormously successful, with attendance of 180 engineers, scientists, and students from industry, government, and academia. The scope of that meeting was intentionally broad and addressed characterization, rheology and plasticity, forming, drying and firing, mechanical properties, and glazing. The proceedings from *Science of Whitewares I*, which sold more than 500 copies, was the top-selling technical title of The American Ceramic Society in 1996.

This new volume gathers the edited proceedings from *Science of Whitewares II*, held May 31–June 2, 1998, at Alfred University. More than 110 members of industry and academia, representing Cameroon, Canada, England, Italy, Mexico, the Netherlands, Peru, Spain, and the United States, attended. This time, the program focused on pre-firing issues of raw materials, polymeric additives, characterization, processing, and forming.

The editors of *Science of Whitewares II* wish to thank the industrial members of the Whiteware Research Center at Alfred University who sponsored the conference and this volume: Buffalo China; The Edward Orton, Jr. Ceramic Foundation; Fusion Ceramics; Georgia Pacific; Hall China; Lenox China; Pfaltzgraff; Sterling China; Syracuse China; Victor Insulators; U. S. Borax; and U.S. Gypsum. We are grateful to the session chairs for their participation, as well: Tom Dombrowski, Nik Ninos (Buffalo China); Tim Stangle (Lenox China); Mark Noirot (U.S. Borax); and Mike Dempsey (Victor Insulators). Special thanks are due to Professor Anne Currier, of the New York State College of Ceramics (NYSCC) School of Art and Design, who presented the banquet address; to NYSCC Dean L. David Pye, who sponsored the conference reception; and to the NYSCC and Alfred University, for their generous provision of conference facilities. We are indebted to Ms. Marlene Wightman, NYSCC Director of Special Programs, and Ms. Victoria Henkes, former Assistant Director of the Whiteware Research Center, for their assistance in planning and coordinating the conference; to Mark Cappadonia and Mike Tindale, who provided audio-visual support; and to several NYSCC graduate and undergraduate students who provided transportation and staffed events. Thanks also to Ms. Jeannette Harris, WRC secretary, for assistance with the proceedings. Finally, our gratitude to Bill Emrick, Director of Conferences and Special Events at Alfred University, for his cool-headedness in the midst of the several inopportune tornado warnings that occurred throughout the meeting.

William M. Carty
Christopher W. Sinton

Raw Materials

THE ORIGINS OF KAOLINITE - IMPLICATIONS FOR UTILIZATION

Tom Dombrowski
21 New Holland Dr.
Cohoes, NY 12047

ABSTRACT

Kaolinite is a mineral that is formed by the alteration of precursor minerals, typically feldspar and mica minerals, in a variety of geologic environments. The term kaolin is used when the amount of kaolinite in a rock is greater than 50%, or when the kaolinite in a rock is exploited commercially. The alteration from feldspar or mica minerals to kaolinite involves the removal of potassium, calcium, sodium and other cations. The alteration process is gradational and the gradational process is responsible for intermediate clay minerals that are often found in kaolin deposits. The gradational alteration to kaolinite is also responsible for the chemical purity and diagenetic maturity of the kaolinite found in a specific kaolin deposit. These factors have a huge impact in the functional properties of kaolinite from a specific kaolin deposit.

Kaolin deposits are classified as either primary or secondary. Primary deposits are those where the kaolinite has formed in place either by residual weathering or hydrothermal alteration, whereas the secondary deposits have been transported by surface processes. Examples of primary kaolin deposits that are used in the ceramic industry include kaolin mines in Cornwall, England; Saxony, Germany; and Maungaparerua, New Zealand. Examples of secondary kaolin deposits familiar to ceramic workers are those in the Southeastern USA and the Amazon region of Brazil. The combination of the features a clay inherits from its source material, and its subsequent geochemical history, impart properties to a kaolinite that determines its applicability for specific applications.

INTRODUCTION

Ceramic engineers are often reluctant to substitute new kaolin into a formulation because they had experiences in which substitute kaolin did not work as well as the original. The geologic history of a specific kaolin deposit controls the different physical and chemical characteristics of a kaolinite from that specific deposit. Therefore, different kaolin deposits, and even various sections of a single deposit, can have large differences in the physical and chemical properties that can make the difference between a kaolinite that works in a specific application and one that does not.

The words "kaolin" and "kaolinite" are often used interchangeably in industrial settings, but the two words have different meanings.[1] Kaolinite is the name of a mineral and therefore refers to a specific alignment of atoms that produce consistent physical properties. Kaolinite has a composition that is expressed as the mineral formula, $Al_2Si_2O_5(OH)_4$, or as the oxide formula $Al_2O_3 \cdot 2SiO_2 \cdot 2H_20$. Halloysite is another kaolin mineral that is part of the kaolin or khandite family of minerals. Halloysite's mineral formula is written as: $Al_2Si_2O_5(OH)_4 \cdot nH_20$. Halloysite in its fully hydrated state contains two water groups that are lightly bound and under normal conditions do not rehydrate. Bailey offers a review of the special nature of halloysite.[2]

Kaolin is a rock term that refers to a material with greater than 50% kaolinite and therefore includes minerals other than kaolinite.[1] This definition has been expanded to include rocks that have a concentration of kaolinite that can be exploited commercially.

No two kaolin deposits, and the kaolinite minerals found within, are exactly the same. The natural processes that form kaolinite and kaolin deposits include variations that are not found in synthetic inorganic chemicals and oxides. Kaolinite is a mineral that is formed by the interaction of water and minerals. The precursor minerals and rocks from which kaolinite is formed are varied in terms of chemical composition and texture. These differences are found in the final material that is mined and processed. For example, kaolinite derived from an iron rich rock will have a larger amount of iron included in the mineral lattice than kaolinite formed from iron poor rocks. This will influence kaolinite properties such as brightness, color, crystal size and rheology. These inherited differences and differences in the post-mineralization history of the deposit also influence the potential utilization of the kaolin deposit.

Formation of Kaolinite

Kaolinite forms by the interaction of water with various precursor minerals such as feldspars, and mica. These are alumino-silicate minerals with different amounts of calcium, sodium, iron, magnesium, potassium, aluminum, and silicon. These minerals are slightly soluble in water at ambient temperature and therefore decompose in the presence of water.[3] The decomposition of the minerals is controlled by the chemistry of the minerals. At ambient temperature surface weathering environments, the alkali and alkali earth elements are more soluble than the transition elements, and aluminum and silicon are the least soluble. Kaolinite is formed when the alkali, alkaline earth and the majority of the transition elements have been removed and the aluminum to silicon ratio is approximately 1:1. The following reaction has been presented for the alteration of potassium feldspar to kaolinite[4]:

$$2KAlSi_3O_8 + H_2O \rightarrow Al_2Si_2O_5(OH)_4 + 4SiO_2 + 2K(OH)$$

Similar equations could be written for other feldspar and mica minerals.

The above equation does not describe the formation of intermediate phases or other additional minerals. SEM microphotographs show kaolinite growing on the faces of feldspar and indicate that that this relationship does occur.

Geologic conditions that control the formation of minerals are variable and as a result the same mineral can be formed by a number of pathways. Also, physical processes exert an influence on the final mineral composition of a kaolin deposit. The specific conditions govern the formation of a mineral and hence the properties and features of the mineral from that location.

Geochemical processes

The alteration process that forms clay minerals including kaolinite is a continuum of transitions from one phase to another. These phases include both minerals and amorphous phases. Mineral stability diagrams show the alteration path from a feldspar-mica source to kaolinite can include mineral assemblages such as: illite and feldspar; smectite; smectite and illite; smectite and kaolinite; kaolinite and silica; and kaolinite and gibbsite.[4] The phases are controlled by the amount of various elements in the system, the solubility of the minerals, the cation saturation of the water, the movement of the water, temperature conditions, and several other factors. The mineralogy is also complicated by the fact that these relationships are reversible and that mixed layer minerals such as illite-smectite and smectite-kaolinite exists. Additionally, the alteration process is not uniform within a deposit.

Geochemical processes continue once kaolinite has been formed. Post-formation alteration can enhance the properties of kaolinite by removing undesirable materials such as iron compounds, or fusing the surfaces of kaolinite crystals. If taken to an extreme the alteration can continue to remove silica and the mineral gibbsite will form. Because alteration reactions are reversible alteration can also proceed to form more chemically complex minerals. Austin has documented the formation of kaolinite from gibbsite.[5] The formation of kaolinite and halloysite can occur in a reversible reaction where kaolinite can be either the precursor or product of the reaction.

Physical influences

The clay composition is also influenced by the primary minerals available for alteration and the fluids that interact with the minerals. In kaolin deposits derived from igneous sources the chemistry is controlled by the magma that formed the rock. This can govern the proportion and composition of the minerals that are available for alteration. The cooling history of the magma controls the crystal size of the minerals and this has a direct relationship to the crystal size of kaolinite. In geologically active areas geothermal water interacts with the rock to alter minerals. The geothermal water is hotter than surface water and can have different water chemistry than groundwater or meteoric water.

Sedimentary processes also control the distribution of the minerals found in secondary kaolin deposits. Stream gradients control the particle size of the minerals transported and also can preferentially include a specific mineral due to shape factors. The abundance of mica in sedimentary kaolin deposits can be attributed to this phenomenon.

VARIATION IN KAOLINITE FROM DIFFERENT SOURCES

The physical and chemical difference among samples of kaolinite from different deposits can have a major impact on the commercial utility of kaolin. In the kaolin district of the Southeastern USA and other regions of North America many of the differences among kaolin deposits can be attributed to different source rocks and alteration histories. These factors control many key properties that can impact the utilization of kaolinite.

The grain size of the precursor minerals has a direct impact upon the particle size of kaolinite. Coarse granite and gneiss give way to coarse-grained kaolinite in Georgia, Idaho, Germany, and England. Fine-grained kaolinite in the Southeastern USA is derived from fine grained metavolcanic rocks and fine-grained halloysite in New Zealand is derived from rhyolite.[6]

The source material also influences the shape of the particles. Pruett showed that igneous and metamorphic rocks produce tabular kaolinite crystals whereas sedimentary kaolins with input from volcanic ash produce spherical halloysite.[7]

Precursor rocks influence the amount of non-constituent oxides present in kaolins. In the Southeastern USA the coarser kaolin deposits show lower amounts of iron oxide, titanium oxide, cobalt and scandium. The iron and titanium oxide content influences the particle size and color characteristics of the kaolinite that is important in ceramic applications. Also, higher iron content is positively correlated to lower Brookfield viscosity, CEC, surface area, and water absorption. The form of iron is important in how it responds to beneficiation treatments. Oxidized iron responds to magnetic separation while reduced iron does not. Differing types of iron can be removed by different froth flotation processes.

Non-kaolinite minerals in kaolin deposits often influence properties important in commercial applications. Montmorillonite is a common contaminant in kaolin deposits and this can influence the rheology, surface area, CEC, and water absorption of kaolin. There is also a slight inverse relationship between crystallinity and montmorillonite. The Mg and Ca that are present in kaolins need to be accounted for in determining the amount of flux needed in a ceramic body. Mica is also a prominent accessory mineral in kaolins. The particle size of mica can have a large impact on the viscosity of kaolin. Large mica is detrimental to viscosity whereas finer mica is not as harmful. The differences in size of mica and their abundance are related to geological processes. Mica from coarse-grained source rocks are found with the coarser kaolins and within a kaolin deposit the mica is generally segregated into the coarser size fractions. In contrast, mica from fine sources are found in most all fractions and depositional processes control the vertical and horizontal distribution of mica. Differences in mica concentration within a deposit can be observed both vertically and horizontally.

Primary kaolin deposits have large amounts of free quartz and feldspar mixed with the kaolinite. In plutonic deposits or coarse-grained metamorphic rocks the grain size is much larger than the kaolinite and the quartz and feldspar are easily removed by water separation. In secondary kaolin deposits the majority of the quartz and feldspar has been removed by aqueous processes. Any remaining quartz in these deposits is easily liberated in a water washed process. In some deposits the quartz and kaolinite have the same particle size distribution and not readily separated by gravitational methods. Tiny crystals of feldspar may be present in kaolin fractions that can have an impact on firing characteristics or abrasion.

The process of post-depositional alteration, or diagenesis, influences the commercial viability of a specific kaolin deposit. White *et al.* studied the influence of post depositional alteration in cores of Middle Georgia kaolins.[8] They found zones of increased alteration are characterized by: more vermicular kaolin crystals, higher crystallinity, less mica and less iron than zones of less intense alteration. Dewu and Durrance developed the concept of a Kaolinization Index to describe the degree of alteration a sample has undergone.[9] They showed enrichments and depletions in various trace elements in zones with increased alteration.

Geochemical alteration dissolves small, non-mineral phases, flattens surfaces of kaolinite crystals, causes growth of vermiform crystals, facilitates the alteration of mica and smectite minerals, oxidizes iron and causes iron to be more soluble so that iron is removed, concentrates titanium dioxide, and can cause the formation of excess Al complexes or gibbsite. Carried to an extreme, Austin shows that gibbsite can be altered to kaolinite by the addition of silica.[5]

These processes change the properties of a kaolin deposit, or more exactly, specific portions of a kaolin deposit. The alteration of montmorillonite to kaolinite will decrease the Brookfield viscosity, CEC, and water absorption. These changes need to be accounted for in industrial processes. The smoothing of crystal surface will cause also reduce viscosity and can change the thermal characteristics of a kaolinite. Elimination of fine particles by Oswald Ripening will change the particle size distribution of deposits and this can also change physical properties of kaolin deposits. The state and amount of mica in kaolin will have an effect on the availability of K_2O as a fluxing agent. More altered mica in a clay may make the potassium in a deposit more effective as a fluxing agent than the same amount present in non-altered mica.

GEOLOGY OF KAOLIN DEPOSITS

Kaolin deposits are classified as either primary or secondary type deposits.[4] Primary deposits are subdivided into residual or hydrothermal types and have had minerals alter to kaolinite *in situ*. The alteration may have occurred by hydrothermal processes, groundwater or surface weathering. Secondary kaolin deposits refer to deposits where the kaolin has been transported and deposited by sedimentary process or sedimentary sediments have been altered to kaolinite.

Primary kaolins

The kaolinite deposits of Saxony, Germany are famous for the Meissen china and represent the longest operating kaolin mines in Europe.[4] These deposits are examples of residual kaolins where the kaolinite was formed in place from feldspar and mica in various types of igneous rocks. Typically these deposits contain quartz, kaolinite, illite-smectite and small amounts of feldspar. The rheology of most residual deposits is high due to the inclusion of swelling clay minerals, salts, amorphous materials, and lack of an adequate particle size distribution. Examples of other commercial residual kaolin deposits are found in Argentina, China, the Czech Republic, Indonesia, Ukraine, and South Africa.[4]

Kaolin deposits in Cornwall, England[10] and Maungaparerua, New Zealand[6] are examples of primary kaolin deposits that were formed by hydrothermal processes and modified by meteoric water. Both these deposits provide very bright kaolinite that is important in ceramic applications.

The kaolin deposits of Cornwall have a very complex history that involves multiple periods of igneous activity followed by alteration of minerals to clays. In these deposits the first alteration was by high temperature hydrothermal water, followed by cycling of heated meteoric water. The final stages involved changes by surface water.

The resulting kaolinite from these deposits is a coarse kaolinite that has a high degree of crystallinity. Impurities include mica, feldspar, and quartz. Differences in the source material and diagenetic history have made it so that some portions of a deposit are ideal for ceramic applications while other areas produce clays that go into paper, paint and rubber applications.

The Maungaparerua deposit of New Zealand is an example of hydrothermal alteration in a fine-grained source rock. Surface weathering modified these deposits. Rhyolitic flow rocks are extremely fined grained and these materials produced kaolinite and halloysite that is very fine. The source rocks are low in iron and the resulting kaolinite and halloysite are low in color impurities. The halloysite associated with these is very pure and is used in whiteware applications throughout the world.

Hydrothermal kaolin deposits are also known in Jingsu Province, China; Sardina, Italy, Japan, Central Mexico, and the western USA.[4] In Utah, some large halloysite deposits have been mined for catalytic applications and there are reports of a new, very pure, halloysite deposit in west central Utah.

Sedimentary kaolins

The kaolin deposits of the Southeastern USA[11] and Brazil[4] are deposits where kaolinite was formed in other locations and transported to local depositional sites. After transportation, post-depositional processes continued to modify the kaolinite.

Different clay deposits were deposited in various portions of the sedimentary basin and therefore have different geometric dimensions. Sedimentary processes have removed much of the quartz, feldspar and mica. Deposits of these types have a very heterogeneous nature due to changes in source rocks, depositional cycles, and alteration histories. The coarse clay in Georgia and South Carolina were derived from granite and gneiss, while the fine clays were derived from fine grained metamorphic rocks. Post depositional changes have caused some clays to be altered to cream or red colors due to oxidation reactions while others have remained reduced and are gray. Also, clays of different maturity have different properties due to the diagenetic changes listed above.

The kaolin deposits of Bavaria are an example of an arkose (feldspar-bearing sandstone) altered to kaolinite.[4] The deposit is rich in quartz, but the rock changes from a kaolinite rich zone to a zone of kaolinite and feldspar to a feldspar rich zone in the direction of groundwater flow.

Other large commercial kaolin sedimentary kaolin deposits are found in Australia, Spain, and Surinam. Worldwide, other smaller types like this are used locally.

SUMMARY

The final quality of a kaolin deposit or even parts of kaolin deposits is controlled by the interaction of the geological and geochemical processes. The gross characteristics of a kaolin deposit are governed by the geology of the source rocks. Many subtle changes that occur as a result of additional alteration can modify the kaolin in ways that may or may not be beneficial to the final application. These differences are difficult to detect and most industries rely on empirical testing to determine the utility of a kaolin deposit.

Companies that sell grade of kaolinite are aware of these factors and try to minimize the differences during processing. Extensive pre-mining analysis and selective mining limits the variability seen in a grade of kaolin. The water washed process causes homogenization enables a consistent product that can be used confidently by the ceramic industry.

REFERENCES

1. H. H. Murray and W. D. Keller, "Kaolins, Kaolins, Kaolins" pp. 1-24 *in Kaolin Genesis and Utilization,* Edited by H. Murray, W. Bundy, and C. Harvey. Clay Minerals Society, Boulder, CO, 1993.
2. S. W. Bailey, "Halloysite-a critical review" pp. 89-99 in *Proceeding of the Ninth International Clay Conference, Strasbourg,* Edited by V. C. Fariner and Y. Tardy. Society of Geological Mem. 86. Washington. 1991
3. G. Millot, *Geology of Clays* p. 429. Springer-Verlag, New York, 1970.
4. H. H. Murray, "Kaolin Minerals: Their Genesis and Occurrences" pp. 67-90, in *Hydrous Phyllosilicates (exclusive of micas) Reviews in Mineralogy, Vol. 19.* Edited by S. W. Bailey. Mineralogical Society of America, Washington, D. C. 1988.
5. R. S. Austin, "Origin of Kaolin in the southeastern US," *Mining Engineering* 50[2], 52-57(1998).
6. R. J. Pruett, and H. H. Murray, "The Mineralogical and Geochemical Controls That Source Rocks Impose on Sedimentary Kaolins" pp. 149-170 in *Kaolin Genesis and Utilization,* Edited by H. Murray, W. Bundy, and C. Harvey. Clay Minerals Society, Boulder, CO, 1993.
7. C. C. Harvey and H. H. Murray, "The Geology, Mineralogy, and Exploitation of Halloysite Clays of Northland New Zealand," pp.233-248 in *Kaolin Genesis and Utilization,* Edited by H. Murray, W. Bundy, and C. Harvey. Clay Minerals Society, Boulder, CO, 1993.
8. W. M. Bundy, "The Diverse Industrial Applications of Kaolin" pp.43-74 in *Kaolin Genesis and Utilization,* Edited by H. Murray, W. Bundy, and C. Harvey. Clay Minerals Society, Boulder, CO, 1993.
9. N. G. White, J. B. Dixon, R. M. Weaver, and A. C. Kunkle, "Recrystallization of Kaolinite in Gray Kaolins," pp.99-116 in *Kaolin Genesis and Utilization,* Edited by H. Murray, W. Bundy, and C. Harvey. Clay Minerals Society, Boulder, CO, 1993.

10. B. B. M. Dewu and E. M. Durrance, "Mobility of U and Granite Kaolinization in Southwest England," pp. 205-220 in *Kaolin Genesis and Utilization,* Edited by H. Murray, W. Bundy, and C. Harvey. Clay Minerals Society, Boulder, CO, 1993.

11. C. M.Bristow, "The Genesis of the China Clays of South-West England-A Mulistage Story" pp. 171-204 in *Kaolin Genesis and Utilization,* Edited by H. Murray, W. Bundy, and C. Harvey. Clay Minerals Society, Boulder, CO, 1993.

12. T. Dombrowski "Theories of Origin for the Georgia Kaolins: A Review," pp.75-98 in *Kaolin Genesis and Utilization,* Edited by H. Murray, W. Bundy, and C. Harvey. Clay Minerals Society, Boulder,CO, 1993.

13. S. M Pickering Jr., V. B. Hurst, J. Elzea, *Mineralogy, Stratigraphy and Origin of Georgia Kaolins, Field Excursion,* 11th International Clay Conference, Ottawa Canada, 1997.

PROCESSING KAOLINS AND BALL CLAYS FOR CERAMIC MARKETS

Haydn H. Murray
Department of Geological Sciences
Indiana University
Bloomington, Indiana 47405

ABSTRACT

There are two basic methods used to process kaolins and ball clays: a dry process in which the clay is dried and sized and a wet process in which the clay is slurried in water and then beneficiated. The wet process is much more expensive and sophisticated and from it a more uniform and higher quality product results.

KAOLIN MINING AND PROCESSING

Kaolin used by ceramic manufacturers can be categorized into two basic types—primary and secondary. Primary kaolins are those that were formed by the alteration of crystalline rocks such as granite and remain in the place where they were formed. The alteration can be caused by weathering or by hydrothermal fluids. Secondary deposits of kaolin are sedimentary and were transported from their place of origin and deposited in beds or lenses associated with other sedimentary rocks such as sands.

The mineralogy of these two kaolin types can be quite different; sedimentary kaolins usually have a higher kaolin content because quartz and other coarse minerals have been removed during transport. In the United States the most widely used kaolins are sedimentary deposits located in Georgia and South Carolina.[1] The majority of kaolins used by the ceramic industry are mined, processed in and shipped from this area (Figure 1).

The kaolin deposits in the Georgia/South Carolina district vary in particle size, plasticity, strength, shrinkage, fired color and PCE (pyrometric cone equivalent). Therefore, it is necessary to maintain good quality control which begins with the mining and extends through the processing. There are two kaolin-bearing stratigraphic units of commercial interest in the district: the Late Cretaceous Buffalo Creek Formation and the Middle Eocene Huber Formation.[1] The Late Cretaceous kaolins are concentrated in the area between Macon and Sandersville that known is Middle Georgia (Figure 1). These kaolins are rather soft and coarse in particle size (averaging sixty-five percent finer than two microns) with a smooth, conchoidal fracture and abundant stack-like crystals packed edge to face (Figure 2a). The Middle Eocene kaolins in the Huber Formation are harder and more dense, considerably finer in particle size (averaging 85% finer than 2 microns), and are dominantly fine plate-shaped crystals which are packed face to face (Figure 2b).

The kaolin deposits occur along what is designated the fall line which is the boundary between the crystalline rocks on the Piedmont Plain and the sedimentary rocks to the southeast on the Coastal Plateau (Figure 1). The so-called kaolin belt averages about fifteen miles wide. The kaolin deposits are located by drilling. Once the kaolin deposit is located and tested, the overburden is stripped off and the kaolin is mined using draglines, shovels or backhoes. The commercial kaolin bodies may be as thick as twelve meters but average closer to three or four meters in thickness. The overburden ratio is normally less than six or seven to one.

Figure 1. Kaolin mining district

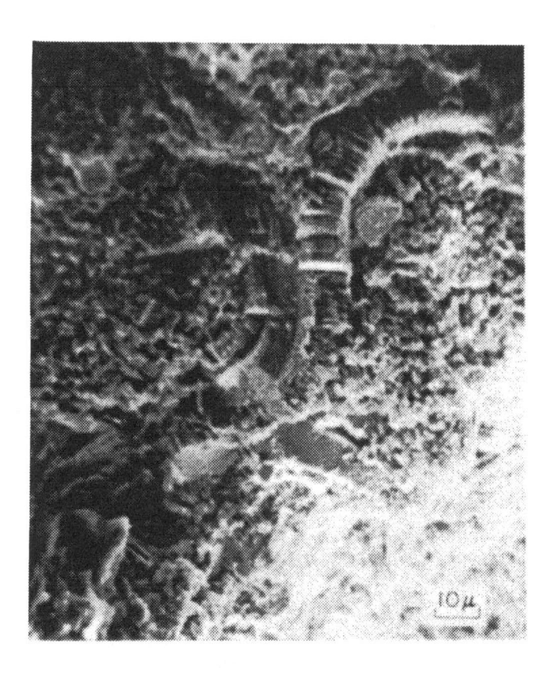

Figure 2a. SEM of soft Cretaceous kaolin

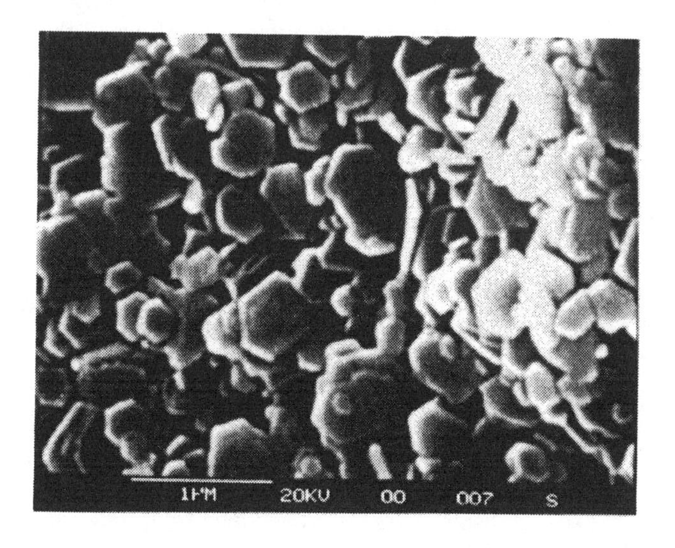

Figure 2b. SEM of Middle Eocene kaolin

Kaolins and ball clays are generally processed using either a dry or a wet method. A typical dry process flow sheet is shown in Figure 3. For the dry process, the kaolin is loaded into trucks after mining and transported to the processing plant.[2] It is important that the physical and chemical properties of the kaolin are known at the kaolin mine face so that the various qualities can be stored in the correct piles or bins at the plant. Once the kaolin is stored in piles or bins according to specified quality parameters, the clay is taken from the pile by a front end loader to a hopper which feeds a clay breaker or shredder. The quality is controlled by blending known qualities of the clay from the various stockpiles. The blending is done by mixing bucket loads from the piles in the feed hopper. The finished product is shipped either as a high-solids slurry or dry. If the clay is to be shipped in slurry form, the crushed clay goes from the shredder directly to a blunger where it is mixed with water and a dispersant such as sodium silicate or sodium polyacrylate. Depending on the viscosity of the slurry, the solids content ranges from sixty-five to seventy percent. After blunging, the kaolin is pumped through screens (normally eighty to one hundred mesh) where the sand, coarse mica, unblunged clay and wood or other trash is screened out. The screened product is then pumped into storage tanks where the quality is checked. If the quality is acceptable, the clay is pumped into tank cars or trucks. If the quality is not acceptable, then other slurry can be added and mixed into it to produce an acceptable product.

If the kaolin product is shipped dry, then the shredded clay is conveyed to a rotary drier. If the customer wants the product in lump form, it can either go directly into the railroad car, be stored in piles or in silos, or it can be bagged. If the customer wants a pulverized air-floated product, then the dried lumps go to a pulverizer and air-separator to produce a product that passes a 200 or 325 mesh screen. The air-separators reject coarse particles such as sand and mica. The pulverized product is either stored, bagged, or loaded directly into hopper cars or trucks in bulk. The moisture is normally six percent or less.

The dry process is less costly than the wet process and the product is usually of lower, more variable quality. However, for many ceramic markets, this is adequate.

The wet process is much more sophisticated and produces a product which is cleaner and normally of higher quality. For the wet process, the mining is similar to that of the dry process. However, the kaolin is either fed directly into a portable blunger or is hauled to a stationary blunger. The kaolin is blunged at about forty percent solids, dispersed in water, and then degritted. The wet process flow sheet is shown in Figure 4. The degritting process is accomplished using

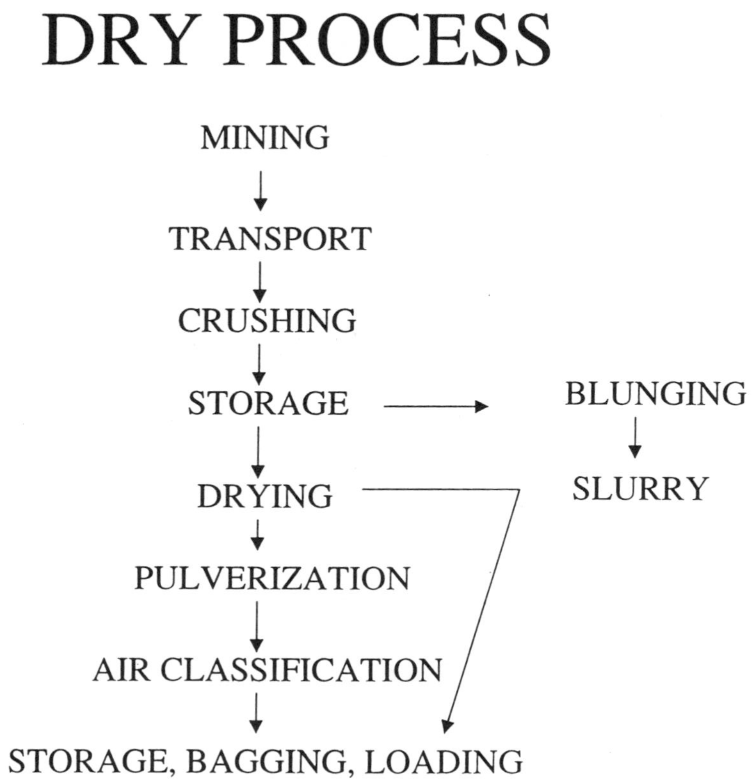

DRY PROCESS

MINING

↓

TRANSPORT

↓

CRUSHING

↓

STORAGE → BLUNGING → SLURRY

↓

DRYING → STORAGE, BAGGING, LOADING

↓

PULVERIZATION

↓

AIR CLASSIFICATION

↓

STORAGE, BAGGING, LOADING

Figure 3. Dry process flow sheet

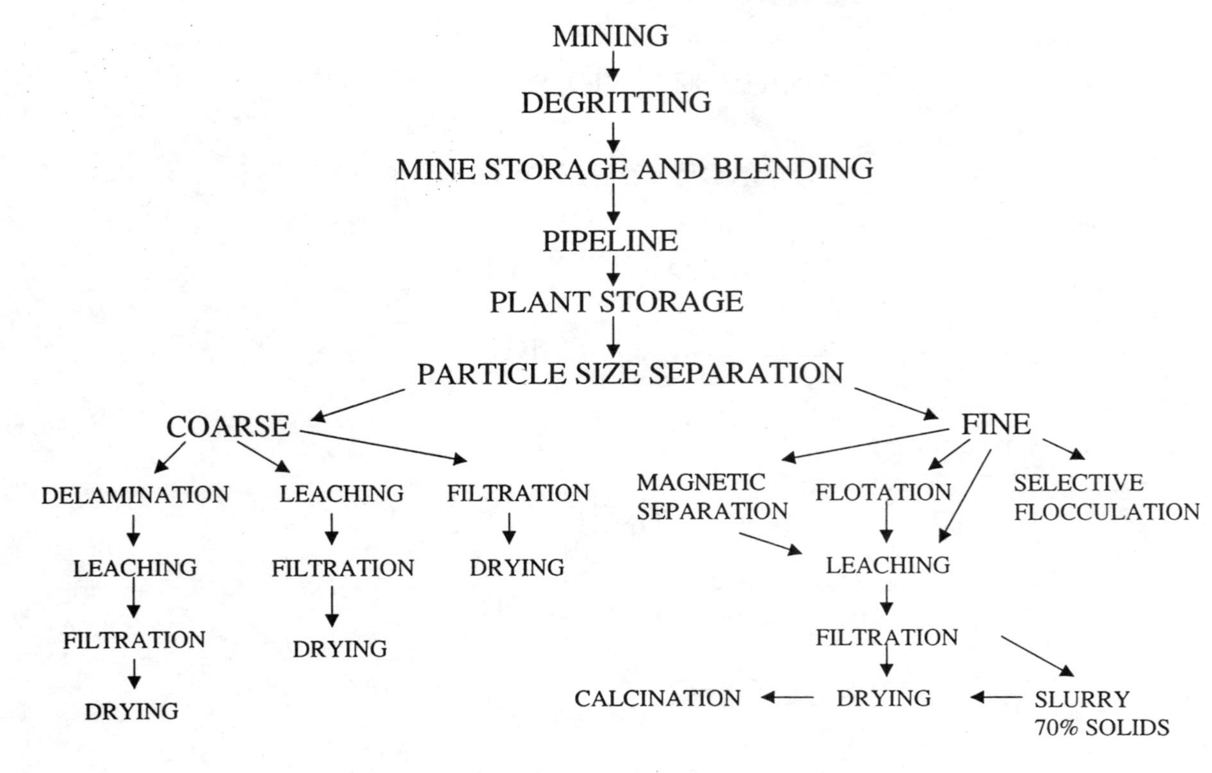

Figure 4. Wet process flow sheet

what are termed sandboxes and screens. The sandboxes are designed to settle out any particles coarser than 44 microns. From the sandboxes the clay is screened to remove coarse mica and other material which does not settle readily. The degritted and screened kaolin is then pumped to storage tanks where the quality is checked. If it is out of specification, it is blended with other clay to bring it up to specification. Once the quality is approved, then the slurry is pumped in pipelines to the plant where it is stored in large terminal tanks. From the terminal tanks the kaolin slurry can go directly to high intensity magnetic separators or directly to centrifuges for particle size separation. Some plants feed all the kaolin slurry to the magnetic separators prior to centrifugation and others wait until after the separation of the coarse and fine particles. Magnetic separation was developed early in the 1970s and is now a standard processing procedure. This method is very effective in removing iron and titanium minerals which discolor the kaolin.[3] Cryogenic magnets that produce field strengths up to 5 Tesla are now commercially available and some are being used currently in the kaolin industry.

After particle size separation in the centrifuges, the coarse clay can go either directly to the filters for dewatering after flocculation with acid, to delaminators to break apart coarse stacks (Figure 5), or to the leaching department where surficial iron stains are removed. Coarse casting clay generally goes directly to the filters after flocculation. After filtering, the coarse kaolin is dried on apron dryers or in rotary dryers to 6 percent moisture or less. The delaminated clays are used primarily by the paper industry, but some are used as fillers in rubber and for some special applications in fine china.

The fine particle size kaolin is processed in several different ways depending on the quality desired and the application. As shown on the flow sheet (Figure 4), the finer kaolin can be floated, selectively flocculated, magnetically separated and leached. The purpose is to improve the brightness and whiteness of the product by removing the iron- and titanium-bearing minerals. After one or more of these processes, the kaolin is filtered and dried or if it is to be shipped in slurry form, the drier is bypassed. For premium quality whiteware, the wet process produces a white firing, high strength product.

Some kaolin is calcined to produce refractory grog. In this process the kaolin is extruded and cut into nodular form, dried, and then calcined in rotary kilns to a temperature of over 1300°C. The product is comprised of dense particles which are crushed and sized for use in refractories to control shrinkage.

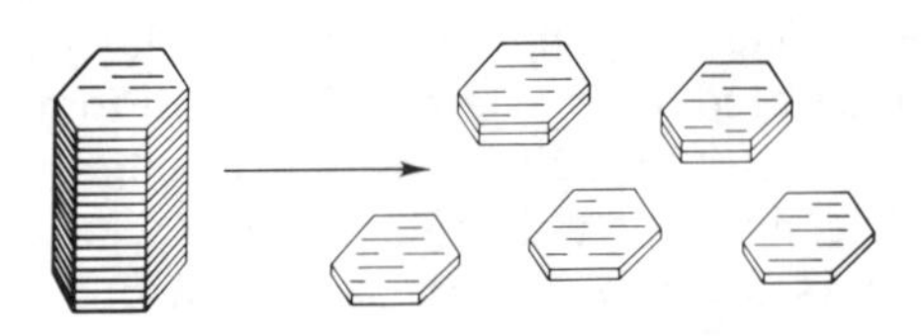

Figure 5. Delamination of stacks to produce plates

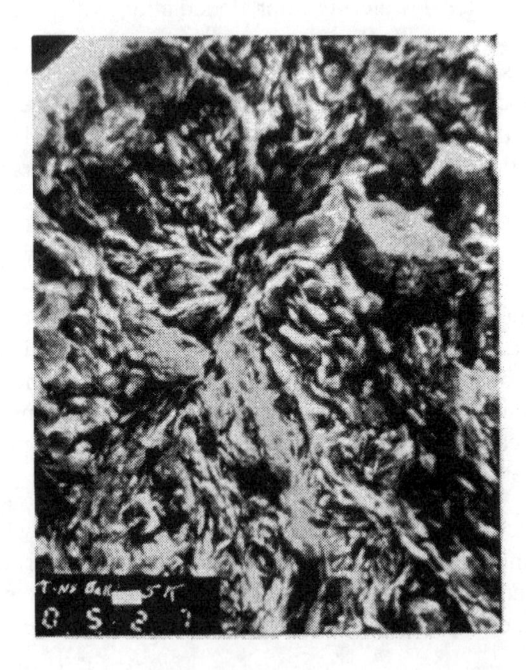

Figure 6. SEM of ball clay from Tennessee

BALL CLAY MINING AND PROCESSING

Ball clays are fine particle size kaolinitic clays (Figure 6) that are plastic and range in color from light gray to nearly back. The dark color is due to the presence of organic matter in the form of lignite or peat. Quartz is the most abundant non-clay mineral in ball clays and ranges from as little as five to as much as 30 percent.[4] Ball clay is used primarily by the ceramic industry. The major ball clay deposits are located in western Tennessee and western Kentucky. There are also some deposits in east Texas. The deposits are Middle Eocene in age and are thought to have been deposited in swamps, river oxbows, and lakes that were present on broad river deltas. The deposits vary greatly in size from a few acres to one hundred acres or more. The deposits also vary greatly in thickness. Some are as thick as ten meters or more, but most deposits are three to six meters thick.

After a ball clay deposit is located by drilling and the quality is determined, the overburden is stripped off, and the clay is mined using a dragline, shovel or back hoe. The clay is loaded into trucks and transported to the plant. The dry process used to process the ball clay is similar to the dry process for kaolin (Figure 3). The quality is checked at the mine face and then again in the stockpiles at the plant. The ball clay is stored in piles according to strength, particle size, fired color, etc. A front-end loader moves the clay from the stockpile to a hopper which feeds a clay breaker or shredder. Quality is controlled by blending bucket loads of clay from different stockpiles according to a predetermined formula.

From the clay breaker or shredder, the ball clay can go in two directions: either to the dry process or to be slurried. If the product is to be shipped as a slurry, the shredded clay is conveyed to a blunger where it is dispersed in water using sodium silicate or sodium polyacrylate at about sixty-two percent solids. The blunged clay is then pumped to screens (usually eighty mesh or slightly finer) where coarse unblunged clay particles, coarse lignite, and other coarse materials are removed. The screened product is pumped to large storage tanks where the clay is tested for various physical and chemical properties to meet the requirements of the customers. If the slurry clay is out of specification, then it can be blended with other clay slurry to bring it into specification. The finished slurry is then pumped into tank cars or trucks for shipment.

If the ball clay is to be shipped dry, then it can either be dried in lump form and shipped, or it can be dried and pulverized to either minus 200 or 325 mesh products. During the pulverization, the clay can be air-separated to remove coarse sand or lignite. The dried clay can be shipped in bulk or in bags.

Ball clay slurry shipments became popular in the 1980s and are primarily used to ship clay to the sanitaryware manufacturers. The quality of the slurry is easier to control and is more uniform than the dry processed ball clay. Slurries can be easily blended to meet specifications. In the dry clay process, the quality is primarily controlled by selective mining and then dry blending the various qualities at the plant.

FUTURE PROCESSING

The dry air-floated process will continue to improve as air classification equipment becomes better-developed to control particle size and purity. The wet processes used by the kaolin industry are highly developed and some of these processes such as centrifugation, magnetic separation, evaporation, filtration, and spray-drying, will eventually be adopted in the ball clay industry. It will then be possible to tailor make high strength or fast-casting products that are more precisely controlled.

In the future, prepared bodies will become available to the ceramic manufacturer.[5] Ball clay, kaolin, talc, feldspar, silica, nepheline syenite, or other mineral constituents can be blended in controlled proportions to meet the requirements of individual users. Automated weigh scales, efficient mixers, and computers will make controlled, prepared bodies possible. Significant cost savings for the ceramic manufacturer can result because of less inventory, reduction of storage silos or bins, and less processing before drying and firing. This will result in fewer rejects and a more uniform high-quality product.

REFERENCES

[1]S. M. Pickering and H.H. Murray, "Kaolin,";pp. 255-277 in *Industrial Minerals and Rocks*, 6th ed. Edited by D.D. Carr. Society for Mining, Metallurgy and Exploration, Littleton, Colorado, 1994.

[2]H.H. Murray, "Dry Processing of Clay and Kaolin," *Interceram*, **31**, 108-110 (1982).

[3]J. Iannicelli, "New Developments in Magnetic Separation," *IEEE Trans. On Magnetics*, V. Mag-12, 436-443 (1976).

[4]W.W. Olive and W.I. Finch, "Stratigraphic and Mineralogic Relations and Ceramic Properties of Clay Deposits of Eocene Age in the Jackson Purchase Region, Kentucky, and Adjacent Parts of Tennessee," *U.S. Geological Survey Bulletin 1282*, 33pp (1969).

[5]H.H. Murray, "Prepared Bodies for Ceramics,"; pp. 173-182 in *Clay Mineralogy and Ceramic Processes and Products*. Edited by F. Veniale. Association Internationale pour l'Etude des Argiles, Milan, 1974.

KAOLINS FROM THE SOUTHWEST OF ENGLAND

Paul Bridgett
ECC International Inc.
100 Mansell Court East
Roswell, GA 30076

ABSTRACT

Kaolins from the Southwest of England have been the primary source of white firing clays in the UK since the start of production of white firing tableware by the likes of Minton and Spode started as early as 1748. Now operated by ECC International, the largest producer of clays in the area with 2.5 million tonnes produced and shipped world-wide each year. The origin, location, processing and control of these clays is outlined along with the unique properties that are used to the advantage of the tableware and sanitaryware markets throughout the world.

INTRODUCTION

Kaolin in the southwest of England was first discovered in 1746 in the far west of Cornwall, near Tregonning Hill, Helston. This discovery was made by the Plymouth Chemist, William Cookworthy.

Two years later, in 1748, Cookworthy discovered extensive kaolin deposits in the St. Austell area. The kaolin was originally worked by Cookworthy to supply the ceramic industry. However, the ceramic industry in the UK soon came down to Cornwall to work their own pits and leading names like Wedgwood, Spode and Minton all carried out kaolin mining in the early days of the industry. However, by about 1840, all the potters had relinquished their leases, leaving the pits to be worked by local Cornish families. By the 1870's there were some 120 separate small enterprises in the area.

Due to mergers between the various enterprises, today there are only three. The most important mergers took place in 1919, when three of the largest producers amalgamated to form English China Clays Limited; and again in 1932, when ECC merged with its two principal competitors, Lovering China Clays Ltd. and H.D. Pochin and Company Ltd., to form English Clays Lovering and Pochin Ltd., today known as ECC International Ltd. Today ECC International Europe produces in excess of 2.5 million tonnes of kaolin each year in the southwest of England.

The southwest kaolins are classified as primary residual and the deposits have been formed as a result of hydrothermal alteration of the granite. Figure 1 shows the principal intrusions of granite in the region, of which only the twin intrusions near St. Austell and the S.W. sector of Dartmoor massif have been extensively kaolinized. The process of hydrothermal alteration occurred some 275 million years ago and kaolinization was a result of the successive effects of high pressure steam and hot acidic vapours and solutions on the feldspars contained in the granite.

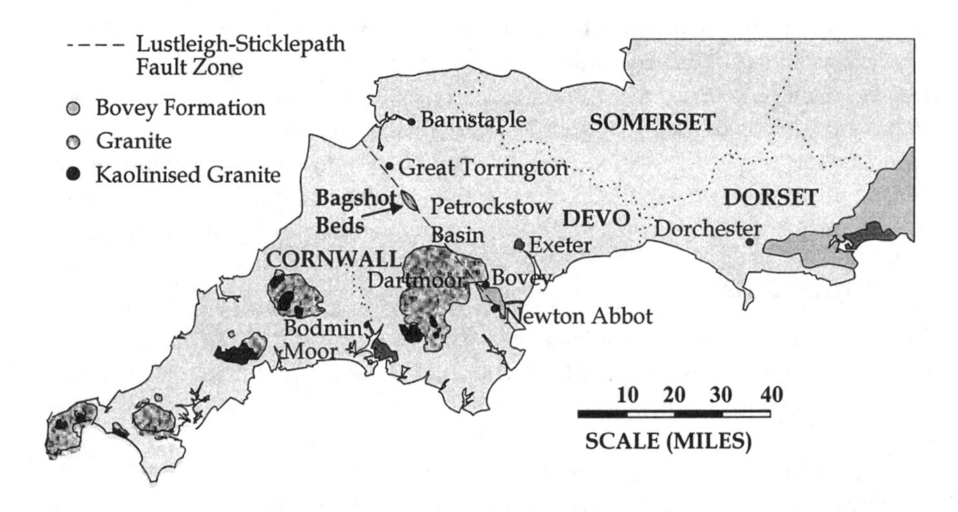

Figure 1. Granite, kaolin, and ball clay deposits in southwest England

Kaolinization extends to depths in excess of 300 meters, the degree of kaolinization and therefore the mineralogical composition of the matrix varies widely. Figure 2 shows that the matrix is an intimate mixture of kaolin, quartz, mica and traces of unaltered feldspar. A typical matrix composition or extraction breakdown is: Rocks 15%, Quartz Sand 70% and Kaolin 15%. Some coarse mica will be present in the sand fraction and some fine mica will be present in the kaolin.

PRODUCTION PROCESS

The ECC international production process divides conveniently into three parts:
(a) Pit Operations
(b) Refining Operations
(c) Blending

All of these are subject to comprehensive quality control procedures which are registered to ISO 9000. Quality control will be reviewed as an overlay of the whole process.

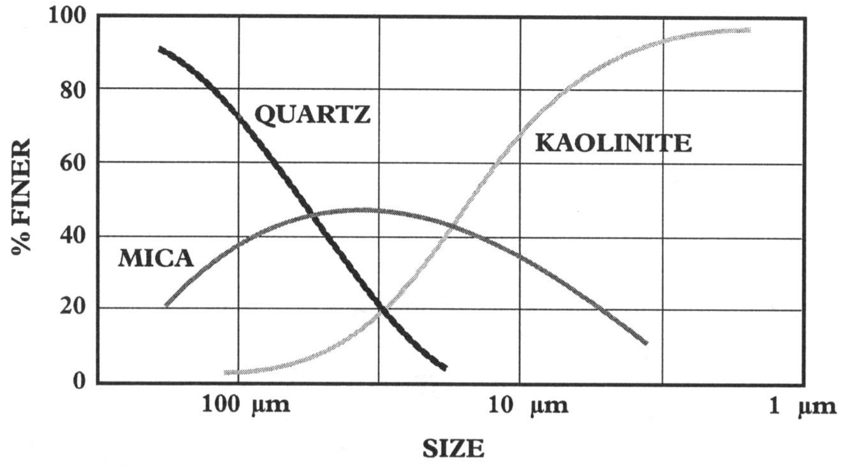

Figure 2. Particle size distribution of the minerals present in the crude kaolin

Pit Operations
Kaolin is extracted using open-cast mining, and operations commence with the stripping of overburden. This overburden consists of soil, peat and a hard rock capping. The exposed kaolin-bearing matrix is loosened by explosive charges and the face is then washed using a high pressure water jet (monitor). During the production process the majority of the other minerals present are removed by particle size separation techniques, the particle size distribution of the minerals present in the crude kaolin are shown in Figure 2. Refining techniques will remove all of the quartz and a large proportion of the mica. The particle size distribution of the latter, however, results in the presence of a significant level of mica in the finished product.

The typical mineralogy of a kaolin product is:

Kaolinite	84%	Mica	14%
Feldspar	1%	Quartz	1%

Figures 3 and 4 show a schematic flow sheet of the production process. The remote controlled water jet (monitor) delivers approximately 2,000 gallons per minute at a pressure of 280 psi. The kaolin wash descends to the bottom of the pit where it is pumped into either spiral or bucket-type sand classifiers. The coarse quartz is removed at this stage and is transferred to the sand tip. The kaolin wash is then pumped to the pit hydroclones which remove all particles greater than 60 microns The residue is mainly coarse mica and some sand which is pumped to the mica dam.

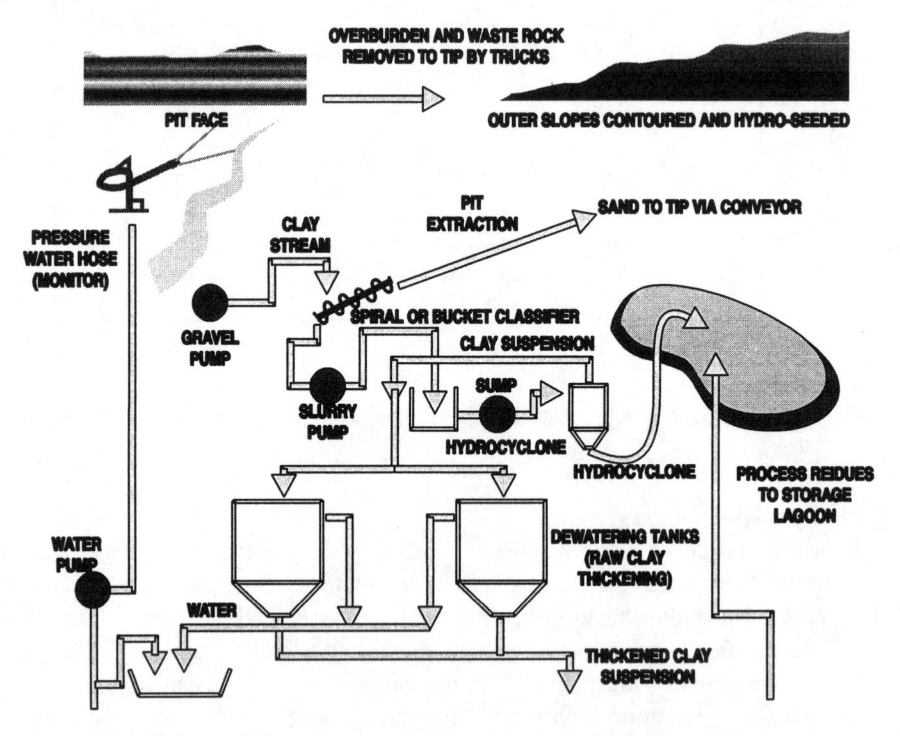

Figure 3. Schematic representation of typical ceramic china clay production

On leaving the pit hydroclones, the kaolin wash is pumped to the surface for dewatering, prior to the refining stage. The pit wash is around 5% solids content and dewaters in 140 ft. diameter tanks to a solids content of 25%. The kaolin is in the natural flocculated state and settles readily. The tanks have a conical base and are fitted with rakes which move very slowly and concentrate the dewatered kaolin to the center of the cone, where it is removed. Each tank holds the equivalent of 1,000 tonnes of dried product.

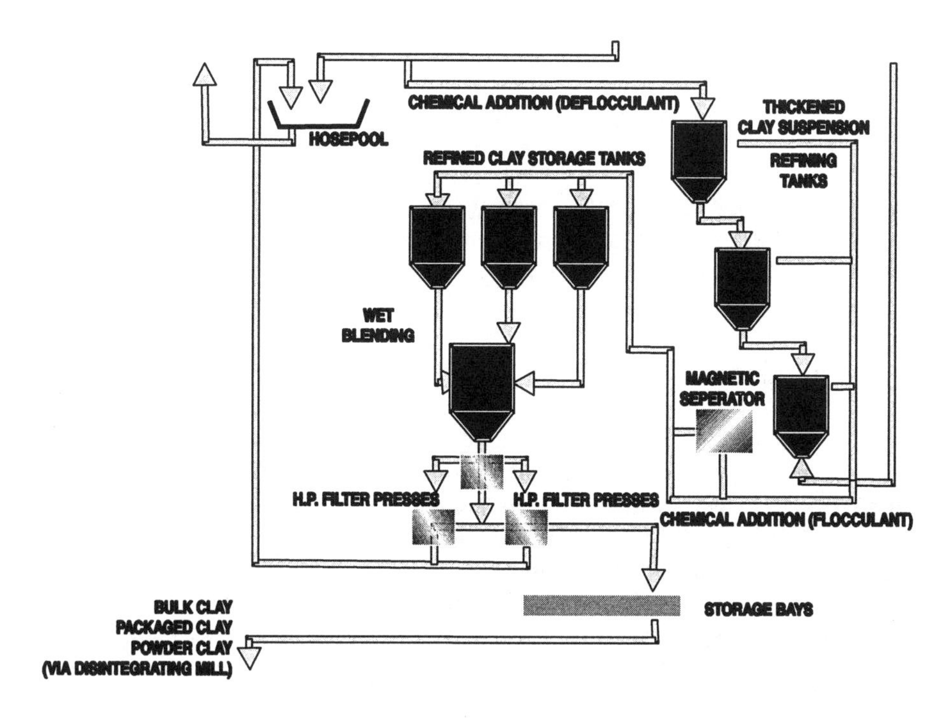

Figure 4. Schematic representation of typical ceramic china clay production

The finished kaolin product will then be made up from a number of kaolin components from different pits, each of which will pass through a refining process. The four major pits in the Cornwall area produce in excess of 20 different kaolin components. The blending process is then used to combine specified components together to produce a final kaolin product.

Refining Operations

Following dewatering, the kaolin passes into the refining plant. Controlled amounts of a mixed deflocculant are added to the slurry, the pH is checked, and the kaolin is screened before passing into a four stage hydroseparator system. At each stage the coarse particle size material settles to the bottom of the tank (underflow) and the fine particle size material (overflow) flows over the tank rim. The underflow material is re-processed and the final rejected underflow material is transported to the mica dam. The hydroseparators refine to 45-50% <2 microns compared with a feed size of 30% <2 microns.

Finer components may then be produced from the hydroseparator products using centrifuges. These operate at a fixed speed, but the desired particle size cut can be adjusted by varying the slurry density and the throughput rate.

After refining the kaolin is subsequently returned to its natural pH by acidification. In addition to refining procedures utilising hydroseparators and centrifuges, a significant level of refining is undertaken using hydroclones.

Blending

After refining, the wet components are blended to give the final wet product. Through the pit operation and refining process the properties of each component are followed very closely by a strict quality control regime. The blending process uses a linear blending programme to smooth out small differences in pit components in order to attain a consistent product. Although the production process is continuous, blending of kaolins is performed on a batch basis. To produce a given tonnage batch of final product, the quality of all suitable components is assessed by computer and the blending programme is then used to specify the blend required to give the final wet properties as shown in Figure 5.

Key Properties	Components					
	PJW	PM	HPSWR	PKT	PLY	Final Grade
K_2O	2.47	2.12	1.74	1.73	1.83	1.97
Fe_2O_3	0.98	1.01	0.94	0.88	1.03	0.94
+10μ	28.00	25.00	24.00	14.00	15.00	23.45
-2 μ	12.00	35.00	39.00	44.00	43.00	31.34
M.O.R.	3.00	4.90	3.00	5.10	5.20	3.38
Casting Rate	15.00	5.00	2.30	2.00	1.80	6.28
Blend (%)	32.00		50.00	18.00		

Figure 5. Blending components needed to produce product grade. Data are from a computer linear blending programme.

Following the blending stage, the wet product is tested to ensure that it is within specification before pumping to the drying plant. Here the slurry is screened at 53 microns prior to filter pressing and drying.

QUALITY CONTROL

The quality control procedures of ECCI Europe have been registered to ISO 9000. The quality control system for kaolins has been developed on the basis of defining the principal operational control parameters that influence quality. For example, in terms of chemical analysis, the day to day variations in Al_2O_3 and SiO_2 contents of a kaolin from a particular area are found to be very small.

The levels of TiO_2, CaO, MgO and Na_2O are also found to be remarkably consistent. However, in terms of quantifying the suitability of a kaolin for a particular sector of the market, the Fe_2O_3 content is extremely important, since the fired colour of a kaolin is directly related to the level of colouring oxides present.

Similarly, the vitrification properties of a kaolin are directly related to the level of fluxing oxides present. Thus % Fe_2O_3 and % K_2O contents provide two important control parameters. Particle size distribution is also a basic control parameter, the particle size distribution of a kaolin will influence strength, rheology, vitrification and shrinkage.

Considering the properties which are important in the final product across all sectors of the market, i.e. fired properties, green strength and casting properties, the important control parameters are: Chemical analysis, particle size distribution, modulus of rupture and casting data. The quality control at each stage of the production process is detailed in Figure 6.

1. BOREHOLE SAMPLES	**FULL EVALUATION**
2. STOPES	**(A) K_2O, Fe_2O_3**
	(B) K_2O, Fe_2O_3 (Al_2O_3) SiO_2 M.O.R.
3. PITWASH	**K_2O, Fe_2O_3, +53u, -2μ**
4. REFINER FEEDS	**(A) K_2O, Fe_2O_3, +53μ**
	(B) K_2O, Fe_2O_3 +53μ, -2μ
5. DORR OLIVER FEEDS	**K_2O, Fe_2O_3, +53μ, -2μ M.O.R.**
6. BLENDING PLANT FEED	**K_2O, Fe_2O_3, +53μ, + 10μ**
	-2μ, pH, M.O.R.
	Al_2O_3, SiO_2, RHEOLOGY
7. FINAL PRODUCT TANK	**K_2O, Fe_2O_3, PSD, M.O.R.**
RHEOLOGY	
8. DAILY DRYING SAMPLES	**RELEVANT PRODUCT DATA**

Figure 6. Quality control stages

Notes on Figure 6

1. Borehole test programs are carried out in advance of pit development and subsequent production. They enable a computer map to be drawn of the pit area in terms of quality and will influence the way in which a pit is developed. The evaluation required at this stage is comprehensive and includes full mineralogical analysis and fired property assessment.

2. Stopes are the sites in the pit at which kaolin is washed, the quality of the wash is monitored on a shift basis.

3. Pitwash passing to the dewatering tanks is tested on a shift basis for chemistry and particle size.

4. Refiner feeds are tested both as received and as laboratory refined.

5. Feeds to the hydroseparators are also tested for modulus of rupture in addition to chemistry and particle size.

6. The component feeds to the blending plant are tested for all parameters.

7. The final wet product tanks are tested for all parameters and approved before drying.

8. The dried product is continuously sampled using automatic samplers which are situated on conveyor belts in the drying plants.

PROPERTIES OF SOUTHWEST KAOLINS

The two major markets for ECCI kaolins are tableware and sanitaryware. 75% of ceramic kaolin production goes to these two sectors of the industry, and for each sector there is a range of special kaolins designed for the particular application.

Tableware Grades

Figure 7 shows typical data for tableware products, Fe_2O_3 contents range from as low as 0.30 and dry strengths cover a range up to 780 psi. The TiO_2 levels in the kaolins from both Devon and Cornwall are the lowest in the world. These low TiO_2 figures enable high translucency values to be achieved in fine porcelain formulations.

Figure 8 shows the relative translucency values of three porcelain bodies. The bodies have been formulated using mixtures of different kaolins, some of which are produced in Europe, the formulations are otherwise identical. The Fe_2O_3

contents of the kaolins and the subsequent formulations are also identical. It can be seen that increasing TiO_2 from the control value of 0.04% (S.W. Kaolin) results in a marked reduction in translucency.

	SUPER STANDARD PORCELAIN	STANDARD PORCELAIN	POLARIS	GROLLEG
% Fe_2O_3	0.39	0.68	0.60	0.75
% TiO_2	0.03	0.02	0.03	0.02
% K_2O	1.20	1.65	2.20	1.85
% + 10μ	1.00	5.00	1.00	8.00
% - 2μ	85.00	70.00	75.00	58.00
M.O.R. MNm^{-2}	5.50	3.20	3.40	2.60
CASTING CONC. (%)	58.00	62.00	59.00	62.00
CASTING RATE (mm^2/min^{-1})	0.35	0.40	0.25	0.50

Figure 7. Typical data for tableware products

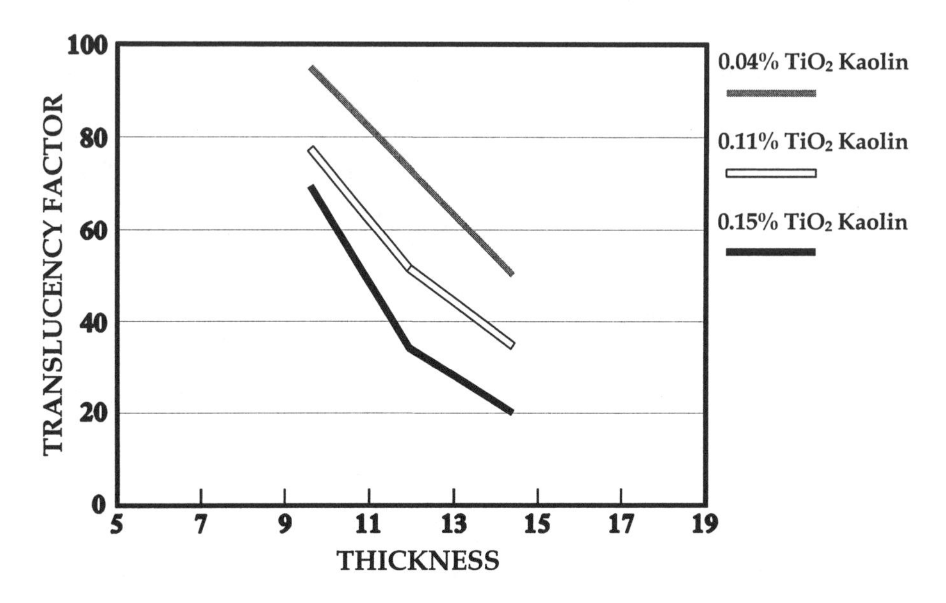

Figure 8. Effect of TiO_2 on relative translucency values of three porcelain bodies

Sanitaryware Grades

The standard range of sanitaryware kaolins is shown in Figure 9. Three principal sanitaryware grades cover the range of requirements, from a fast-casting grade developed especially for pressure casting to a stronger, more plastic grade for bench and battery casting. A number of permutations is possible in order to obtain specific values of both casting rate and strength.

	NSC	REMBLEND	KERNICK	SUPERCAST
% K_2O	2.00	2.00	2.10	2.00
% +10μ	19.00	20.00	25.00	20.00
% - 2μ	40.00	40.00	28.00	40.00
M.O.R. MNm^{-2}	1.30	1.10	0.60	0.90
% CASTING CONC.	69.00	66.00	64.00	60.50
% CASTING RATE (mm^2/min^{-1})	0.90	1.75	8.00	6.50
% DEFLOC. DEMAND	0.50	0.48	0.30	0.37

Figure 9. Standard range of sanitaryware kaolins.

SUMMARY

The kaolins of the southwest of England are unique and therefore important materials for the ceramic industry. The chemistry and mineralogy of these kaolins is particularly suitable for the tableware and sanitaryware sections of the industry, where low levels of colouring oxides, good plasticity and consistent rheology are prime requirements. In some areas of the industry these criteria are so important that even with the high cost of transportation demands that they are used throughout the world. In the whiteware industry especially, the whiteness and plasticity of these clays allows for the production of white tableware which would otherwise have non plastics levels so high as to produce severe manufacturing limitations and high loss.

FELDSPAR EXTRACTION AND PROCESSING

David Marek
PO Box 2004
Big Bear City, California 92314-2004

Christopher W. Sinton
New York State College of Ceramics at Alfred University
Alfred, NY 14802

INTRODUCTION

Feldspars are one of the most ubiquitous minerals in the earth's crust. The major elements found in feldspars are Na, K, Ca, Al, and Si that are coordinated with oxygen to make a framework crystal structure. Based on composition, the feldspars can be divided into two types: plagioclase feldspars and alkali feldspars. The plagioclase series is a complete solid solution between anorthite ($CaAl_2SiO_8$) and albite ($NaAlSi_3O_8$). The alkali series is a complete solid solution between albite and orthoclase/microcline ($KAlSi_3O_4$). The compositions of feldspars are most conveniently displayed on a ternary diagram (Figure 1). Limited substitution of Ba, Ti, Fe^{2+}, Fe^{3+}, Mg, and Sr may occur in natural feldspars.[1]

Although feldspars are a common rock-forming mineral, there are few deposits in which the mineral can be mined economically. Economic deposits are associated with pegmatites (rare igneous intrusions with meter-long crystals), a type of granite called alaskite (composed of feldspar and quartz), and in some beach sands. Pegmatites and alaskite are the last products of crystallization of a cooling granitic magma, at which point most of the Fe-bearing minerals have already crystallized. This produces feldspar deposits that have relatively low concentrations of iron. Nevertheless, mica and garnet still need to be removed from the ore during beneficiation.

In this paper, we describe the process of exploring for profitable feldspar deposits, the extraction of the material and subsequent processing that results in a commercially useful product. We use the Spruce Pine District of North Carolina, the most productive alkali feldspar deposit in the United States, as an example.

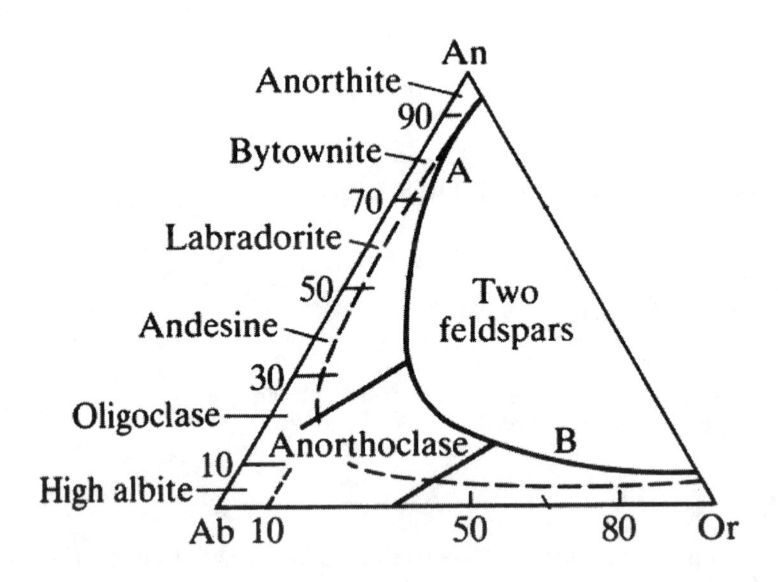

Figure 1. Ternary diagram showing anorthite ($CaAl_2Si_2O_8$), albite ($NaAlSi_3O_8$), and orthoclase ($KAlSi_3O_8$).[1]

USES OF FELDSPAR

Feldspars are widely used in the glass and ceramic industry as a source of soda, potash, and alumina. Feldspars are also used as fillers for other products such as abrasives and paints. When blended into ceramic bodies and glass batches, feldspar decreases the firing temperature (due to the fluxing effect of Na and K), increases the strength and durability of the final product (due to the Al), and cements crystalline phases of other ingredients. Feldspar accounts for 10-35% of the raw material for ceramic bodies and 30-35% of the glaze. In glass, approximately 110 lbs. of feldspar is used to produce a ton of container glass and 100 lbs. is used to make a ton of float glass. Because Fe is undesirable in most glass and whiteware applications, it is necessary that mined feldspar deposits contain very low amounts of Fe.

DEPOSIT LOCATIONS

Feldspar deposits are found around the world. Because it is a bulk raw material, many manufacturers, particularly those in less developed countries, rely on local deposits. While some of these operations may be primitive and the feldspar may be of relatively low purity, processing of local feldspars can often be more cost efficient than importing a higher purity product.
North Carolina is the primary producer of feldspar in the United States. The Spruce Pine Mineral District of North Carolina is mined by three companies and supplies 65-70% of the feldspar consumed in North America. The district is composed of up to twelve intrusions occurring over an area of approximately 0.5 x 1 mile. The extent of the deposit below the surface is unknown.

Originally, feldspars were taken from the pegmatite deposits that contain individual feldspar crystals up to four feet in length. In addition, large muscovite mica crystals were mined for use as isinglass or as insulators. Currently, feldspar is mined from both pegmatite and alaskite deposits. The alaskite is made of: 35% albite; 30% orthoclase; 24% quartz; 10% muscovite mica.

EVALUATION AND DEVELOPMENT OF A FELDSPAR MINE

After the initial discovery of a promising feldspar deposit, it can take up to three years to bring it into production. Before this can occur, the deposit must first be assessed to determine whether or not it can be economically mined. Hand samples of the rock are collected and then brought into the lab. The sample is broken up in a mill and the different minerals are separated and analyzed for composition. If the samples are of sufficient purity, then the process continues.

The next step is drilling the area with diamond drills, similar to methods that are used in gold and silver exploration, to assess the size, homogeneity, and quality of the deposit. This method of drilling is expensive and can cost $20-30$/foot, but the recovered cores are critical in obtaining accurate depth information. With the drill cores and a survey of the area, the dimensions of the deposit can be accurately assessed.

When the deposit has been thoroughly evaluated, development drilling begins. Here a 12 foot by 12 foot pattern of six inch holes are drilled to a depth of 65 feet. The holes are filled with explosives and detonated. Each blast produces 20,000-25,000 tons of material that takes three to four weeks to remove.

The blasting progresses to develop a series of benches or steps that allow safe access within the mine. Each bench is 65 feet high and a good mine plan allows several benches to be mined simultaneously. On a bench, a front end loader loads the blasted rock to a truck which hauls the material to a jaw crusher. The opening at the bottom of the jaws, which can be set from 3-15 inches, determines the size of the resulting aggregate. At this point water is sprayed on the crushed rock to control dust.

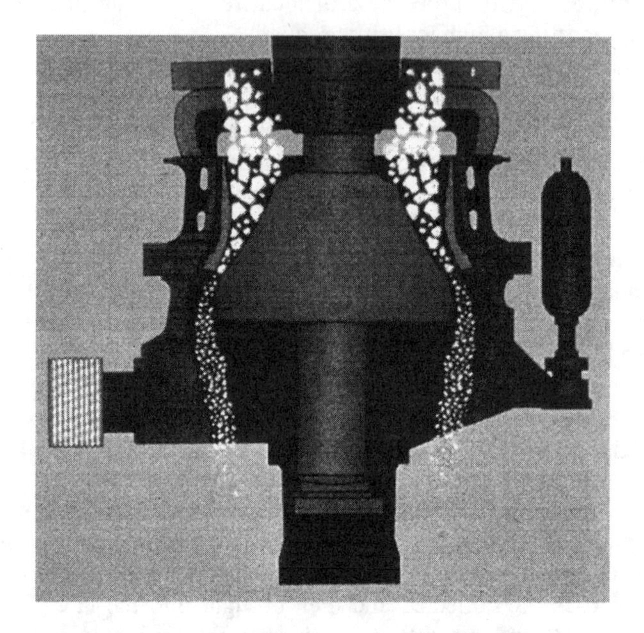

Figure 2. Schematic diagram of a cone crusher.

The crushed material is then conveyed to through screens and into a cone crusher (Figure 2). The cone crusher consists of two inverted four foot diameter rotating steel bowls. After crushing, the material is stockpiled, loaded onto trucks, and taken to a central processing facility that is fed by several nearby mines.

PROCESSING OF FELDSPAR

At the processing facility, the crushed material and water is fed into a rod mill to further reduce the size of the pieces. The mill consists of a drum filled with ~100 steel rods and can process up to 70 tons of material per hour (Figure 3). After the ore leaves the rod mill, it is sent through a hydrosizer. Material that is less than 30 mesh moves on in the process while the larger pieces are fed back to the rod mill.

At this point, the grains are small enough that the individual grains can be separated by four banks of four to ten flotation cells (Figure 4). The flotation process is the dominant mineral separation method used for any mineral and non-mineral (e.g., coal). It is based on the dissimilar surficial behavior of different mineral types in a solution. A desired mineral will become hydrophobic when coated with a specific organic compound. The mineral grain will then attach to air bubbles that rise through the tank and can then be skimmed off. Uncoated minerals sink to the bottom of the cell and the process is repeated with a different reagent mix until all mineral phases are separated.

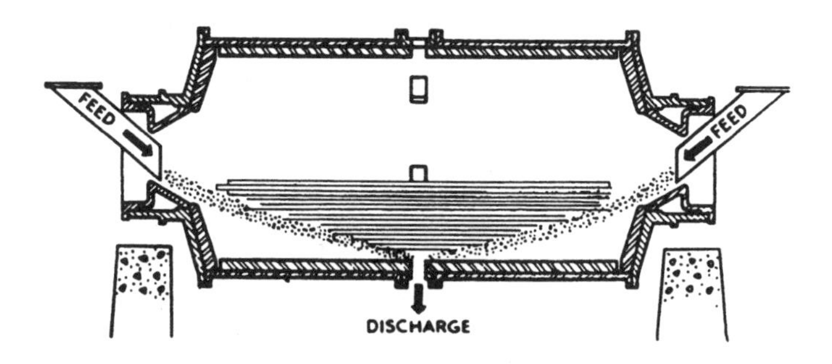

Figure 3. Schematic diagram of a rod mill

At the Spruce Pine facility, there are three flotation steps to separate mica, Fe-bearing minerals, and feldspar. The ore powder is first placed in conditioner cells where the material is scrubbed and retained prior to flotation. The first cell floats the mica in a mix of amine acetate and sulfuric acid. The sulfuric acid helps sink the feldspar and quartz. The floated mica is paddled off and drained. The mica can be used in the manufacture of wallboard and joint compound.

The next minerals to be removed are Fe-bearing garnet and biotite using promoters and frothers. Again, sulfuric acid is used to sink the quartz and feldspar. This step also removes iron and steel contaminants from the rod milling.

At this point, the material is composed of feldspar and quartz. It is easier to float the feldspar, which can be done in a solution of amine and hydrofluoric acid (which helps sink the quartz). The quartz is of high purity and is presently being sold to the semiconductor industry. Earlier, the sand was sold to regional golf courses for sand traps.

After flotation, the feldspar is passed over a vacuum filter belt to de-water and lower the moisture content to 6% prior to feeding the feldspar to a fluid bed dryer. Once dried, the feldspar is now a marketable product named F20 (20 mesh feldspar). It can be loaded into trucks, rail cars, or it is packed in bags.

Value-added products are made from F20 by passing the dried feldspar over rare earth magnets to produce a magnetically separated material. This low-iron feldspar is fed to pebble mills and ground to 325 mesh, 250 mesh, and 200 mesh products. The fine grind material is pneumatically separated and loaded into bulk rail, bulk truck, and bags for shipping.

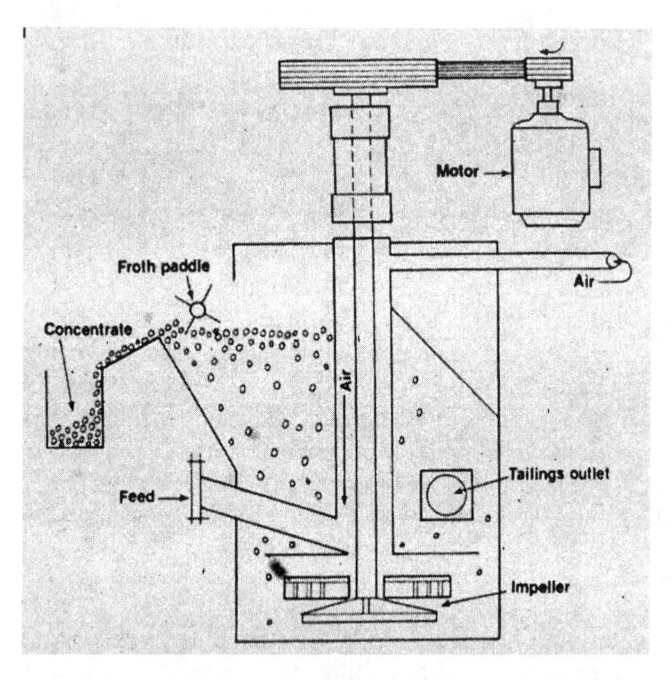

Figure 4. Schematic diagram of the froth flotation process.

The feldspar is tested at the facility to ensure quality control. Analyses performed include full chemical analysis, XRD, fired color, and surface area.

ENVIRONMENTAL IMPACT

The separation and processing of feldspar generates substantial waste that needs to be managed. The Spruce Pine facility uses 3.5-4.0 million gallons of water each day. Some of this water has been contaminated with hydrofluoric acid and therefore must be treated prior to discharge or re-use.

Fine particles must be removed from the process water through clarifying tanks. This material is hauled away to landfill sites in the neighboring community. The cleaned process water is re-circulated back to the plants for continued use.

REFERENCES

[1] Deer, W.A., Howie, R.A., and Zussman, J., *An Introduction to the Rock-Forming Minerals*, Longman, Essex UK (1992).

Polymeric Additives

POLY(VINYL ALCOHOL): IMPORTANT PROPERTIES RELATED TO ITS USE AS A TEMPORARY BINDER IN CERAMIC PROCESSING

Eric H. Klingenberg
Air Products and Chemicals, Inc.
7201 Hamilton Blvd.
Allentown, Pa 18195-1501

ABSTRACT

Poly(vinyl alcohol) (PVA) is one of the most commonly used polymers as a temporary binder in ceramics processing. Because of the wide range of standard and specialty grades available, it is important to know the differences between grades and how to choose the appropriate one for a given application. Therefore, the synthesis and general properties of PVA are discussed and the plasticization and thermal degradation of PVA are reviewed. An understanding of these properties allows for the effective use of PVA.

INTRODUCTION

Increasingly, synthetic polymeric processing aids are being utilized in ceramics processes as rheology modifiers, dispersants, and temporary binders. Incorporation of these materials has greatly improved process consistency and product quality. Although the number of different polymeric processing aids available and their use has increased over the years, a need to fundamentally understand how the polymer properties effect overall performance in a given application still exists. For instance, knowing how the glass transition temperature (T_g) of a particular polymer system is effected by changes in moisture content, plasticizer type, and plasticizer level, can impact how it is used in processes such as dry pressing of spray dried powders. Therefore, to more effectively utilize organic polymers in these applications, it is necessary to understand how the polymer's physical and chemical properties are manifest in the application. In this paper we focus on PVA and the important properties related to its use as a temporary binder.

Because of its water solubility, ease of plasticization, high binding strength, and clean burn-out, PVA is one of the most utilized polymers as a temporary binder in ceramic processing.[1,2] However, it is important to realize that there are a number of different grades of PVA that can be used in these applications. The different grades of PVA vary in molecular weight and acetate level, which largely determine the polymer's physical properties. The general structure of PVA is shown in Figure 1. The goal of this paper is to improve the general understanding of PVA, how it is synthesized, and how the polymer properties change in accordance with the chemical structure of the polymer.[3] These general polymer properties will then be related to using PVA in ceramic processing applications.

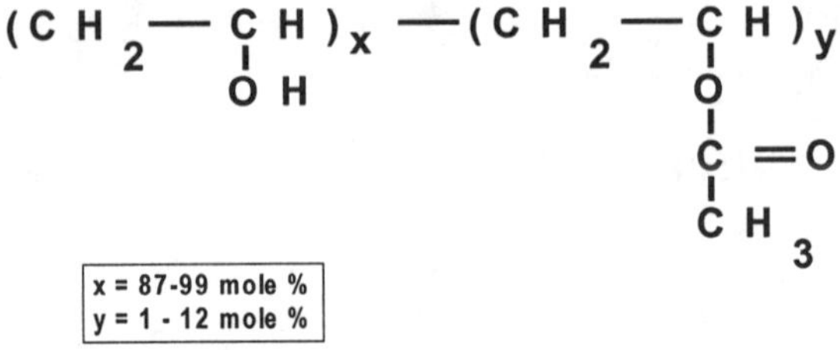

$$(CH_2-CH(OH))_x-(CH_2-CH(O-C(=O)-CH_3))_y$$

x = 87-99 mole %
y = 1 - 12 mole %

Figure 1. General structure of poly(vinyl alcohol).

GENERAL PVA PROPERTIES

PVA is one of the world's largest volume synthetic polymers. It is commercially produced in a two step process (Figure 2). In the first step, vinyl acetate is polymerized in methanol by a free radical mechanism to yield poly(vinyl acetate). The poly(vinyl acetate) is then converted to PVA through a transesterification reaction in the second step of the process. By controlling process variables such as time, temperature, concentration of initiators for polymerization, and catalyst concentration and type for the transesterification reaction, a variety of different grades of PVA can be produced. These different grades of PVA are then characterized by their molecular weight and hydrolysis level.

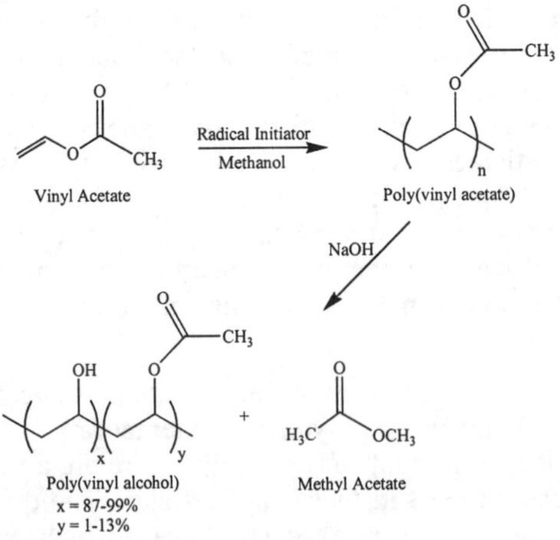

Vinyl Acetate
Radical Initiator
Methanol
Poly(vinyl acetate)
NaOH
Poly(vinyl alcohol)
x = 87-99%
y = 1-13%
Methyl Acetate

Figure 2. Synthesis of poly(vinyl alcohol).

The molecular weight of PVA is usually expressed as the 4% aqueous solution viscosity. The higher the 4% aqueous solution viscosity the higher the molecular weight. In general, higher molecular weight PVA materials are used in applications which require higher viscosities, high tensile strength, and reduced water sensitivity. On the other hand, lower molecular weight materials provide increased water solubility and improved flexibility. Since PVA is commercially available in a wide range of molecular weights, care should be taken in choosing the correct molecular weight grade of PVA with the appropriate hydrolysis level.

Hydrolysis level refers to the mole percent of vinyl alcohol units present along a polymer chain. In general the hydrolysis level varies from partially hydrolyzed (87-89% alcohol functionality) to super hydrolyzed grades (99.3+% alcohol functionality). Because PVA contains a large number of hydroxyl groups which are capable of forming very strong hydrogen bonded networks, these polymers are semicrystalline materials. The degree of crystallinity is dependent on the hydrolysis level, because the presence of acetate groups inhibits the formation of the hydrogen bonds. This results in materials with lower degrees of crystallinity (Figure 3). The crystallinity dramatically effects the properties of these materials in terms of water solubility, tensile strength, and adhesion to different substrates.

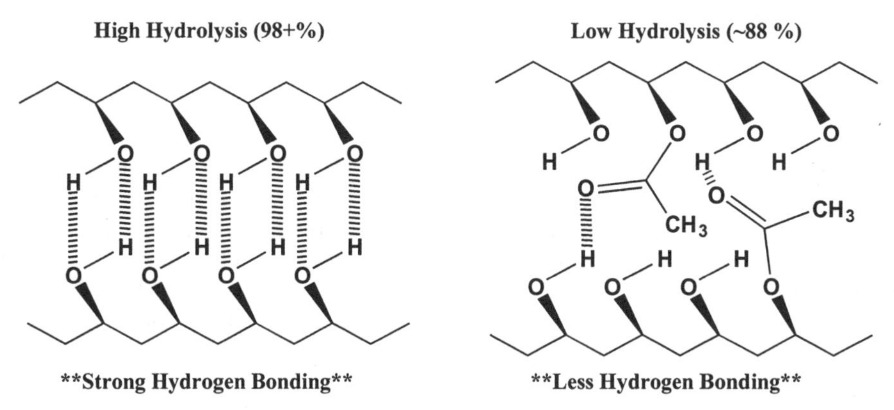

Figure 3. Illustration of acetate groups interfering with hydrogen bonding.

Because the properties of PVA are largely determined by the molecular weight and hydrolysis level, it is important to consider both features when selecting a particular grade of PVA for a given application. Figure 4 indicates the typical trends in polymer properties as they are related to molecular weight and hydrolysis level. The properties listed on the top half of Figure 4 represent those properties which are affected by the molecular weight of the polymer. The lower half of figure 4 indicates properties which are effected by hydrolysis level. Increasing molecular weight and hydrolysis level decreases the water solubility but improves the tensile strength and the adhesion of the material to hydrophilic substrates. Lower molecular weight and lower hydrolysis levels reverse these trends, providing improved water solubility, flexibility, and adhesion to more hydrophobic substances. Since there are a large number of different commercial grades of PVA available, it is possible to choose a grade which has a unique balance of properties to provide optimal performance for a specific application.

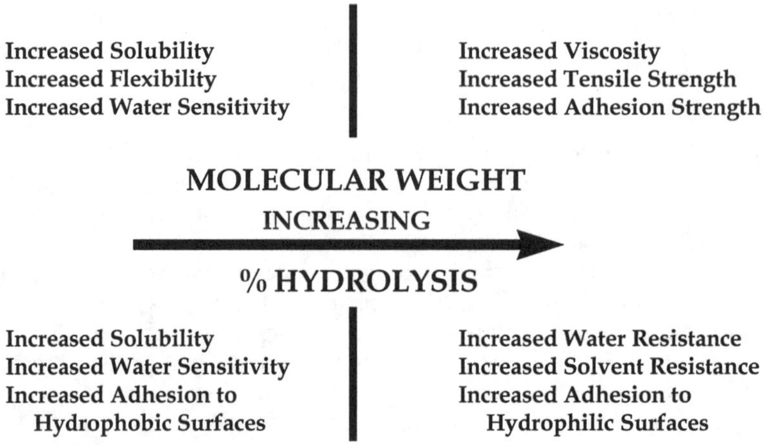

Increased Solubility
Increased Flexibility
Increased Water Sensitivity

Increased Viscosity
Increased Tensile Strength
Increased Adhesion Strength

MOLECULAR WEIGHT

INCREASING

% HYDROLYSIS

Increased Solubility
Increased Water Sensitivity
Increased Adhesion to Hydrophobic Surfaces

Increased Water Resistance
Increased Solvent Resistance
Increased Adhesion to Hydrophilic Surfaces

Figure 4. Typical trends in polymer properties as a function of molecular weight and hydrolysis level.

PROPERTIES OF PVA RELATED TO FORMING AND BURN-OUT

PVA is used as a temporary binder in a number of different ceramic processes to provide strength to the unfired ceramic part. These processes include slip casting, injection molding, extrusion, and dry pressing. In each of these processes, the required properties of the binder may be slightly different. Two of the most important binder properties related to ceramics processes are the effect the polymer has on the flow behavior of the system (Forming) and the thermal degradation of the binder system (Binder Burn-out).

Forming

The effect PVA has on the flow behavior in ceramics processing is not only related to the grade of PVA as discussed above, but also on the solids loading of the system, amount of binder used, and plasticization. Depending on the process, it is possible to adjust these parameters to achieve the desired rheological characteristics of the system. Because ceramic systems are generally molded above the T_g of the binder but near room temperature, PVA (T_g = ~60°C) performs best when plasticized. There are a number of different plasticizers available for use with PVA. These plasticizers include glycerine, low molecular weight poly(ethylene glycol) (PEG), and water. It is necessary to understand how these plasticizers interact with PVA to determine the best plasticizer type and level for a given process.

As with any water soluble polymer, water is one of the most effective plasticizers for PVA. Water can be present in the system either by conscious addition or by adsorption from the atmosphere. The amount of water present in the system determines the viscosity of a dissolved polymer or the T_g of the undissolved system. In applications such as dry pressing of spray dried powders, the T_g of the binder system plays an important role in determining the effectiveness of the binder in providing strength to the green part and minimizing the defects formed during the pressing process. Because PVA can adsorb moisture from the atmosphere, it is useful to know how environmental conditions such as relative humidity effect the T_g of PVA. In this paper we have limited the discussion to a low molecular weight (4% aqueous solution viscosity of 5 cps), 88% hydrolyzed grade of PVA. Although the discussion is limited to one grade of PVA, the results described are very similar between different grades of PVA.

The water content of PVA is directly affected by the relative humidity of the environment in which the material is stored. Figure 5 illustrates the effect relative humidity has on the amount of moisture adsorbed by PVA. The higher the relative humidity, the more moisture is adsorbed by PVA. Increasing the water content of the PVA lowers the T_g of the polymer as shown in Figure 6. Because water is such an efficient plasticizer for PVA, the effect of humidity on T_g should be a consideration when formulating.

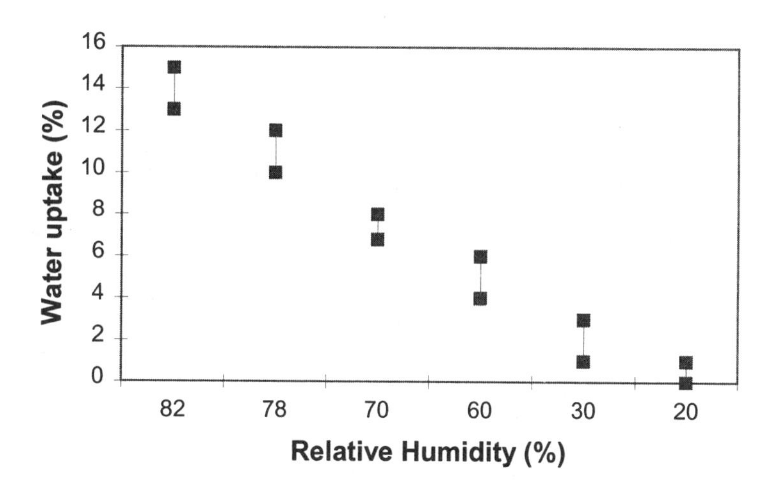

Figure 5. Effect of relative humidity on water content of PVA.

In addition to water, it may be necessary to utilize additional plasticizers with PVA. The level and type of plasticizer used depends on the desired pressing pressure, required green strength, and required green density. Understanding how the plasticizer can effect these properties is especially important in a dry pressing process where it is necessary to achieve some level of polymer flow during compaction. The ability of the polymer to flow during the compaction allows for the formation of a connective network which enhances the green strength of the body by acting as the glue that holds the particles together. However, the addition of too much plasticizer can result in the polymer being to flowable and thus deficient in mechanical strength. In this case, the green strength of the green body is reduced. Therefore, it is important to understand the interactions between the plasticizers and the binder used.

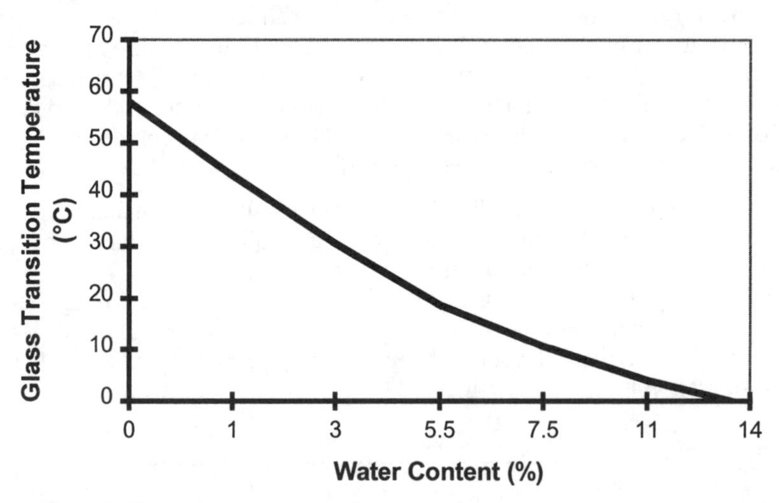

Figure 6. Effect of water content on the glass transition temperature (T_g) of PVA.

The interactions between different plasticizers and binders can be examined using Dynamic Mechanical Analysis (DMA). This type of test is used to examine the mechanical properties of the binder film. Specifically, it can be used to measure both loss (E') and storage modulus (E") of the plasticized polymer films as a function of temperature. The ratio of the loss modulus to the storage modulus is the tan(δ) (E"/E'). The tan(δ) is a measure of the viscoelastic state of the material. For example, an elastic solid has a tan(δ) = 0, and a liquid has a tan(δ) = ∞. When measuring the mechanical properties of a film against temperature, a peak in the tan(δ) represents a transition from one mechanical state to another. In the examples illustrated here, the peaks in the tan(δ) verses temperature plots are the T_g values of the systems. Analysis by this technique provides information on both the T_g of the polymer systems and on the mechanical integrity of the polymers. These properties can then be correlated to the binder's ability to provide green strength to the green body. This information is very helpful when choosing the best plasticizer type and level for a given process and binder system.

Two of the most common plasticizers for PVA in ceramic manufacturing processes are glycerin and PEG-400.[4-6] However, both of these materials interact very differently with PVA. Figure 7 shows a series of DMA plots for PVA by itself and plasticized with 20 wt.% glycerin or PEG-400 at 50% relative humidity. Examination of the midpoint of the peak in the tan(δ) versus temperature curves indicates that at equal weight percentages, glycerin is a more effective plasticizer for PVA. Glycerin reduces the T_g by 50 °C while PEG-400 only reduces the Tg by 35 °C. The difference in the effectiveness of each of these plasticizers is related to the chemical compatibility of these materials with PVA. Because of the differing chemical nature between PVA and PEG, these systems are somewhat incompatible and can phase separate upon drying, reducing the effectiveness of PEG-400 as a plasticizer for PVA. This is evident by the breadth of the peak in the DMA analysis. The PEG-400/PVA system shows the broadest peak which even extends above 50 °C. This suggests that there are regions of unplasticized PVA in the film. Therefore, it takes more PEG-400 by weight to decrease the T_g of PVA. Figure 8 shows the effect of the amount of plasticizer on the midpoint T_g of the PVA systems. As a result of these differences in plasticization, pellets pressed from spray dried powders prepared using equal weight percentages of glycerin and PEG-400 differ in green strength (Figure 9). However, it is also important to consider other properties of the plasticizer/polymer system such as the volatility of the plasticizer for spray dry operations, green density effects, and the potential for die sticking to occur when choosing the best plasticizer for a process.

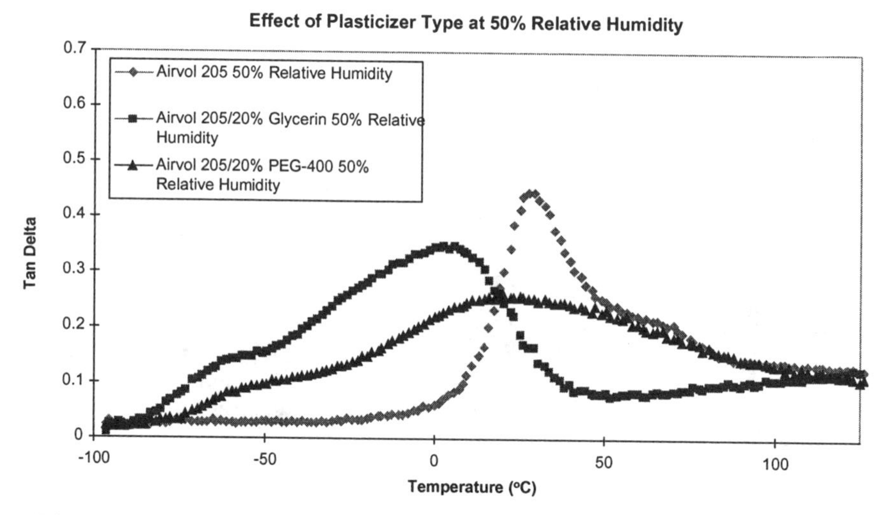

Figure 7. DMA results indicating the partial incompatibility of PEG-400 and PVA.

As a general rule, to provide maximum green strength and green density for an operation the T_g of the binder system should be below the pressing temperature. The desired T_g range will also depend on the pressing pressure and green density requirements of a given system. However, it is important to note that even though the T_g of the plasticized system is below the pressing pressure, it is possible to over plasticize the binder and reduce the green strength. Figure 10 illustrates the effect of too much plasticizer on the green strength of an alumina system that contains PVA as a binder. The addition of too much glycerin in this system has resulted in a significant loss of green strength. This is a direct result of a loss in the mechanical strength of the PVA film with plasticization.

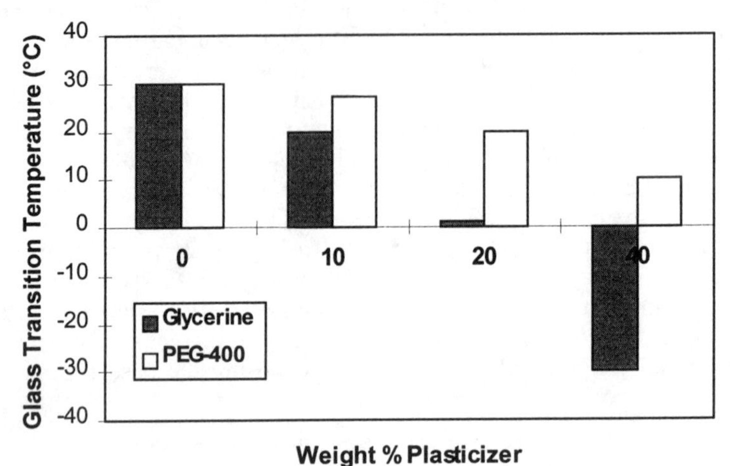

Figure 8. Effect of plasticizer type and level on the T_g of PVA at 50% relative humidity

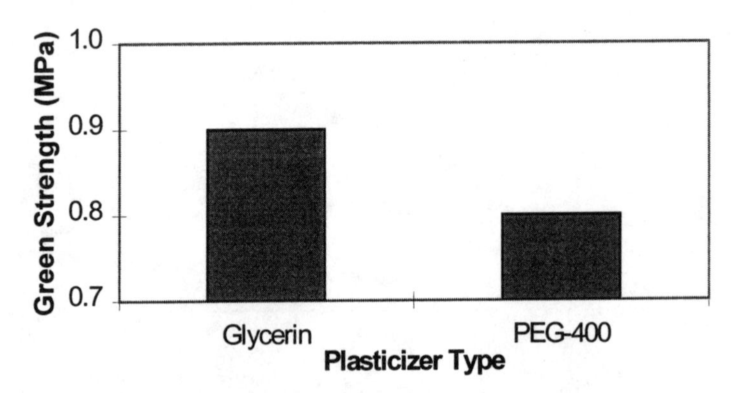

Figure 9. Effect of plasticizer type on green strength. Samples pressed at 10,000 psi, 50% Relative Humidity, 20 wt % plasticizer, and based on alumina A-16 S.G. (Alcoa).

Analysis by DMA of PVA films with increasing levels of glycerin reveals that not only is the T_g of the PVA films reduced with increasing plasticizer levels but the film breaks in the test at lower temperatures with increasing plasticizer content (Figure 11). This is seen as the rapid rise in the tan(δ) values above 100 °C which end abruptly because of the polymer film breaking. These results indicate that it is not only important to consider the level of plasticizer in the system but also the type of plasticizer.

By understanding the interactions between binders and plasticizers, it is possible to design a system which provides the maximum all around performance. This includes choosing the right plasticizer type and level to obtain the appropriate green strength and green density for a given process.

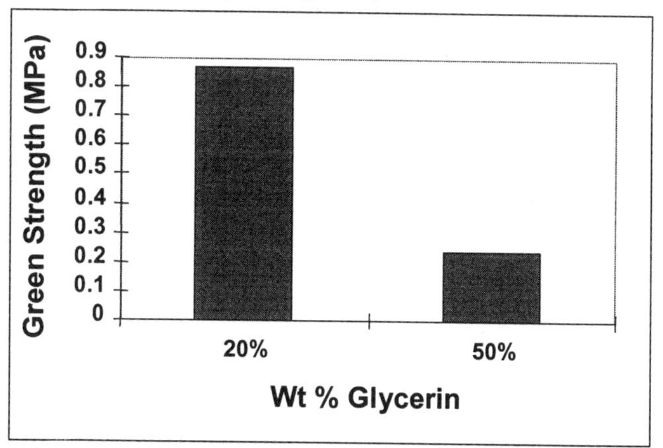

Figure 10. Effect of plasticizer level on green strength. Samples conditioned at 50% relative humidity, 2 wt % binder on alumina, and pressed at 10,000 psi.

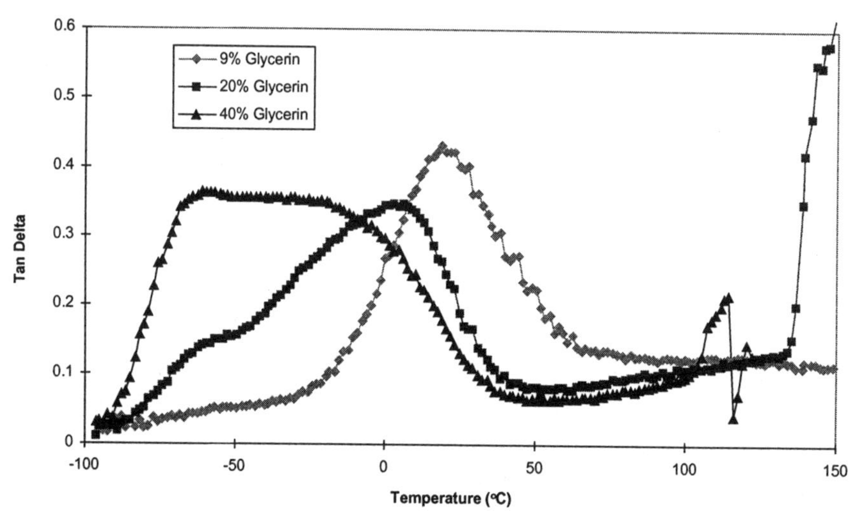

Figure 11. DMA analysis of effect of plasticizer level on the mechanical strength of PVA

Binder Burn-out

The removal of the organic materials in the pre-sintering phase of the firing process is often referred to as the burn-out stage. The thermal degradation of the binder during this stage of the firing process is extremely important. The binder degradation not only has to be controllable, but the binder has to burn-out completely leaving the ceramic part defect free. To complicate matters, this burn-out must occur in a variety of atmospheres depending on the type of ceramic part being manufactured. For instance, metallized ceramic materials may have to undergo binder removal under reducing atmospheres, protecting them from oxidation.

PVA burns-out in air and oxygen deficient atmospheres, such as nitrogen, leaving very little residual material in the ceramic part. This is illustrated in the Thermal Gravimetric Analysis (TGA) trace shown in Figure 12. It can be seen from this TGA trace that in nitrogen, the weight loss occurs at a fast rate initially then becomes more gradual. In air, however, the weight loss is slower than that observed for the nitrogen atmosphere initially, but the complete burn-out is accomplished much earlier. The reason for this behavior is that the mechanism and by-products of the thermal degradation depend on the amount of oxygen present.[7-9]

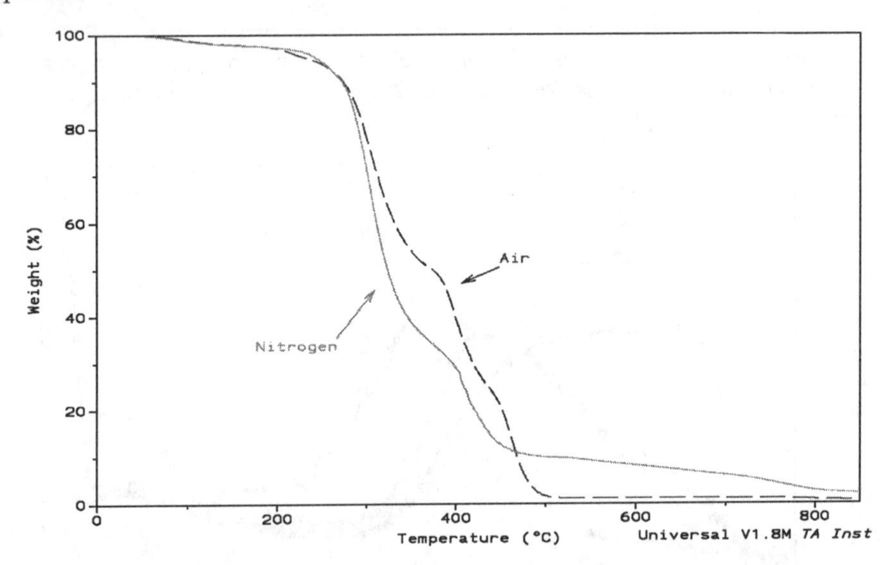

Figure 12. Thermal gravimetric analysis trace of PVA (Airvol 205) in air and nitrogen, (2°C/min to 850°C).

In air PVA, Airvol® 205, burns out cleanly to carbon dioxide and water. A simplified chemical mechanism is shown in Figure 13. This mechanism is observed experimentally by analyzing the volatile components from the TGA experiment by Fourier Transform Infrared Spectroscopy. Figure 14 indicates the volatile components as a function of time and temperature for the experiment by monitoring absorbances specific to the organic materials. This indicates that for the majority of the experiments the by-products are only carbon dioxide and water. The small amount of organics observed can by eliminated by running the experiment at slower heating rates. However, in low oxygen environments PVA, like all organic polymers, degrades to a mixture of different small molecule organic compounds. The majority of the organic compounds produced in these instances by PVA are aldehydes, in the form of crotonaldehyde and acetaldehyde (Figure 15). The formation of these compounds must be accounted for in the process to minimize potential release of these substances into the environment.

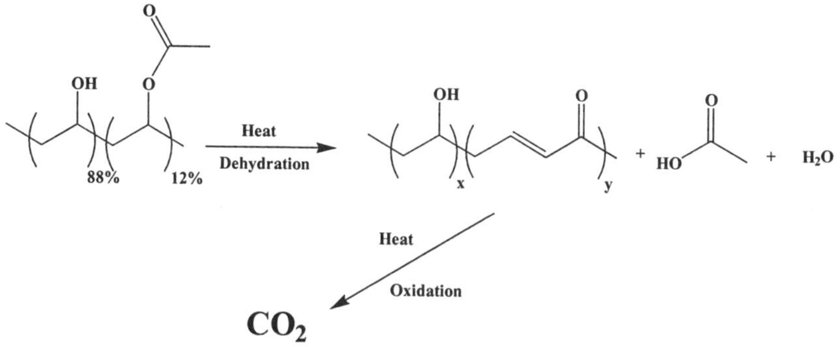

OH O 88% 12% —Heat / Dehydration→ OH O x y + HO(C=O)CH3 + H_2O

—Heat / Oxidation→ CO_2

Figure 13. Simplified chemical mechanism for the thermal degradation of PVA in air.

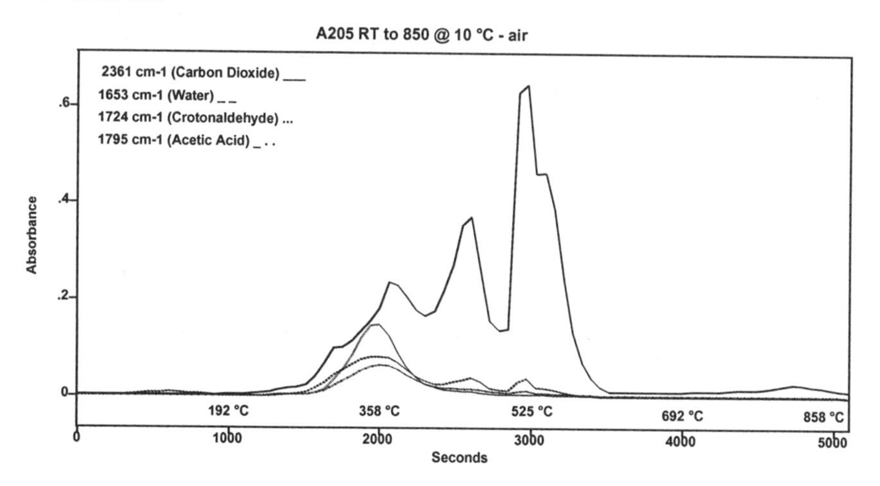

Figure 14. IR absorbances for the volatile organic components from the thermal degradation of PVA in air as a function of time and temperature.

Knowing how PVA degrades thermally under different atmospheric conditions, it is possible to adjust the processing conditions to allow for the efficient and total removal of the polymer from the ceramic part. It is also possible to degrade PVA cleanly by providing enough oxygen to the system to promote the formation of carbon dioxide and water directly from PVA. This is an important consideration for decreasing industrial emissions.

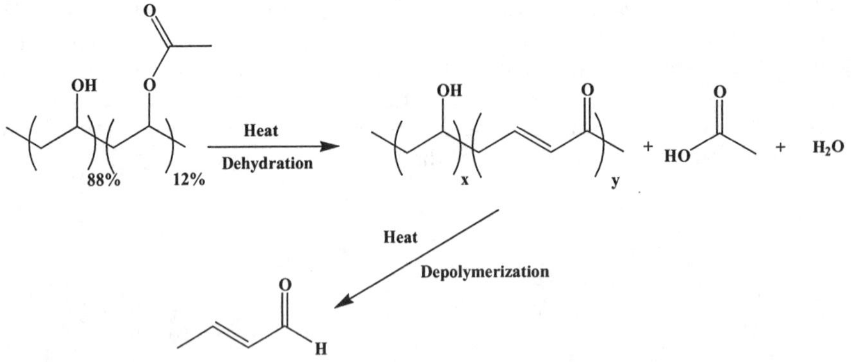

Figure 15. Simplified chemical mechanism of thermal degradation in nitrogen.

SUMMARY

All of the characteristics described above determine the binders usefulness in a given process and must be taken into account when choosing the binder or binder system. As discussed, it is possible to choose between a wide variety of different grades of PVA to obtain the necessary balance of properties required for the different processes. Combining the proper choice of PVA grade, plasticizer, and process conditions, it is possible to obtain the maximum performance from the polymer for a variety of applications.

ACKNOWLEDGMENTS

The author would like to thank Amir Famili, Sherri Bassner, Prof. William Carty, Prof. Gary Messing, and Prof. Dan Shanefield for useful discussions. The author also gratefully acknowledges the collaborative efforts of Menas Vratsanos and Robert Berner for the DMA analysis, and Terry Slager and Frank Prozonic for the TGA/IR analysis.

REFERENCES

1. D. S. Shanefield, *Organic Additives and Ceramic Processing*, 2nd ed., Kulwer Academic Publishers, Norwell, MA, 1996; and references within.
2. J. S. Reed, *Principles of Ceramic Processing*, 2nd ed., John Wiley and Sons, Inc., NY, 1995; and references within.
3. *Encyclopedia of Polymer Science and Engineering*, Volume 17, 2nd ed., pp 167-198, F. L. Martin, John Wiley and Sons, Inc., NY, 1989; and references within.
4. R. A. DiMilia and J. S. Reed, "Dependence of Compaction on the Glass Transition Temperature of the Binder Phase," *Amer. Ceramic Soc. Bull.*, **62** [4] 484-86 (1983).
5. C. W. Nies and G. L. Messing, "Effect of Glass Transition Temperature of Polyethylene Glycol-Plasticized Polyvinyl Alcohol on Granule Compaction," *J. Amer. Ceramic Soc.*, **67** [4] 301-04 (1984).
6. W. J. Walker, J. S. Reed, and S. K. Verma, "Polyethylene Glycol Binders For Advanced Ceramics," *Ceram. Eng. Sci. Proc.*, **14** [11-12] 58-79 (1993).
7. Y. Tsuchiya and K. Sumi, "Thermal Decomposition Products of Poly(vinyl alcohol)," *J. of Polymer Sci.: Part A*, **7** 3151-58 (1969).
8. C. Vasile, C. N. Cascaval, and P. Barbu, "Thermal Degradation of Vinyl Alcohol/Vinyl Acetate Copolymers. III. Study of the Reaction Products," *J. of Poly. Sci.: Polym. Chem. Ed.*, **19** 907-16 (1981).
9. C. Vasile, L. Odochian, S. F. Patachia, and M. Popoutanu, "Thermal Degradation of Vinyl Alcohol/Vinyl Acetate Copolymers. V. Study of the reaction Products of Statistical Copolymers," *J. of Poly. Sci.: Polym. Chem. Ed.*, **23** 2579-87 (1985).

PHYSICAL AND CHEMICAL PROPERTIES OF CELLULOSE ETHERS

Michael R. Smith
Dow Chemical Company
P.O. Box 400 Bldg. 2512
Plaquemine, LA 70765-0400

ABSTRACT

Cellulose ethers consist of a wide variety of water-soluble and organo-soluble polymers, with an equally wide variety of properties. These properties are a function of the type and amount of substitution of hydroxyl groups in the original cellulose backbone, and the molecular weight of the finished product. These polymers can provide water retention, lubricity, green strength, and binding during extrusion and forming. They also help reduce surface defects during drying and firing.

INTRODUCTION

Many of the basic properties of cellulose ethers are determined by the structure of cellulose itself. Cellulose is a polymer (degree of polymerization **n**) consisting at the simplest level of two anhydroglucose rings joined by a β-linkage between C-1 of one ring and the C-4 oxygen of the next (cellobiose, Figure 1).

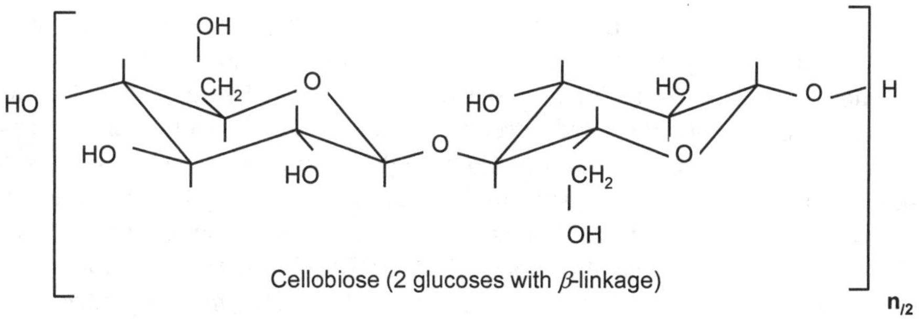

Cellobiose (2 glucoses with β-linkage)

Figure 1. Structure of cellulose.

This β-linkage causes the polymer chains to lie in a straight line, with a slight twist, thereby allowing adjacent chains to form many hydrogen bonds with one another. This hydrogen bonding makes cellulose with a degree of polymerization greater than about seven insoluble in water. Thus the minimum requirement for a substituent to make a water-soluble polymer from cellulose is to 1) disrupt the hydrogen-bonding of the chains enough to make the dissolved state more favorable than the crystalline state in terms of free energy, or 2) to generate a greater free energy by solvating a powerfully hydrophilic substituent such as a carboxylate salt. One of the most critical properties of cellulose derivatives is the degree of polymerization, which can be reduced, but not augmented, during the derivatization reactions. The degree of polymerization is a function of the source of cellulose (cotton, softwood, or hardwood), the type of pulping and bleaching processes, and incidental or deliberate mechanical and chemical degradation during the reaction or finishing processes.

TYPES OF CELLULOSE ETHERS

The other determinants of the properties of the polymer are the specific type of substituent, the average number of hydroxyl groups substituted per anhydroglucose ring, and the uniformity of distribution of substituents. The average number of substituents per ring is referred to as the Degree of Substitution (DS) for substituents which do not replace the ring hydroxyl with a hydroxyl of their own, or Molar Substitution (MS) for substituents which do contain a hydroxyl capable of further reaction with other molecules of the same or other reactant. The maximum DS is 3.0, since there are 3 hydroxyls per ring, while the maximum MS is theoretically unlimited.

The most common non-hydroxyl-containing derivatives are methoxyl, ethoxyl, and carboxymethoxyl cellulose (commonly referred to as methyl cellulose(MC), ethyl cellulose (EC), and carboxymethyl cellulose (CMC), since the oxygen in the ether linkage was present in the cellulose before derivatization). The most common hydroxyl-containing substituents are hydroxyethoxyl and hydroxypropoxyyl (called hydroxyethyl cellulose (HEC) and hydroxypropyl cellulose (HPC). Small amounts of C6 or higher chains are sometimes added to modify the rheology of hydroxyethyl cellulose via hydrophobic interactions between the solvated chains (HmHEC for hydrophobically modified HEC). All of these substituents or combinations of them (HPMC and HEMC are the most common) result in non-ionic polymers except for the carboxymethyl products.

BASIC MANUFACTURING PROCESSES

Regardless of the exact derivative being manufactured, all processes require an activation step to prepare the cellulose to be reacted. For most products this is done by treating the ground or sheeted cellulose with concentrated NaOH. For derivatives where the NaOH is a stoichiometric reagent, the amount of NaOH will depend on the target substitution. For derivatives where the NaOH is only catalytic (principally HEC), the cellulose is activated with NaOH, most of which is then neutralized before adding the reagent (ethylene oxide), in order to maximize the yield of the oxide. The non-hydroxyl-containing substituents are added by reacting the alkali-cellulose with methyl chloride, ethyl chloride, or sodium chloroacetate. Hydroxyl-containing substituents are formed by adding ethylene or propylene oxide.

After reaction, most products are purified to different degrees by washing with hot water or organics to remove the salt and organic by-products. Carboxymethyl cellulose, on the other hand, is sold mostly as the crude reaction product. The purified products are then dried and ground to achieve the desired particle size. The products are sold in granular, fine granular, or powder form.

PROPERTIES OF CELLULOSE ETHERS

All of the water-soluble, non-ionic derivatives except hydroxyethyl cellulose exhibit inverse temperature solubility, being soluble in cold water and insoluble in hot water. The gelation or precipitation temperature is a function of the type and amount of the substituent. Methyl cellulose has the lowest gelation temperature of the soluble derivatives, ethyl cellulose is only organo-soluble, and combinations of methyl, ethyl, hydroxyethyl, and hydroxypropyl ethers gel or precipitate at various temperatures, depending on the type and amount of each substituent. The strength of the gel formed, if any, and the temperature of its formation are key properties of many cellulose ethers, particularly with respect to ceramic applications. The effect of various substituents on the water and organic solubility and the gelation or precipitation temperature is shown in Figure 2.

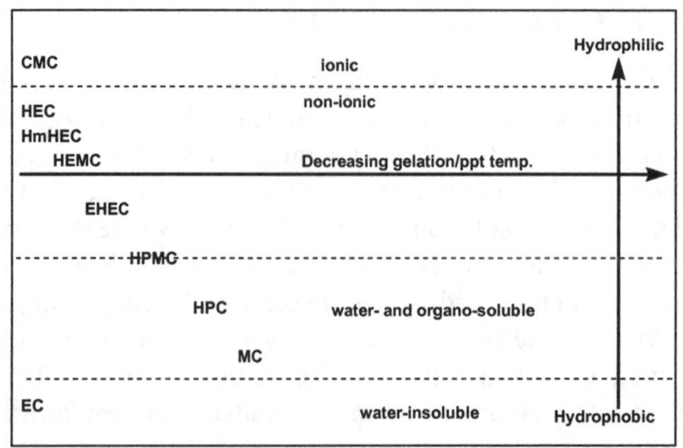

Figure 2. Effect of substituent on solubility and gelation temperature.

The numerous possible products that can be made by mixing the various types of substituents can be grouped into four basic categories: 1) the organo-soluble only (EC), 2) the water- and organo-soluble, thermally-gelling or precipitating derivatives (HPC, MC, HPMC, HEMC, and EHEC), 3) the water-soluble non-gelling types (HEC and HmHEC), and the ionic types (CMC). The most common commercially available products are described in Table I, which shows typical substitution levels for the various derivatives, which contain one or two types of substitutions.

Table I. Typical substitution levels of commercially available cellulose ethers.[1]

Product	R1 Molar Substitution	R2 Molar Substitution	Key Properties
CMC	0.4-0.8	-	Hydrophilic; low cost
HEC	1.6-3.0	-	Hydrophilic; non-gel
EHEC	0.8-1.0	0.6-0.9	Mixed properties
HmHEC	<0.5	1.6-3.0	Hydrophobic interaction; non-gel
HEMC	0.1-0.4	1.7-2.1	Hot-water insoluble; low gel strength
HPMC	0.05-1.5	1.3-2.1	HWI; weak to moderate gel strength
HPC	3.5-4.2	-	HWI; no gel strength; organosoluble
MC	1.8-2.3	-	HWI; Strongest gel; Lowest gel temp
EC	1.5-2.5	-	Organosoluble only

The particular combination of substituents determines the balance of properties present in that product, which determines what sort of applications the product is suited for. The key characteristics of each polymer are given in the last column of the table. These are by no means the only characteristics, but they are the ones which usually determine whether that product will be used in a given application. (An equally important characteristic is the solution viscosity or molecular weight, a broad range of which is available for all the different products).

FUNCTIONS OF CELLULOSE ETHERS IN CERAMICS

Cellulose ethers exhibit a number of properties which are useful in the manufacture of ceramic articles, depending on the type and amount of substituent added to the cellulose backbone. All cellulosics provide binding, water retention, pseudoplastic rheology, suspension or emulsification, and lubricity to varying degrees. Cellulosics containing hydrophobic groups such as methyl, ethyl, or hydroxypropyl can behave as surfactants, lowering the surface tension of formulations. The methyl-containing ethers have a unique function in ceramics in their ability to form a gel with significant yield strength at various temperatures(shown as an increase in mixing torque in Figure 3). The main utility of this function is to add green strength during drying. Formation of the gel

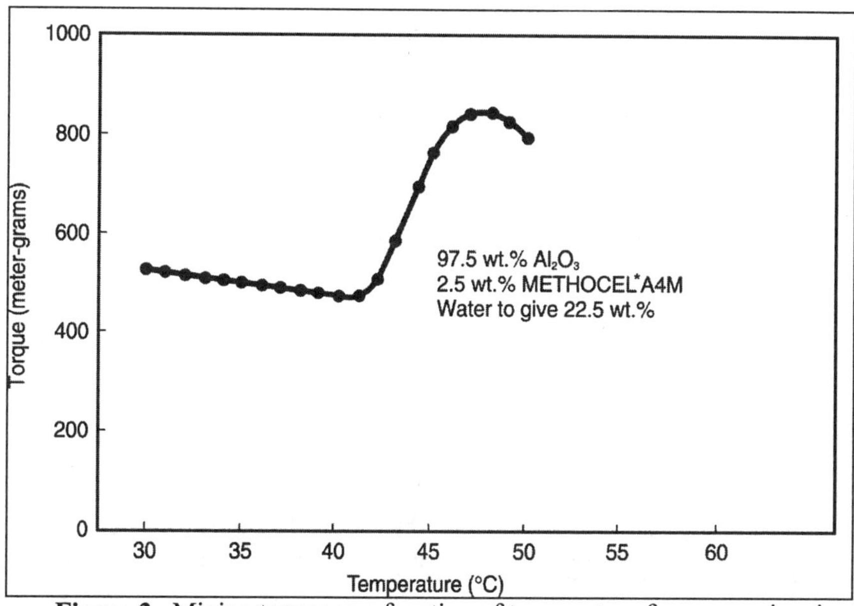

Figure 3. Mixing torque as a function of temperature for a ceramic mix containing MC.[2]

during heating also helps moderate water loss, thereby reducing cracking and blistering. As a result of these properties, MC and HPMC have found extensive use in extruded ceramics such as automotive catalytic converter honeycomb substrates.

All non-ionic cellulose ethers also provide clean burn-out during firing. The rate of burn-out depends on the temperature, as shown by the time to 50% wt. loss of cellulosic during a differential thermal analysis in Figure 4. Actual rate of burn-out will depend on the particular ceramic formulation.

---

*METHOCEL is a trademark of The Dow Chemical Company. A4M is the designation for MC with a 2% aqueous solution viscosity of 4,000cp.

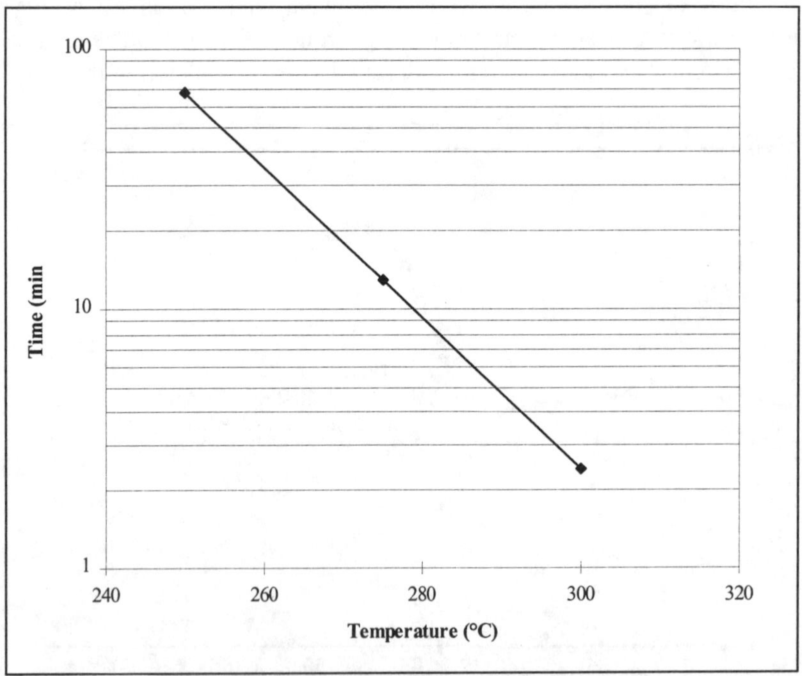

Figure 4. Differential thermal analysis of METHOCEL.

The balance of properties provided by cellulose ethers renders these polymers useful in a variety of related ceramic applications, in addition to extrusion. These applications are listed, along with the types of properties provided by MC and HPMC, in Table II.

Table II. Benefits of MC and HPMC in Ceramic Applications.[2]

Application	Typical Benefits
Ram Extrusion	Precise control of viscosity and rheology increases stability of mixes. Permits broader operating ranges.
Screw Extrusion	Lubricity reduces extruder and die wear and promotes smoother surfaces. Thermal gelation permits extrusion of extremely delicate, thin-walled shapes without sagging or deformation.
Glazes	Improved control of viscosity and rheology.
Porcelain Enamel	Glazes bond tightly to greenware. Smoother finishes. Unfired glazes are tougher. Fires out completely in kiln.
Injection Molding	Uniform and rapid mold filling. Smoother finishes. Higher green densities and better green strength.
Dry, Semi-Dry, and Isostatic Pressing	Provides optimum grain lubrication for tighter, more uniform packing. More predictable green densities, less shrinkage during firing, higher green strength. Reduces die wall friction. Fewer rejects due to warping, distortion, or cracking during firing. Improved workability and application properties. Low ionic salt residues.
Refractory Mortars	Improved workability and application properties. Low ionic salt residues.
Tape Casting	Provides excellent control over viscosity and inhibits settling in slips. Lubricity and surface activity permit tighter particle compaction for greater green strength and less shrinkage during firing. Thermal gelation provides added strength and stability during drying and early stage firing.

SUMMARY

Cellulose ethers provide many useful properties with applications in ceramics. The choice of cellulose ether depends on which balance of properties is the best fit for the application being developed. Key attributes to be considered are molecular weight or viscosity, ionic or non-ionic, water or organic solubility, and thermal gelation or high temperature solubility. All non-ionic ethers provide good burnout, but the best drying control and green strength are provided by ethers containing the methyl substituent.

REFERENCES

[1]Data in this table was compiled from the *Handbook of Water-Soluble Polymers*, edited by R. L. Davidson. McGraw-Hill, New York, NY, 1980.

[2]"METHOCEL Cellulose Ethers, Binders and Processing Aids for Ceramics" bulletin, Form No. 192-00997-396GW, The Dow Chemical Company, March 1996.

COMPARISON BETWEEN POLY(VINYL ACETATE) AND POLY(VINYL ALCOHOL) AS BINDERS FOR WHITEWARE BODY COMPOSITIONS

F. Andreola, M. Romagnoli
University of Modena
Via Campi, 183
41100 Modena-Italy

W. M. Carty
NYS CACT-Whiteware Research Center
New York State College of Ceramics
Alfred University, 2 Pine street, Alfred, NY 14802

ABSTRACT

Binders are used to allow easy handling of green bodies and to avoid spring-back. This work compares the effects of one industrial poly(vinyl acetate) and three poly(vinyl alcohol)s, of different molecular weights, on the suspension rheology, pressing behavior, ease of scrap recycling, and thermal decomposition in a typical porcelain dinnerware body.

INTRODUCTION

Organic additives have long been used in the processing of ceramics as deflocculants, dispersants, binders, etc. Organic binders are often polymers whose main role is to improve the green body strength to allow the handling of the product before firing. This operation can be extremely delicate in the case of large ceramic tiles (60x60 cm or more). The addition of a binder is also necessary to avoid a specific phenomenon of pressing, commonly known as "spring-back," or the elastic re-expansion of pressed bodies soon after shaping.[1] The economically driven need for increased production rates of pressed bodies has naturally made spring-back an increasing problem. The binder acts in opposition to the elastic recovery that is caused by two mechanisms: the sudden release of the elastic energy stored in the solid materials, and the re-expansion of the compressed air trapped in the pores.[1,2] The immediate expansion is particularly dangerous because it leads to micro-cracking, which weakens the body, potentially leading to cracking and warping during the firing stage.

Two particular cases where spring-back is of concern are the production of dinnerware by isostatic pressing and the uniaxial pressing of large tiles. Automated isostatic pressing is a relatively new forming technology, but is becoming more and more important for the efficient manufacturing of high quality tableware and tile products.[3] Compared with the traditional forming methods, which themselves have reached a high level of sophistication, isostatic pressing shows strong economic and technological advantages.[4]

Poly(vinyl acetate) (PVAc) and poly(vinyl alcohol) (PVOH) are among the most popular binders used in the traditional ceramics industry. This paper compares the performance of PVOH to a commonly used commercial PVAc. The evaluation is carried out from several points of view: the influence of the binder on the rheological characteristics of the slurry; the tendency to crack during pressing; the potential and ease of scrap recycling; the binder decomposition on heating; and the residual porosity in the fired body.

BINDER CHARACTERIZATION

PVAc is obtained by the polymerization of the vinyl acetate monomer, while the PVOH is produced by hydrolysis of the polyvinyl acetate because monomeric vinyl alcohol cannot be obtained in quantities and purity sufficient to make polymerization of PVOH feasible.[5] Both binders possess a moderate affinity for oxide particles in water.[6] PVOH has a lower percentage of acetyl groups and the degree of hydrolysis is decisive for the application and technological properties.[7,8] In practice, approximately 12 to 20% of the –OH side groups are replaced by acetate groups to make it soluble in water.[9] The solubility of PVAc is high below 60°C and under strong agitation.

During pressing, the binders improve the strength of the as-formed product by forming a polymer-polymer bonded film between the particles.[6] High green strength is achieved by either the wetting of the grain surface by the surface-active behavior of this binder type, or by the binder migration to the surface of the pressed body.[10] Among the disadvantages of binder migration is increased granule hardness that may prevent rupture and flow during the pressing operation leaving voids and density gradients. Moreover, the PVOH and PVAc tend to foam during milling, often requiring the use of de-foaming additives.

EXPERIMENTAL PROCEDURE

The base slurries were prepared using a typical white porcelain tableware formulation. A low- (degree of polymerization: 800); medium- (degree of polymerization: 1600); high- (degree of polymerization: 2000) viscosity grade partially hydrolyzed (85-90% molar) PVOH and an aqueous industrial PVAc solution were used as binders. They were added to the slurry in quantities corresponding to 0.5, 1.0, and 1.5 % d.w.b. (dry weight basis) together with 0.05% (d.w.b.) poly(ethylene glycol) plasticizer. Sodium metasilicate was used as a deflocculant, consistent with industrial practice. The rheology measurements were performed using a controlled shear rate rotational viscometer (Haake VT550) with a coaxial cylinder measuring system (Searle). The slurries were spray-dried in a semi industrial counter-current spray-dryer. The granulate size distribution, determined using several sieves, ranged between 180-500 μm and the moisture content was about 2 wt% (d.w.b.).

For each slurry/binder mixture, 100 plates (30 cm diameter and ~610 g) were prepared by isostatic pressing under industrial conditions. Three-point breaking load testing (Gabrielli) was performed following the European standard EN 100. The possibility to recycle the flash derived from the isostatic pressing process and damaged material before firing was determined by preparing water suspensions of the rejected material at a 50 wt% level. After grinding in a ball mill for 20 and 40 minutes, the dry residue on a 180 μm sieve was measured. The carbon content was determined by elemental analysis (CE instruments) on samples heated to 200, 350, 450, and 700°C. The porosity of the samples was measured using a mercury porosimeter (Micromeritics).

RHEOLOGICAL BEHAVIOR

To reduce the energy cost during the spray-drying of the suspensions, it is necessary to maintain a high solids loading; for pumping, a low viscosity is necessary, requiring the use of dispersants. In the production of tableware and tiles, sodium metasilicate is commonly used. It allows the preparation of suspensions with a solid content up to 65-67 wt% that are flowable. The addition of PVOH or PVAc increases the viscosity of the slurry compared to the suspension containing only dispersant, as illustrated in Figure 1. The suspension exhibit shear rate dependence with apparent viscosity decreasing with increasing shear rate (i.e., shear rate thinning). PVOH or PVAc increase shear rate thinning behavior.

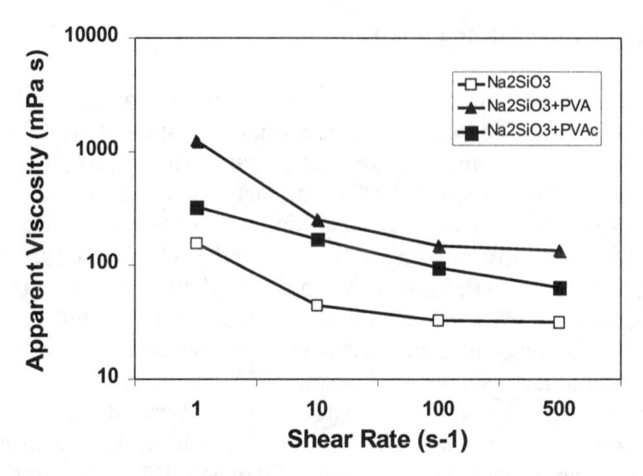

Figure 1. Apparent viscosity versus shear rate for the slurries with and without binders.

The increase in viscosity can be interpreted as the interaction between binder and the suspended powders. The binder can adsorb onto the surface of the particles, forming bridges and entanglements among the polymer chains on two near solid particles. Based on this classical model, the larger the degree of complexity of the network and the intensity of interaction forces, the higher the apparent viscosity. The decrease of the apparent viscosity with the increase in shear rate is caused by the flow conditions inside the viscometer that breaks the continuity of the network. At high shear rates, the polymer chains tend to align with the shear direction, thus reducing the viscosity. When the shear is reduced or stopped, the polymer chains get back a random distribution with a consequent more increasing of the viscosity with respect to the slurry without binder.

The higher viscosity of the slurry containing PVOH, compared to the slurry containing PVAc, can be explained with a higher amount of poly(vinyl alcohol) adsorbed on the particle surface. In fact, the amount of adsorbed polymer is proportional to the degree of hydrolysis. The molecular weight of the binder molecules influences the rheological properties of the slurries, as shown in Figure 2. The flow curves (shear stress vs. shear rate), relative to the slurries prepared with the same percentage of poly(vinyl alcohol), show an increase in the apparent viscosity with an increase of the molecular weight.[10] A longer molecular chain allows loops and tails of greater sizes so a more intense phenomenon of the "bridging" takes place. An analogous effect can be observed with the increase of the binder concentration.

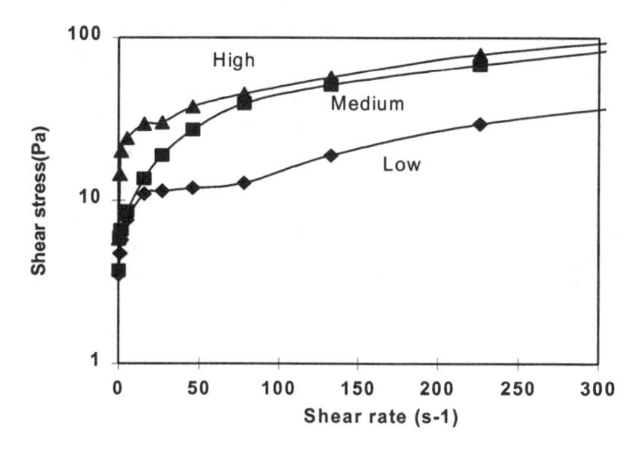

Figure 2. Effect of the poly(vinyl alcohol) molecular weight on the apparent viscosity.

PRESSING EFFICIENCY

In order to estimate the efficiency of the binders, the number of samples obtained per minute without fracture was determined, as shown in Figure 3, using an industrial isostatic press. Statistics were obtained on 100 pressed bodies for each binder and at each binder concentration. The binders at medium and high viscosity grade showed a lower efficiency at a concentration of 1.5%. They resulted in, respectively, a production rate of 6 and 5 pieces per minute with a 20% damage rate. For the other binders, the result was 7.2 pieces per minute without damage. Figure 4 shows the data obtained by the breaking load test.

The modulus of rupture shows a close dependence on the binder percentage, the binder formula, and the molecular weight. The values for bodies containing PVOH are higher than those prepared with PVAc. There is also an observed increase of the mechanical proprieties of the pressed bodies with an increase in binder content. The results demonstrate that 0.5 wt% of low molecular weight PVOH is sufficient to permit handling during the process.

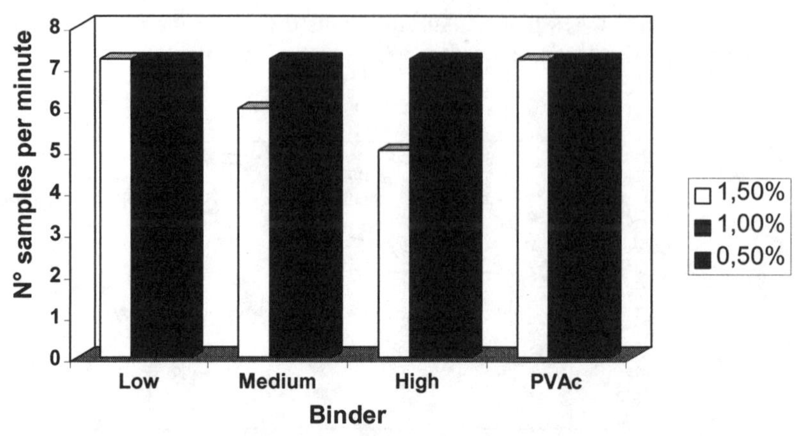

Figure 3. Industrial test: number of pressed samples per minute.

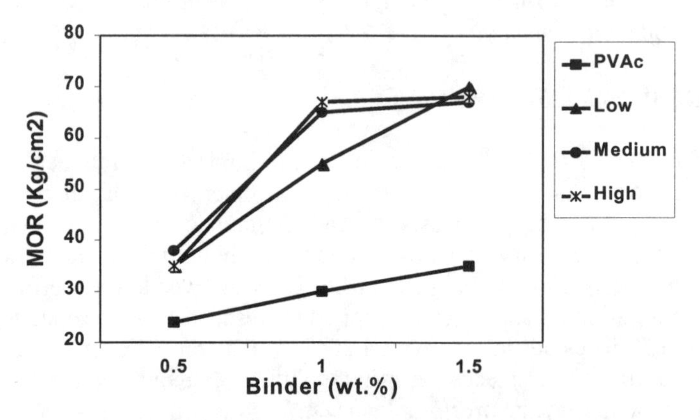

Figure 4. Modulus of rupture for the green bodies.

The high number of broken samples produced from binders with medium and high molecular weights can be explained by the elevated hardness of the spray-dried powders. A portion of the granulates are not destroyed and when the pressure decreases, releasing stored elastic energy and leading to breakage. The phenomenon increases with increasing binder percentage, so that the concentration of 1.0 wt% represents a limit value for a production rate of about 7 pieces per minute. Another phenomenon that contributes to the breaking of the pieces during the pressing operation is represented by the adhesion of the sample surface to the pressing die. The adhesion increases with the presence of the binder on the surface of the pressed body. For this reason, the increase of the polymer concentration contributes to an increase in the number of broken samples. In the case of 1.5 wt% PVOH with low molecular weight, a smaller adhesive ability with respect to the mould can be assumed.

RECYCLING TEST

During the industrial process, about 15-20 wt% of green pressed material is rejected due to the finishing stage and from material handling issues prior to firing. This high rate of defective units is costly and generates a large amount of waste. Recycling the defective units can reduce the cost associated with loss and scrap generation. The re-introduction of the material into the slurry during the milling step and its reconstitution as a suspension is probably the best method of recycling. In this step, the binder must not be a handicap to the dispersion of pressed body and spray-dried granules. In order to check the recycling potential of bodies prepared with different additives, a simple laboratory test was used and the results are reported in Figure 5. With the lowest viscosity-grade PVOH, the pressed body disintegrated easily and the residue after ball milling for 20 minutes is strongly reduced while after 40 minutes the residue is essentially eliminated. The increase of binder chain length worsens the grinding behavior. The binder with the highest degree of polymerization shows a minor disruption and therefore prevents recycling. Samples with PVAc show a behavior between the binder with the lowest viscosity-grade and the other two ones. For all the binders, the increase of up to 1.5 wt% inhibits recycling efforts.[11]

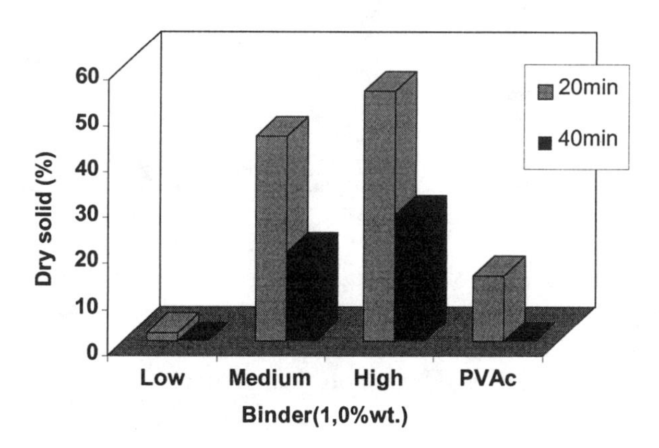

Figure 5. Results of the recycle test on samples with 1.0% of binder.

FIRED PRODUCTS

The determination of the elemental carbon percentage was assumed to be an indirect measure of the amount of binder (or products of its burnout) within a sample.[12] Using PVOH it was possible to observe a relationship between the thermal decomposition and molecular weight, as it is shown in Figure 6. With the increased backbone length, thermal decomposition requires greater temperature. This phenomenon is caused by diffusion limiting the oxidation of organic matter within the porous body structure. In an industrial cycle, with a heating rate higher than that used in the laboratory tests, the decomposition range shifted to higher temperatures. The PVOH with the lowest molecular weight shows efficient burnout, very similar to the PVAc. The total open porosity of the fired bodies is influenced by the molecular weight as well. As illustrated in Figure 7, porosity increases with degree of polymerization.

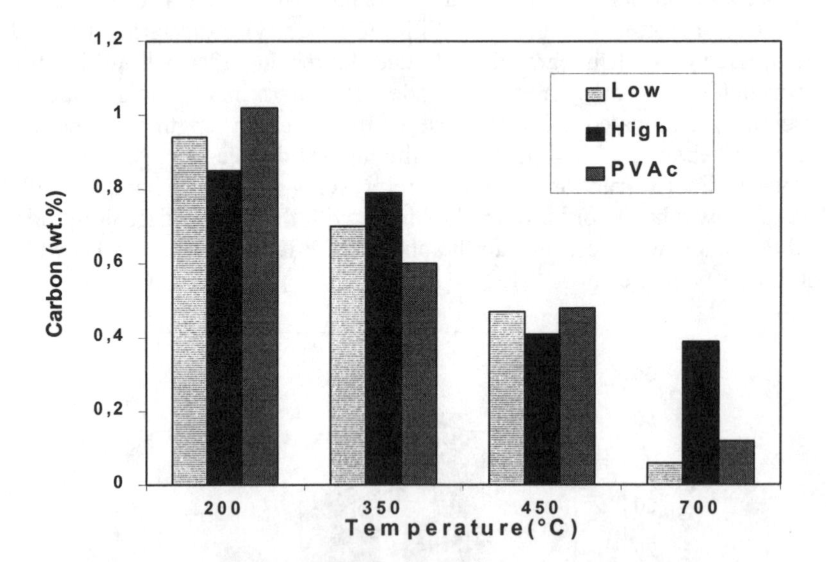

Figure 6. Carbon content at different temperatures for the samples with poly(vinyl acetate) and the two poly(vinyl alcohol)s at low and high molecular weight.

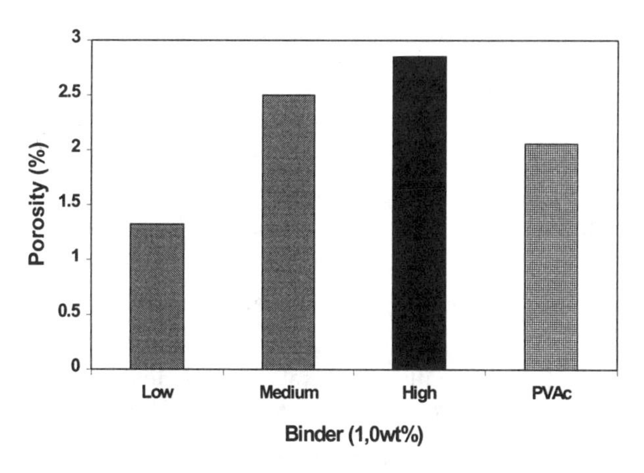

Figure 7. Open porosity of samples fired at 1240°C.

CONCLUSIONS

The addition of binders to a whiteware composition is necessary to prevent the phenomenon of "spring-back " and to increase the green body strength. Poly(vinyl acetate) is one of the binders traditionally used in industry. During the production cycle, a large amount of material produced in the phase that precedes the firing must be recycled. In this phase, the poly(vinyl acetate) can cause some disadvantages, such as the formation of spray-dried granulates hard enough to avoid recycling inside the slurry. A partially hydrolyzed poly(vinyl alcohol) is a good alternative. It increases the slurry viscosity more than poly(vinyl acetate), but permits a more efficient re-dispersion of the scraps, increases the mechanical properties of the pressed bodies and improves the burnout. The industrial tests have indicated the percentage of 0.5 wt% of poly(vinyl alcohol) with low molecular weight as the best concentration and binder.

REFERENCES

1. R. Sladek, "Die problematik des Teller-stapelbrandes im gluhbrand bei isostatisch geprebten tellern," *DKG*, **3**, 67 (1990).
2. A. Geigle, K. Hauswurz, and S. Mager, "Optimization of spray-dried granulates for isostatic pressing of tableware," *Interceram*, **42** [1], (1993).
3. V. Ramakrishnan, "Isostatic dry pressing of tableware," *Interceram*, 35 [3], 48-51, (1986).
4. R. Bartusch and W. Schulle, "Special process features of isostatic pressing of whiteware bodies," *Interceram*, **41** [4], (1992).
5. K. Othmer, *Encyclopedia of Chemical Technology*, 4th Ed., **24**, pp.980, John Wiley & Sons.
6. J. S. Reed, *Principles of Ceramics Processing*, 2nd Ed., Chapter 11, John Wiley & Sons, 1995.
7. R. Bast, "Organic Additives for Dry Pressing," *Interceram,* **39** [6], 1990.
8. E. H. Klingenberg, "Poly(Vinyl Alcohol): Important Properties Related to its Use as a Temporary Binder in Ceramic Processing," this volume.
9. E. Carlstrom, "Surface Chemistry in Dry Pressing," in *Surface and Colloid Chemistry in Advanced Ceramics Processing*, Marcel Dekker p.258, 1994.
10. X. K. Wu, D. I. Arnold, W. L. Kaufell, and D. W. Whitman " Comparison study of Polyvinyl alcohol, Polyethylene glycol and Polyacrylic binders for dry pressing ceramics," presented at the *95th Annual Meeting of the American Ceramic Society*, 1994.
11. F. Andreola, M. Romagnoli, *et al.*, "Organic Additives in Body Compositions for the Production of White Tableware by Isostatic Pressing," *Fourth Euro Ceramics,* **10**, 21-29, Faenza Editrice, 1995.
12. F. Andreola, M. Romagnoli, *et al.*, "Sintering and Modification of Porous Structure Caused by Binders Added to Whiteware Body Composition," *Ceram. Eng.Sci.Proc.*, **18**[2] (1997).

Characterization

MINERALOGY OF CERAMIC GRADE KAOLINS FROM THE SOUTHEASTERN UNITED STATES

J. Elzea [1], J. Bruns [2] and S.Rice [3]

[1] Thiele Kaolin Company, 520 Kaolin Road, Sandersville, GA 31082
[2] Albion Kaolin Company, 1 Albion Road, Hephzibah, GA 30815
[3] McCrone Associates, 850 Pasquinelli Drive, Westmont, IL 60559

ABSTRACT

Kaolin is added to ceramic compositions to control slip rheology, casting rate, final product porosity and fired body color. It is also used in glazes to impart stability and improve adhesion. The mineral component primarily responsible for these properties is kaolinite, which is an alumino-silicate formed from the decomposition of feldspathic rocks. Much of the white-firing, non-plastic kaolin used in ceramic manufacture in the United States comes from an extensive series of deposits located in South Carolina and Georgia. One of the most important factors determining the ceramic properties of these kaolins is their mineral composition. Chemistry, particle shape, particle size, surface area and impurity mineral content all profoundly influence quality.

With the availability of sophisticated analytical tools including X-ray diffraction (XRD), transmission electron microscopy (TEM), BET surface area analysis and laser diffraction particle size analysis, raw material suppliers now have the ability to carefully characterize their clays and closely monitor the effects of processing on fundamental properties. This paper presents a detailed mineralogical profile of a series of ceramic grade kaolin products and raw materials from the Southeastern United States. The objective will be to 1) discuss the mineralogical properties of these kaolins, 2) demonstrate variability in mineralogy and 3) relate mineralogy to ceramic properties.

INTRODUCTION

The kaolin mining district in Georgia and South Carolina is part of a northeast-southwest trending belt of deposits that closely parallels the fall line which marks the boundary between older igneous and metamorphic rocks to the northwest and younger coastal plain sediments to the southeast (Figure 1). The kaolins occur in Cretaceous and Tertiary age strata as lenses and 3 to 15 m thick horizontal beds that are a few hundred meters to several kilometers in length. They were derived from kaolinite-rich detritus weathered from feldspathic crystalline rocks and transported to the coastal plain.[1] After deposition, the sediments underwent additional cycles of weathering. Weathering was largely responsible for producing essentially monomineralic commercial-grade kaolins.

This clay type has undergone in-situ recrystallization, which resulted in the formation of large vermiforms and the coarsening of individual kaolinite crystallites. Hard kaolins are restricted to Early to Middle Eocene age marine-strata and appear to have formed as a result of flocculation in salt water which caused tightly packed face-to-face association between submicron sized clay platelets (Figure 3). Vermiforms are scarce in the hard kaolins, which have undergone little in-situ recrystallization, at least on a macroscopic scale. TEM reveals micro-scale recrystallization where there has been repeated or continuous long-term oxic weathering.[2] A common recrystallization sequence is the transformation of halloysite, a kaolinite polymorph, to kaolinite (Figure 4).

Figure 2. Scanning electron micrograph of the coarse fraction of Middle Georgia type crude showing kaolinite stacks.

MINERALOGY OF COMMERCIAL GRADE KAOLINS FROM GEORGIA

Kaolin is a light colored clay comprised mostly of kaolinite. The term derives from the Chinese word, *Kauling*, which means high ridge, and refers to the site where this mineral was first mined over 3000 years ago.[3] Originally

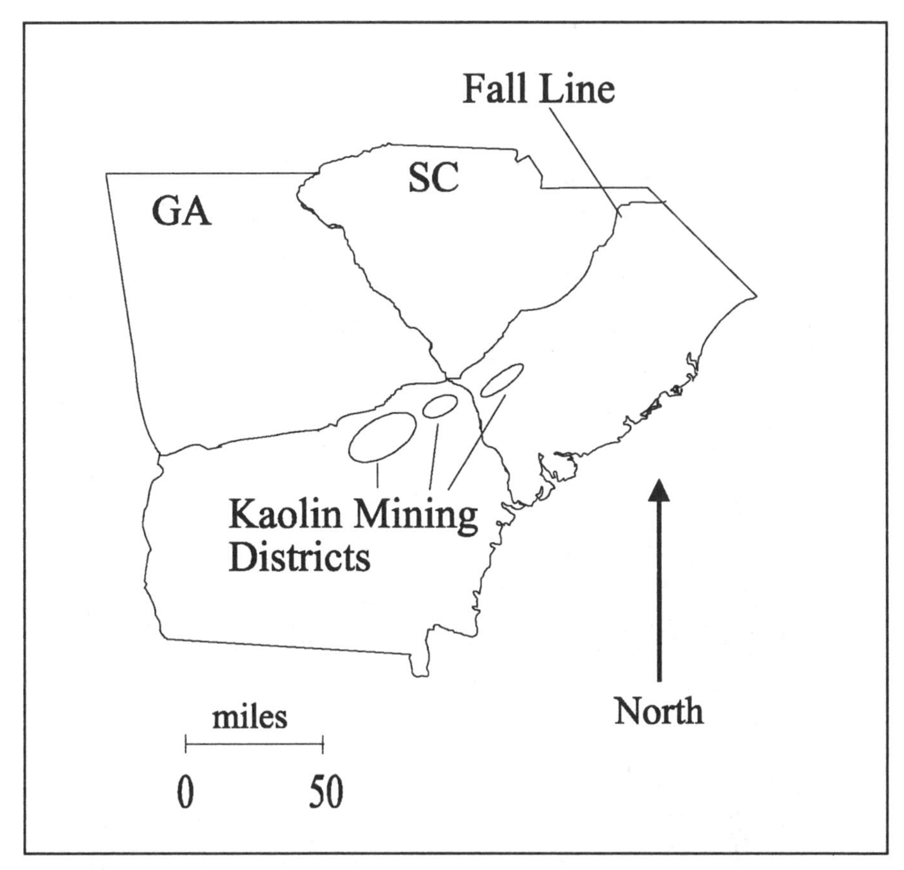

Figure 1. Map showing location of major kaolin mining districts in Georgia and South Carolina.

There are three types of kaolin in Georgia: hard, soft and flint. The soft kaolins are relatively coarse, have a soapy texture, and break easily with a conchoidal fracture. The hard kaolins are finer grained, have an earthy texture, and are more difficult to break. The fracture surface of hard kaolins is typically hackly. Flint kaolins are harder than the hard kaolins and closely resemble flint in appearance and texture. The flint-like quality of these clays is due to silica. These siliceous kaolins occur in very localized areas, often as small (1 to 2 m) pod-shaped bodies at the top of beds of hard kaolin.

In general, the soft kaolins are found in Cretaceous and Paleocene age strata. These kaolins are best described as mixtures of interlocking fine to coarse euhedral kaolinite platelets and larger kaolinite stacks or vermiforms (Figure 2).

Figure 3. Scanning electron micrograph of an East Georgia type kaolin showing face-to-face packing of kaolinite crystallites.

Figure 4. Transmission electron micrograph of the recrystallization of halloysite tubes (left) to elongate kaolinite crystals with sawtooth edges (right).

kaolin was primarily used as a raw material for porcelain. Now it has many industrial applications. Paper coating accounts for the largest portion of the domestic market. Approximately 75% of the kaolin produced in Georgia is sold as a raw material for paper coating and filler compared to the approximately 10%

that goes into ceramics. Ceramic grade kaolins are mined from soft and hard kaolins.

Kaolinite

Kaolinite is comprised of stacked sheets of silica tetrahedra linked to alumina octahedra in a one-to-one ratio. The ideal chemical composition is 39.5% Al_2O_3, 46.5% SiO_2 and 14.0% H_2O. Small amounts of iron and titanium substitute for aluminum in the octahedral sheet. Aluminum also substitutes for silicon in the tetrahedral sheet. These substitutions result in crystal imperfections and create a negative surface charge, which gives kaolinite a small cation exchange capacity.

Two major types of kaolinite are recognized in the Georgia deposits. These are low defect and high defect types (Figure 5). The low defect kaolinite is also referred to as high Hinckley Index kaolinite, or well-crystallized kaolinite, and is typically found in the soft kaolins. This type of kaolinite is low in structural iron content, tends to be coarser in particle size, and often occurs as large vermiforms. The high defect kaolinite, which is also known as low Hinckley index kaolinite or poorly crystallized kaolinite, is finer in particle size and has a higher structural iron content. High defect kaolinite is the principal component of the hard kaolins.

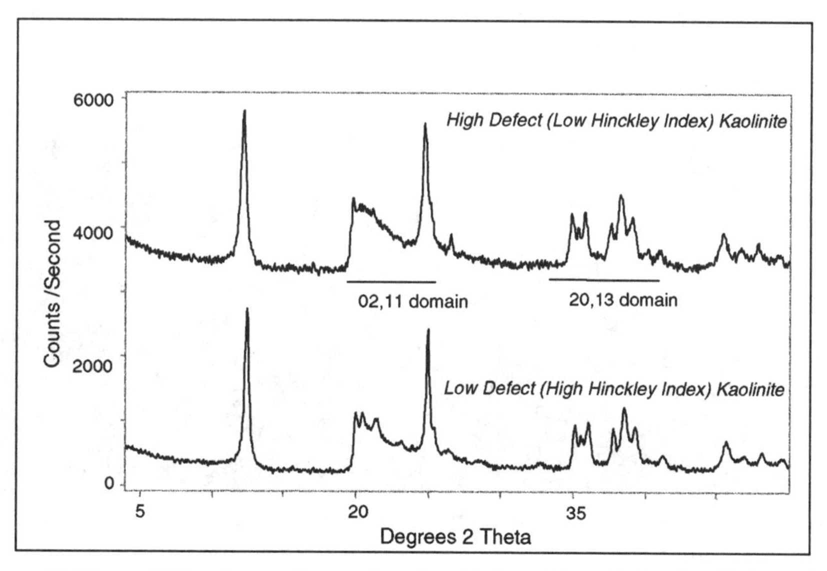

Figure 5. X-ray diffraction patterns showing high and low defect kaolinite. Note the differences in the 02,11 domains of the two types of kaolinite.

Common Impurity Minerals

The Georgia kaolin deposits contain trace quantities of impurity minerals that, despite their low concentration, strongly impact quality. Many of these minerals are removed during processing or are avoided by selective mining. Some impurity minerals, such as smectite, are considered advantageous for certain ceramic applications.

Mica/Illite: Mica/illite is ubiquitous, occurs in all size fractions and is the major source of potassium in the kaolins. The coarse muscovite and the finer illite found in these deposits are predominantly dioctahedral $2M_1$ polytypes. The illite, which is a degradation product of coarser detrital mica, has a chemical composition between muscovite and dioctahedral vermiculite and contains 2 to 8% K_2O [4]. Quantities of these minerals range from 5 to 25%.

Smectite: Smectite is found in both the hard and soft commercial-grade kaolins in quantities ranging from less than 5% up to approximately 40%. The smectite is best characterized as a 15 to 17Å montmorillonite with Ca and Mg as the principal exchangeable cations.[5] This montmorillonite has a high layer charge and is often concentrated in the fine fraction of the kaolins. In some cases it is attached to edge or basal surfaces of kaolinite (Figure 6). It also occurs as discrete particles forming a matrix around kaolinite. This mineral increases plasticity and improves green strength.

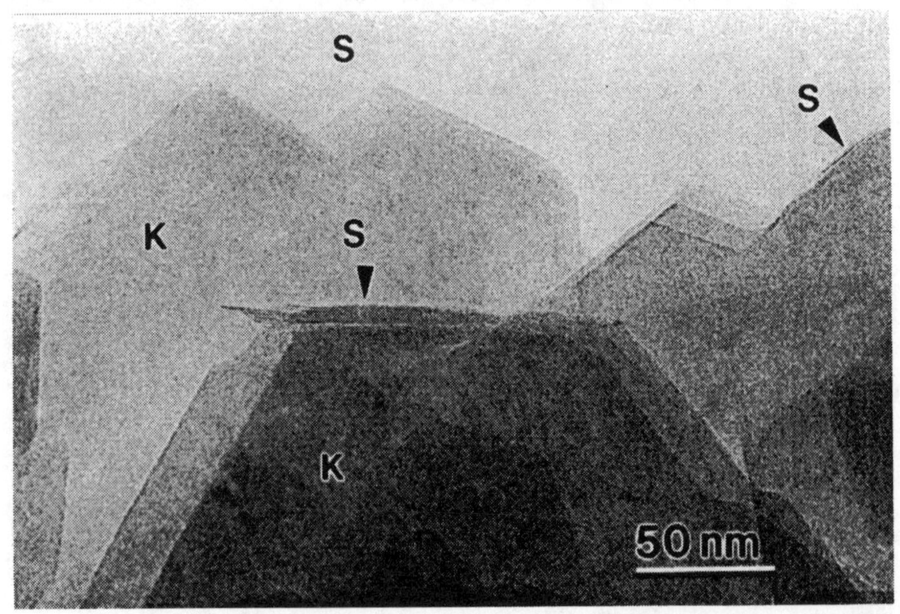

Figure 6. Transmission electron micrograph showing smectite (S) on the edge of kaolinite (K).

Figure 7. Transmission electron micrograph depicting stellate growth habit of goethite.

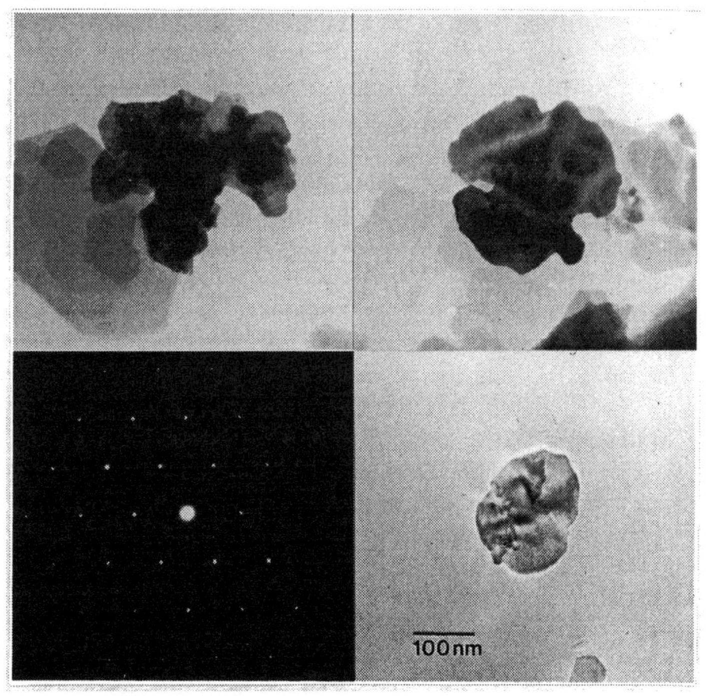

Figure 8. Transmission electron micrographs of hematite.

Fe and Ti Oxide Minerals: Iron oxide minerals account for approximately 2 to 30% of the total iron in Georgia kaolins. The remaining iron is in the kaolinite structure. These minerals discolor the raw material as well as the fired body. Goethite (FeOOH) is the most common iron oxide mineral in the commercial grade kaolins. It occurs as coatings and discrete colloidal crystallites that typically form single laths or stellate intergrowths (Figure 7). Hematite (Fe_2O_3) is less abundant. It tends to be coarser than goethite and most often forms subhedral to anhedral pseudohexagonal crystallites that range from 0.05 to 0.2 microns in diameter (Figure 8).

Both hematite and goethite contain trace quantities of transition metals including Ti and V. Goethite causes a yellow discoloration and hematite causes a pink discoloration in the crude kaolin. Because goethite is thermally unstable, it converts to hematite when fired resulting in a pink product. This transformation takes place at approximately 350^0 C.

The TiO_2 minerals, which are discolored due to trace quantities of Fe, comprise 1 to 2% of the soft kaolins and 1 to 3% of the hard kaolins. These minerals may also control fired color. Anatase, like goethite, is ubiquitous in the kaolins, and is generally submicron in size. Besides Fe, anatase contains trace quantities of Cr, Zr, and Nb.[6] Typical crystal morphology is submicron euhedral pseudocubic crystals and coarser equant cube-shaped skeletal crystals (Figure 9). Rutile is less abundant and occurs as much larger detrital grains.

Iron Sulfide Minerals: Pyrite and marcasite are most prevalent in hard kaolins. The pyrite exhibits a range of morphologies from very small, disseminated, typically octahedral crystals and framboids to large masses of cubo-octahedral to cubic crystals.

Heavy Minerals: Heavy minerals reported in the kaolins include ilmenite, tourmaline, zircon, epidote and kyanite. These minerals are inherited from the original feldspathic source rocks that altered to form the kaolins. These heavy minerals occur in trace quantities that are typically below XRD detection limits.

Opal Minerals: Opal-CT is found in flint kaolins as well as occasionally in the commercial grade kaolins. The opal-CT cementing the flint kaolins has been studied in detail by XRD and high resolution TEM.[7] It is recognized by a broad X-ray diffraction maximum centered at 4.1Å and is poorly ordered due to interlayered tridymite-like stacking sequences.

Organic Matter: Commercial grade kaolins contain from 0.01 to 0.08% organic matter. In general, the hard kaolins are more organic-rich than the soft kaolins. Little is known about the nature of the organic matter in these clays.

PHYSICOCHEMICAL PROPERTIES

Critical qualities monitored for kaolin used in whiteware manufacture are casting properties and fired color. In order to demonstrate the relationship between these properties and mineralogy, five ceramic products derived from Georgia kaolin were analyzed using a variety of techniques (Table I).

Table I. Physicochemical characterization of Georgia kaolins (A = anatase, G = goethite, H = hematite, M = mica, S = smectite)

Properties	Product 1	Product 2	Product 3	Product 4	Product 5
wt% Fe_2O_3	0.61	0.57	0.26	0.32	0.20
wt% TiO_2	0.32	1.59	1.63	1.68	1.35
wt% K_2O	0.28	0.15	0.07	0.07	0.07
%<10 micron	86.1	91.1	92.4	90.7	97.1
%< 2 micron	52.5	65.3	58.4	57.7	74.1
% < 1 micron	42.9	55.4	45.8	45.6	60.5
Aspect Ratio (D/t)	21.2	16.7	13.3	14.5	9.3
BET Surface Area (m^2/g)	20.2	16.8	10.5	9.1	33.5
MBI	4.1	4.4	2.3	1.9	13.2
Fired Color:					
Brightness	77.4	76.1	79.8	76.5	84.1
L	92.0	91.7	93.2	92.2	90.3
a	1.7	1.3	1.3	1.8	0.2
b	5.9	6.6	5.8	7.0	5.0
Casting Rate (g)	79	na	119	123	na
Deflocculant Demand	0.55	na	0.58	0.61	na
Impurity Minerals	A,G,H,M	A,G,M	A,G	A,G	A,G,S

Casting Properties

Two tests used to determine the casting properties of kaolin slips are casting rate and deflocculant demand. Casting rate measures how quickly the slip forms a cast on the wall of a plaster mold and is determined by pouring the slip into a mold and allowing it to drain for 30 minutes. Excess liquid is then poured off and the cast is allowed to drain for another 15 minutes before measuring the weight of the cast. The final cast weight is reported as the casting rate. Deflocculant demand is the quantity of deflocculant required to fully disperse the slip. These rheological properties are controlled by a number of parameters including particle size distribution, particle shape, surface area, and the presence of viscosity modifying impurity minerals such as smectite and degraded mica.

Figure 9. Transmission electron micrograph of cube shaped (left) and skeletal (right) anatase.

For example, finer particle size kaolins tend to have lower casting rates because the finer particles blind the surface of the mold and form low void volume slips, which inhibits dewatering. In some cases, the relationship between casting rate and particle size distribution is obscured by the presence of high surface area minerals such as smectite and degraded mica. Product 1 is an example of a relatively coarse product with a high surface area and low casting rate. The XRD pattern of this product shows trace quantities of degraded mica (Figure 10). This degraded mica most likely causes the low casting rate by altering the flocculation structure of the slurry.

Aspect ratio also controls casting rate. High aspect ratio clays, which are comprised largely of broad, thin plates, tend to blind the surface of the mold in much the same way as fine particles.

Fired Color

In the Georgia kaolins the minerals that control fired color are hematite, goethite and iron anatase. Typically clays containing goethite will fire pink as the result of the conversion of goethite to hematite. A decrease in Hunter b-value and an increase in Hunter a-value reflect this mineral transformation (Table II). Clays with large quantities of iron anatase may also fire pink due to the exsolution of Fe and the subsequent formation of Fe oxide.[8] It is therefore critical to determine how much iron is present and in what form. Total iron gives a gross indication of fired color but more detailed mineralogical characterization is required to accurately predict how the fired product will absorb light in the visible region of the spectrum (Figure 11).

Organic matter, which imparts a gray color to the unfired kaolin, also influences fired color. Clays with higher organic carbon content tend to fire whiter than clays that contain smaller amounts of organic matter. This shift in color is due to the thermal destruction of the organic matter, which results in an increase in L-value.

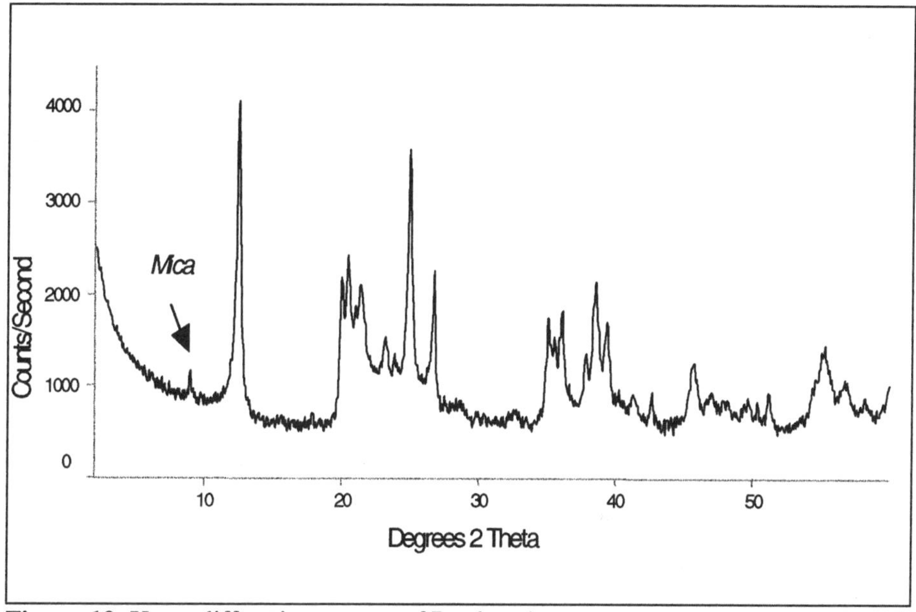

Figure 10. X-ray diffraction pattern of Product 1.

Table II. Hunter a- and b-value before and after firing

Product	Fe Oxides	a-value		b-value		L-value	
#	(raw clay)	before	after	before	after	before	after
1	H and G	1.28	1.71	7.02	5.92	81.87	91.99
2	Goethite	0.28	1.27	6.01	6.63	89.25	91.74
3	Goethite	0.04	1.30	7.17	5.81	93.12	93.23
4	Goethite	-0.04	1.81	8.29	6.96	92.67	92.23
5	Goethite	0.16	0.17	6.58	4.96	90.31	95.00

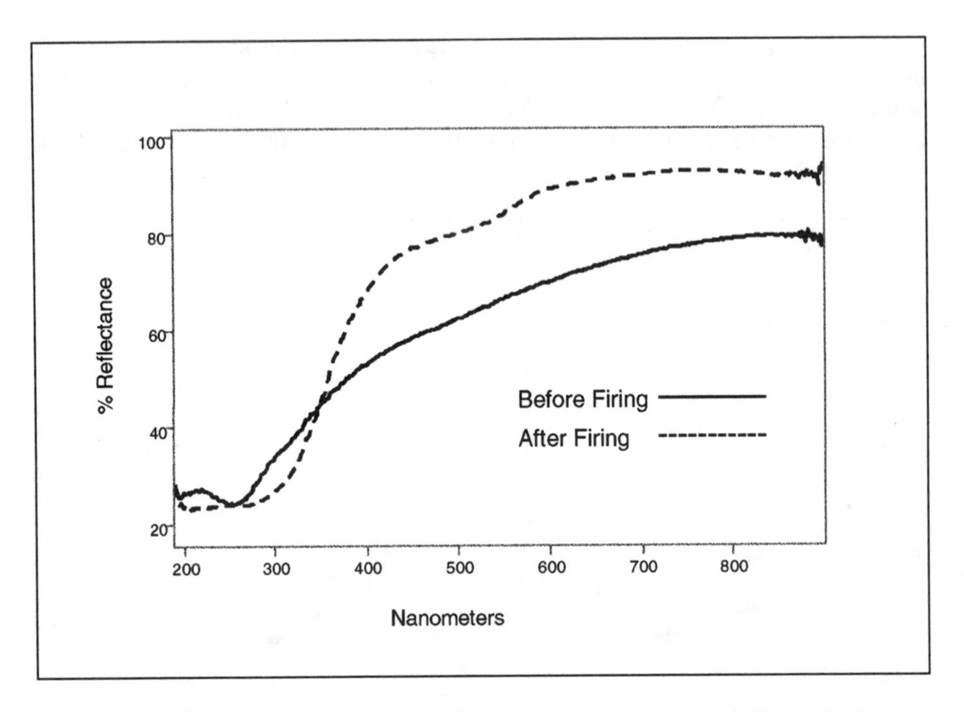

Figure 11. Diffuse reflectance spectra of Product 1 showing the effect of firing.

SUMMARY

Kaolins mined for ceramics in the Southeastern US are comprised predominantly of well-ordered kaolinite. Differences in the quality of these clay deposits are attributable to differences in kaolinite particle morphology and particle size distribution as well as impurity mineral content. Particle size and shape strongly influence properties related to whiteware applications and therefore must be controlled. Particle size is controlled by selective mining and can be manipulated through blending or classification. Aspect ratio is an inherent property of the raw material and is not normally altered in the processing of ceramic grade kaolins.

Certain impurity minerals, even in trace quantities, profoundly influence ceramic application. Two groups of impurity minerals are of concern to the ceramic manufacturer. These are impurity minerals that control rheology including smectite and degraded illite and impurity minerals that control color including the Ti and Fe oxide minerals. Abundance of these minerals is controlled through selective mining practices and, to a lesser degree, by processing.

Because the Georgia kaolins exhibit a range of properties due to sometimes-subtle variations in mineralogy, these deposits provide new opportunities for developing improved products tailored to meet customer specifications. Creating new products can only be accomplished through detailed characterization of the raw material coupled with an understanding of how various mineralogical and physicochemical attributes control key ceramic properties. With the increasing availability of research grade analytical instruments, such as high resolution TEM and laser diffraction particle size analysis, it is now more feasible to carefully characterize the mineralogy of these kaolins on a scale that was not possible until very recently.

REFERENCES

[1]R.S. Austin, "Origin of Kaolin in the Southeastern US," *Mining Engineering*, **50** 52-57 (1998)

[2]V.J. Hurst and S.M. Pickering, "Origin and Classification of Coastal Plain Kaolins, Southeastern USA, and the Role of Groundwater and Microbial Action," *Clays and Clay Minerals*, **45** 274-285 (1997)

[3]S.M. Pickering and H.H. Murray, "Kaolin," pp. 255-277 *in Industrial Minerals and Rocks*, 6th ed. Edited by D.D. Carr. Society for Mining Metallurgy and Exploration, Inc. Colorado, 1995.

[4]O.C. Malcom, "Illite of the Commercial Grade Kaolins of the Middle Georgia Kaolin District," *Clay Minerals Society 31st Annual Meeting, Meeting Program and Abstracts*, 115 (1994).

[5]R.A. Lowe, P.B. Malla, and J.L. Jordan, "Characterization of Smectites in Tertiary and Cretaceous Kaolins of Georgia," *Clay Minerals Society 30th Annual Meeting, Meeting Program and Abstracts,* 146 (1993)

[6]R.J. Pruett, J. Yuan, L.P. Keller, and J. Bradley, "Trace Element Chemistry of Anatase in Georgia Kaolin," *Clay Minerals Society 32nd Annual Meeting, Meeting Program and Abstracts,* 133 (1995)

[7]J.M. Elzea and S.B. Rice, "TEM and X-Ray Diffraction Evidence for Cristobalite and Tridymite Stacking Sequences in Opal," *Clays and Clay Minerals*, **44** 492-500 (1995)

[8]U.Schwertmann, J. Friedl, G. Pfab, and A. Gehring, "Iron Substitution in Soil and Synthetic Anatase," *Clays and Clay Minerals*, **43** 599-606 (1994)

X-RAY CHARACTERIZATION OF CLAYS

Michele M. Hluchy
Geology and Environmental Studies
Alfred University
Alfred, NY 14802

ABSTRACT

X-ray diffraction is a primary tool used to characterize clay minerals, however samples must be prepared and treated appropriately to obtain valid and useful data. Identification of clay mineral phases is usually done using oriented sample mounts, prepared specifically to enhance the basal (00*l*) reflections. Several different methods are employed to obtain oriented sample mounts, including settling of clay-water suspensions onto flat substrates such as glass slides. Random mounts may also be used, particularly for determining *a* or *b* crystallographic axis dimensions, order or disorder in stacking sequences, or to obtain true peak intensities for quantitative analysis. Samples are often chemically or thermally treated prior to X-ray diffraction in order to differentiate between phases giving similar diffraction patterns. Phase identification is accomplished by using the d-values of the peaks, which reflect the unit cell's *c*-axis spacing. The chemical composition and diffracting domain size also influence the nature of the X-ray diffraction pattern. Computer models developed in recent years have proven to be powerful aids to the identification and characterization of clay minerals from their X-ray diffraction data.

INTRODUCTION

Characterization of clay minerals using X-ray diffraction (XRD) is a broad topic that has been addressed in several publications devoted entirely to the subject, including an excellent text entitled *X-ray Diffraction and the Identification and Analysis of Clay Minerals* written by Moore and Reynolds[1] and the monograph *Crystal Structures of Clay Minerals and their X-ray Identification* edited by Brindley and Brown.[2] For more details on the principles discussed here, the reader is referred to those publications as well as others referenced in the text. This paper will concentrate on sample treatments and preparation techniques, basic analysis of XRD data, and a brief introduction to the use of computer models to help characterize clays. Much of the information presented comes from Moore and Reynolds,[1] Bish and Reynolds,[3] Eslinger and Pevear[4] and several chapters in the monograph edited by Brindley and Brown.[2]

SAMPLE PREPARATION

Sample Mounts

Identification of clay minerals and partial characterization of phyllosilicate mineral structures is usually done using oriented aggregates of crystallites. The unit cell parameters in the X and Y directions (indicating *a* and *b* axis spacings) for many of the phyllosilicates are similar to each other, making the *c* axis spacings diagnostic for most clay minerals. To maximize the intensity of XRD reflections which give *c*-axis spacings (the 00*l* series), it is desirable to prepare a sample with as many crystallites as possible oriented with their X-Y planes parallel to the sample surface,[3(p.89)] keeping in mind that this type of sample preparation will change the relative intensities of all of the XRD peaks. Bish and Reynolds[3 (p.89-90)] list four conditions which maximize preferred orientation of samples: (1) sample crystallites should have a good platy morphology, (2) samples should be well-dispersed in aqueous suspension, (3) non-platy minerals should be removed from the sample, and (4) a smooth sample substrate should be used.

Preparation of an oriented sample mount from aqueous dispersed samples is accomplished using a variety of methods. Perhaps the simplest method is to use an eye dropper to place some of the clay-water suspension onto a clean glass microscope slide, adding enough sample to cover the entire surface of the slide yet not so much that the liquid overflows the surface. The slide is then air-dried at room temperature or in a 90°C oven until the water evaporates, leaving behind a clay film on the surface of the glass. A second method that is used to prepare oriented sample mounts is to smear a thin layer of clay-water paste onto a glass slide with a spatula. A third method, described first by Kinter and Diamond,[5] involves centrifuging clay-water suspension onto an unglazed ceramic tile. This method produces sample mounts with a high degree of preferred orientation, but results in particle size segregation of the sample, which makes the XRD data unsuitable to use for quantitative analysis.[1,3] A fourth method, originally suggested by Drever,[6] uses a vacuum filtration system to draw the clay-water suspension through a filter. The filter and the clay film remaining on it are then inverted onto a glass microscope slide and allowed to dry almost to completion, at which time the filter is quickly peeled off, leaving the film of clay on the glass slide.

Each of these sample preparation methods has its advantages and disadvantages. As mentioned above, the centrifuge porous plate method provides samples with very good orientation but which suffer from particle size segregation. Samples prepared using the "eye dropper" method also have some particle size segregation and are generally too thin to obtain accurate diffraction intensities at high diffraction angles,[1 (p.214)] but they are easy to prepare and are adequate for qualitative analysis. The "smear" method does not result in samples with particle size segregation[1] but preferred orientation of the crystallites is not maximized. Samples prepared using the "filter peel" method do not have significant particle size segregation, have reasonable crystallite orientation, and can be used for quantitative analysis,[1,3] although it takes some practice to develop the skill to prepare these slides well.

Random powder mounts are also used for clay studies, particularly for detailed characterization of the mineralogy. It is crucial that an XRD pattern be obtained from a random powder mount for complete crystal structure determination, because the true relative intensities of all of the (*hkl*) reflections must be measured. For example, the octahedral occupancy of clay minerals (dioctahedral vs. trioctahedral) is determined by measuring the *d* value of the 060 diffraction peak, because the *b* unit cell dimension is sensitive to the size of cations and to octahedral site occupancy.[1 (p.245)] For most micas and clay minerals, *d*(060) values generally range from 1.49Å to 1.56Å.[1,7,8] Identification of the mineral's polytype (systematic displacements in the stacking of the layers within the crystal) also requires the use of random powder mounts.

Achieving completely random orientation of the particles in a sample mount is difficult when the particles have a platy morphology, as do the clay minerals and other phyllosilicates. A variety of methods for obtaining random orientation of platy minerals have been described in the literature[1,3,9], however, to prepare a random mount, powdered samples are commonly packed into cavity mounts using either a side-loading or back-loading technique. Spray-drying or freeze-drying the sample prior to packing and using a powder with a small, uniform particle size will help to maximize the randomness of the particle orientation.[1,3]

Sample Treatments

Identification of clay minerals by X-ray analysis usually requires that two or more XRD patterns be collected from the same sample exposed to different chemical or thermal treatments. This is necessary because the basal spacings of the clay minerals are variable and the minerals respond differently to a change in chemical and/or thermal conditions, in some cases resulting in a change in *c*-axis spacings. In fact, the behavior of the minerals when exposed to these chemical and/or thermal treatments is used as a means of identifying many clays. All clay samples should be analyzed *at least* in the air-dried state (dried from suspension at temperatures less than 100 °C) and after solvation with ethylene glycol.[1] Thermal treatments are also often required.

Solvation with ethylene glycol is done as a diagnostic test for the presence of smectites, which adsorb the reagent in their interlayer region and swell, increasing the *d*(001)value from that of the mineral in the air-dried state (see original work by Bradley.[10]) Ethylene glycol solvation can be accomplished by placing the sample mount (glass slide or ceramic tile) into a closed container containing ethylene glycol and placing the container into an oven at 60 °C for at least 8 hours. At this temperature, the reagent will evaporate, so the atmosphere in the container will be rich in ethylene glycol vapors. These vapors will be adsorbed by any smectite in the sample. The sample should be analyzed immediately after removal from the ethylene glycol vapor because prolonged exposure to normal atmospheric conditions will cause the reagent to desorb from the smectite structure.

Thermal treatments are often also necessary for complete mineral identification and these are usually done to the sample once it is mounted on a slide or tile. These treatments are used particularly (but not exclusively) for samples containing expandable clays such as smectites or vermiculite. The sample mount is placed into an oven at an appropriate temperature (usually either 300 °C or 550 °C) for 1-12 hours. The exact temperature used and the amount of time required is mineral-specific. For example, heating a non-hydroxy-interlayered vermiculite to 300 °C for one hour will cause the $d(001)$ value to change from approximately 14Å -15Å to 10Å as the mineral is dehydrated.[11] Heating samples containing chlorite to 550 °C for at least one hour will cause a change in the basal spacings and confirm the presence of the mineral because the chlorite structure will dehydroxylate at that temperature.

Other chemical treatments may be used to identify specific minerals. These treatments may include cation saturation (commonly K^+, Mg^{+2}, Ca^{+2}, or Li^+), exposure to acids, or solvation with other reagents (e.g. formamide to distinguish kaolinite from halloysite,[12] dimethylsulfoxide to differentiate kaolinite from chlorite and serpentine,[13] or glycerol to distinguish vermiculite from smectite.[11]) An excellent guide to the identification of clay minerals describing specific chemical and thermal treatments on a mineral-by -mineral basis is given in Moore and Reynolds.[1]

ANALYSIS OF XRD DATA

X-ray diffraction patterns contain a wealth of information about clay minerals. Peak positions reflect spacings between planes of atoms in the crystal structure and are thus used for phase identification. Peak intensities are used to determine chemical compositions and positions of atoms in the unit cells. Peak breadths yield information about the thickness of crystallites in the sample.

Phase Identification

The first step in identifying mineral phases using XRD data is to assign d values to all of the peaks. Most of the peaks that are used for phase identification of clay minerals are found at (two-theta) diffraction angles of 40° or less. If the sample has been prepared to maximize preferred orientation of the crystallites (as described previously), most of the more intense peaks in the diffraction pattern will be from the $(00l)$ or basal series. An $(00l)$ series of reflections from one mineral can be identified by looking for a set of d values that are related to a common d value by division by an integer. For example, if the pattern contains peaks with d values of 10Å (10/1), 5Å (10/2), 3.33Å (10/3), 2.5Å (10/4) and 2Å (10/5), then it is reasonable to assume that these are the $(00l)$ series for a mineral with a 10Å basal spacing. These inferred basal spacings can then be compared to published values of known c-axis spacings to determine phase identities. Table I lists $d(001)$ values for some common clays or groups of clay minerals for which the $(00l)$ series is diagnostic. The $d(060)$ values, referred to in the previous section as an indicator of octahedral occupancy, are also given. Figure 1 shows two calculated XRD patterns for common clay minerals, (a) kaolinite and (b) illite followed by the calculated XRD pattern for a (c) 50% – 50% mixture of the two minerals to illustrate the relationships between peaks in the basal series for each mineral and how a typical XRD pattern with more than one phase might appear. All XRD patterns that appear in this paper were calculated using the computer program NEWMOD© written by R.C. Reynolds, Jr.[14]

Table I: *d* values for common clay minerals and phyllosilicates. (after Brown and Brindley[9 (p.323)])

Mineral or Group	d(001) in Å	d(060) in Å
Kaolinite group	7.15 – 7.20	1.489
Mg-serpentine	7.25 – 7.35	1.536 – 1.540
Fe-serpentine, Berthierine	7.04	1.555
Pyrophyllite	9.20	1.493
Talc	9.35	1.527
Muscovite (and Illite)	10.0 – 10.05	1.499
Biotite	10.0	1.530
Vermiculite (trioctahedral)	14.3	1.541
Chlorite (magnesian)	14.15 – 14.35	1.549
Chlorite (iron-rich)	14.10 – 14.25	1.560
Smectite group	variable	variable

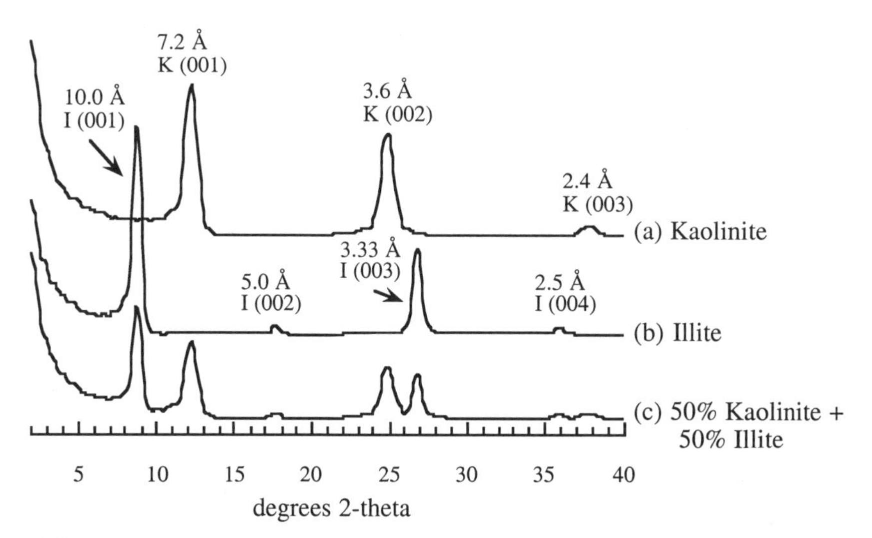

Figure 1: Computer-calculated XRD patterns for (a) pure kaolinite, (b) pure illite, and (c) a mixture of kaolinite and illite.

The Use of Sample Treatments

One of the major problems that arises when identifying clay minerals from XRD data is that of superimposed peaks or two or more minerals having peaks with the same *d* values. One common example of this is the XRD pattern from a sample containing both chlorite and kaolinite. The (001) value for chlorite is 14.10–14.35Å (see Table I) and the *d*(002) value is half of that, or 7.05–7.18Å. The *d*(001) value for minerals in the kaolinite group is 7.15–7.20-Å, so the 001 peak of kaolinite is found at the same location on the XRD pattern as the 002 peak of chlorite. Similarly, the 002 kaolinite peak is superimposed on the 004 chlorite peak. The kaolinite peaks can therefore be hidden by the chlorite peaks, causing it to be difficult to detect the presence of kaolinite in the sample. This is a situation where sample treatment is necessary, followed by another XRD analysis. The sample treatment in this particular case could be heating, boiling in acid, or chemical treatment using formamide or dimethylsulfoxide.[1, 13, 15] Each treatment has an effect on the *d*(001) of either the chlorite or the kaolinite, causing a change in the XRD pattern.

Another example of a mineral mixture that is difficult to characterize without sample treatments is vermiculite and montmorillonite (a member of the smectite group). The XRD pattern of the sample mixture analyzed in the air dried state may contain a large, sometimes very broad peak with a *d* value near 14Å and several very small peaks at higher diffraction angles (see Figure 2a). It is difficult to determine the mineralogy of that sample using only the XRD data from the air-dried state. However, after ethylene glycol solvation, the pattern changes considerably (Figure 2b). The pattern now contains *two* peaks in the low diffraction angle region, clearly indicating that montmorillonite, which swells with ethylene glycol solvation, is present.

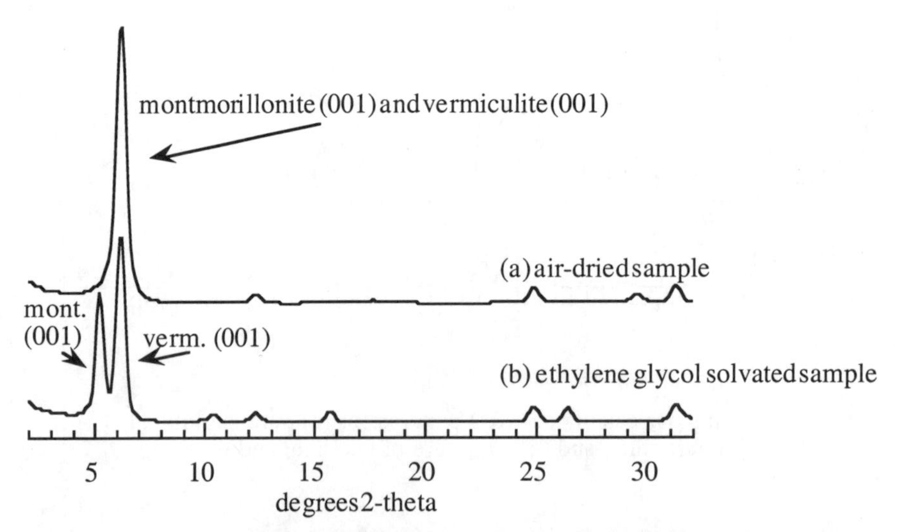

Figure 2: Computer-calculated XRD patterns for a 50%-50% mixture of montmorillonite and vermiculite in the (a) air dried state and (b) after ethylene glycol solvation.

Another complication that is often encountered occurs when a single mineral has variable $d(001)$ values due to the material that occupies the interlayer region – e.g. hydrated cations, non-hydrated cations, organic materials, etc. Identification of minerals in the smectite group may initially be difficult because of this problem. Figure 3 demonstrates how the XRD pattern of a smectite changes with degree of interlayer hydration and intercalation with ethylene glycol. This is another situation where sample treatments are necessary. In the case of members of the smectite group, the treatment involves saturating the sample with a known cation. This is followed by ethylene glycol solvation and/or heat treatments. XRD data is collected after each treatment, so the response of the mineral to each treatment can be determined. For some clay minerals, complete characterization of the mineralogy may not be possible until several (five or six) XRD patterns are collected!

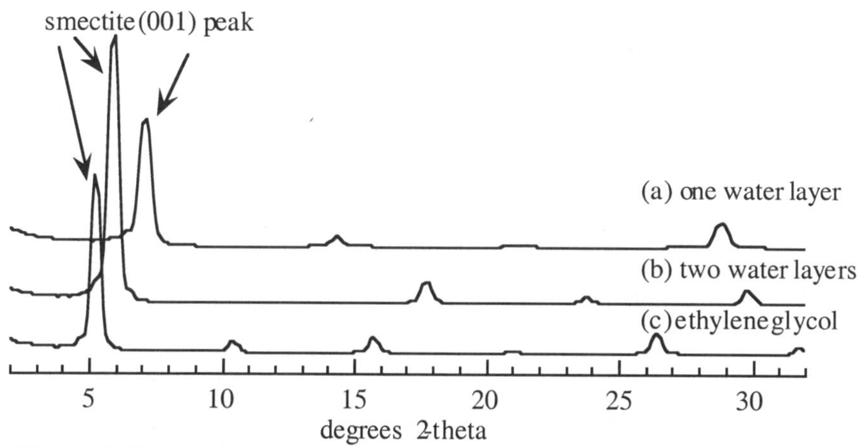

Figure 3: Computer calculated XRD patterns of a dioctahedral smectite with (a) one layer of water, (b) two layers of water, and (c) ethylene glycol, in the interlayer region.

The Effect of Chemical Composition on XRD Data

Substitution of one ion for another into clay mineral structures is common, so two samples of the same mineral may have slightly different chemical compositions. At the same time, the scattering amplitude of an X-ray beam, and therefore the intensity of an XRD peak, is a function of the chemical composition of the mineral and where individual atoms are situated in the mineral structure. This means that two samples of the same mineral with slightly different chemical compositions will have slightly different XRD patterns. In general, the positions of the XRD peaks do not change appreciably with ionic substitutions, but the *intensities* of the peaks may change, especially if the ion that substitutes into the mineral structure has a vastly different scattering power than that of the ion that is removed. A good example of this is phenomenon can be seen in XRD patterns of chlorite. Chlorite has two octahedral sheets in its structure, and the octahedral sites can be occupied by relatively heavy metals (Fe, Co, Cr, Mn, Ni) or light metals (Mg or Al). The number of heavy or light atoms in the octahedral sites, *and which octahedral sheet they are situated in*, has such a profound effect on the relative intensities of the (00*l*) reflections that those intensities can be used to estimate the heavy atom content of a chlorite sample.[16]

Peak Breadths

XRD peak breadths also contain information about the sample. As the diffracting domain or crystallite size increases, the peak breadth decreases. XRD peaks from clay minerals are often broader than peaks from non-clay minerals because the crystallite or diffracting domain size of clay minerals tends to be smaller than that of non-clays. Peak broadening also occurs when the mineral is an interstratification of two or more clays, and, in fact, the peak breadth can be used to detect and quantify small amounts of interstratification in a variety of minerals.[1,17]

COMPUTER MODELING

Because the physics and mathematics of XRD are well understood, computer modeling of XRD data is possible if the structure and chemical composition of the mineral are known. Computer models that are used for phase identification are very valuable in clay research because they allow the researcher to make changes to an input mineral structure or chemical composition and recalculate the XRD pattern as many times as is necessary until a good match with experimental data is obtained. A good computer model must contain (either as built-in components or input variables) the necessary algorithms to simulate the interaction of X-rays with an array of atoms in a mineral structure, instrumental parameters which allow the model to simulate the response of an individual diffractometer, and a complete description of the mineral structure and composition.[1] One program that is used widely by researchers and is commercially available is NEWMOD© by R.C. Reynolds, Jr. [14] This program allows the user to specify a mineral type, some aspects of the chemical composition, the interlayer complex (if appropriate), instrumental parameters (radiation wavelength, goniometer radius, etc.), and sample parameters (degree of preferred orientation, crystallite size, etc.). Once these are designated, one-dimensional diffraction patterns are calculated and can be used to aid in phase identification.

Computers are also used for a variety of other purposes in clay analyses, including refining crystal structures using the Reitveld method,[18] determining particle thicknesses[19] and deconvolving instrumental signatures from XRD peak profiles.[20] For more details on these applications, the reader is referred to the publication *Computer Applications to X-ray Powder Diffraction Analysis of Clay Minerals,* edited by Reynolds and Walker.[21]

SUMMARY

Attention to sample preparation (mounting and treatment) is crucial when characterizing clays by XRD. For routine phase identification, oriented sample mounts are often used, but random powder mounts are necessary for complete structural characterization. Ethylene glycol solvation, specific cation saturation, and exposure to elevated temperatures are the most common treatments employed during clay analyses, but specialized treatments for individual minerals are also sometimes required. The XRD data is then analyzed with the mineral structures and mineral behavior in mind. Computer modeling of diffraction phenomena is an extremely useful way to confirm or refine phase identifications or to predict XRD patterns for clay minerals, and these techniques are widely used in clay mineral research.

REFERENCES

[1]D.M. Moore and R.C. Reynolds, Jr., *X-Ray Diffraction and the Identification and Analysis of Clay Minerals*, 2nd. ed., Oxford University Press, New York, 1997.

[2]*Crystal Structures of Clay Minerals and Their X-ray Identification*, Edited by G.W. Brindley and G. Brown, Mineralogical Society Monograph 5, Mineralogical Society, London, 1980.

[3]D.L. Bish and R.C. Reynolds, Jr., "Sample Preparation for X-ray Diffraction"; pp. 73-00 in *Modern Powder Diffraction*, Edited by D.L. Bish and J.E. Post, Reviews in Mineralogy vol 20, Mineralogical Society of America, Washington, D.C., 1989.

[4]E. Eslinger and D. Pevear, *Clay Minerals for Petroleum Geologists and Engineers*, SEPM Short Course Notes No. 22, Society of Economic Peleontologists and Mineralogists, Tulsa, 1988.

[5]E.G. Kinter and S. Diamond, "A New Method for Preparation and Treatment of Oriented-Aggregate Specimens of Soil Clays for X-ray Diffraction Analysis", *Soil Science*, **81**, 111-120 (1956).

[6]J.I. Drever, "The Preparation of Oriented Clay Mineral Specimens for X-ray Diffraction Analysis by a Filter-Membrane Peel Technique", *American Mineralogist*, **58**, 553-554 (1973).

[7]S.W. Bailey, "Structures of Layer Silicates"; pp. 1-123 in *Crystal Structures of Clay Minerals and Their X-ray Identification*, Edited by G.W. Brindley and G. Brown, Mineralogical Society Monograph 5, Mineralogical Society, London, 1980.

[8]G.W. Brindley, "Order-Disorder in Clay Mineral Structures"; pp. 125-195 in *Crystal Structures of Clay Minerals and Their X-ray Identification*, Edited by G.W. Brindley and G. Brown, Mineralogical Society Monograph 5, Mineralogical Society, London, 1980.

[9]G. Brown and G.W. Brindley, "X-ray Diffraction Procedures for Clay Mineral Identification"; pp. 305-359 in *Crystal Structures of Clay Minerals and Their X-ray Identification*, Edited by G.W. Brindley and G. Brown, Mineralogical Society Monograph 5, Mineralogical Society, London, 1980.

[10]W.F. Bradley, "Molecular Associations Between Montmorillonite and Some Polyfunctional Organic Liquids", *Journal of the American Chemical Society*, **67**, 975-981 (1945).

[11]G.F. Walker, "Reactions of Expanding Lattice Minerals with Glycerol and Ethylene Glycol", *Clay Mineral Bulletin*, **3**, 302-313 (1958).

[12]G.J. Churchman, J.S. Whitton, G.G.C. Claridge, and B.K.G. Theng, "Intercalation Method for Differentiating Halloysite from Kaolinite", *Clays and Clay Minerals*, **32** [4] 241-248 (1984).

[13]C.S. Calvert, "Simplified, Complete CsCl-hydrozine-dimethylsulfoxide Intercalation of Kaolinite", *Clays and Clay Minerals*, **32** [2] 125-130 (1984).

[14]R.C. Reynolds, Jr., *NEWMOD©: a Computer Program for the Calculation of One-Dimensional Diffraction Patterns of Mixed-Layered Clays*, R.C.Reynolds, Jr., 8 Brook Road, Hanover, NH (1985).

[15]D.M.C. MacEwan and M.J. Wilson, "Interlayer and Intercalation Complexes of Clay Minerals"; pp. 197-248 in *Crystal Structures of Clay Minerals and Their X-ray Identification*, Edited by G.W. Brindley and G. Brown, Mineralogical Society Monograph 5, Mineralogical Society, London, 1980.

[16]J.R. Walker, M.M. Hluchy, and R.C.Reynolds, Jr., "Estimation of Heavy Atom Content and Distribution in Chlorite Using Corrected X-ray Powder Diffraction Intensities", *Clays and Clay Minerals*, **36** [4] 359-364 (1988).

[17]R. C. Reynolds, Jr., M.P. DiStefano, R.W.Lahann, "Randomly Interstratified Serpentine/Chlorite: Its Detection and Quantification by Powder X-ray Diffraction Methods", *Clays and Clay Minerals*, **40** [3] 262-267 (1992).

[18]D.L. Bish, "Studies of Clays and Clay Minerals Using X-ray Powder Diffraction and the Reitveld Method"; pp. 79-121 in *Computer Applications to X-ray Powder Diffraction Analysis of Clay Minerals*, Edited by R.C. Reynolds, Jr. and J.R. Walker, The Clay Minerals Society, Boulder, 1993.

[19]D.D. Eberl and A. Blum, "Illite Crystallite Thickness by X-ray Diffraction"; pp. 123-153 in *Computer Applications to X-ray Powder Diffraction Analysis of Clay Minerals*, Edited by R.C. Reynolds, Jr. and J.R. Walker, The Clay Minerals Society, Boulder, 1993.

[20]R.C. Jones and H.U. Malik, "A Computer Technique for Rapid Decomposition of X-ray Diffraction Instrumental Aberrations from Mineral Line Profiles"; pp. 156-171 in *Computer Applications to X-ray Powder Diffraction Analysis of Clay Minerals*, Edited by R.C. Reynolds, Jr. and J.R. Walker, The Clay Minerals Society, Boulder, 1993.

[21]*Computer Applications to X-ray Powder Diffraction Analysis of Clay Minerals*, Edited by R.C. Reynolds, Jr. and J.R. Walker, The Clay Minerals Society, Boulder, 1993.

THE COLLOIDAL NATURE OF KAOLINITE

William M. Carty
NYS Center for Advanced Ceramic Technology—Whiteware Research Center
School of Ceramic Engineering and Materials Science
New York State College of Ceramics at Alfred University
Alfred, NY 14802

ABSTRACT

The commonly held view of kaolinite particles in water, that of negatively-charged basal plane surfaces with positive edges, is inconsistent with the mineralogy and dispersion behavior. Kaolinite is a 1:1 sheet silicate, one side silica-like and other aluminum hydroxide-like, with a variable edge nature. Therefore, based on the colloidal theory and the generation of surface potential in an aqueous environment, it is proposed that the silica-like surface will be negatively charged and the aluminum hydroxide-like surface will be positively charged over a broad pH range. The edges of the particle will adopt a net charged (either negative or positive) as a function of the suspension pH. The dispersant demand for kaolinite particles is consistent with roughly one-half as much dispersant necessary to reach the minimum in the viscosity curve, compared to the dispersant demand for colloidal alumina, further supporting the dual basal plane surface model.

INTRODUCTION

Kaolinitic clays are the foundation of the whiteware and traditional ceramics industry. Clays account for nominally 95% of the specific surface area of a typical whiteware batch (composed of 45% kaolin and ball clays, 25% quartz, and 30% feldspar) and thus dominate the batch rheology. Since the production of whitewares begins most frequently in the slurry stage, understanding the colloidal nature of kaolinite is critical to developing a well-controlled, repeatable process. The commonly held view that kaolinite possesses negatively charged basal plane surfaces and positively charged edges is inconsistent with the mineralogy of kaolinite – a 1:1 sheet silicate composed of a $[Si_2O_5]^{-2}$ tetrahedral layer and an $[Al_2(OH)_4]^{+2}$ octahedral layer. It is proposed instead that kaolinite particles possess a dual basal plane surface nature and establishes a basis for the evaluation of observed dispersant and rheology data for kaolin suspensions. Based on the colloidal behavior of silicas (isoelectric point of 2.0-3.5) and aluminas (isoelectric point 8.5-10.4),[1] in the pH range of 3.5 to 8.5 the silica-like basal plane surface of a kaolinite particle must be negatively charged and the alumina-like basal plane surface must be positively charged.

GENERAL INTRODUCTION TO KAOLINITE AND KAOLINS

Kaolinite and other clay minerals are formed by the decomposition of feldspars via geological processes.[2] The term kaolin refers to a rock, or in this case, a clay powder, composed of at least 50% of the mineral kaolinite.[3] Kaolin deposits are either primary (i.e., deposited where the weathering process occurs) or secondary (i.e., transported a considerable distance from the source rock). In commercial kaolin deposits, the most common mineral impurities include quartz, micaceous minerals (micas), smectites (also commonly referred to as montmorillonites), titania (rutile or anatase), and iron oxide.[3,4] Beneficiation of the raw kaolin frequently eliminates all but a few percent of the impurity minerals.

In the United States, the major ball clays deposits are located on the eastern side of the Appalachian Mountains in western Kentucky and Tennessee; the major kaolin deposits are located on the southeastern side of the mountains in central Georgia and South Carolina. All commercial kaolins and ball clays in the U.S. are mined from secondary deposits. Kaolins and ball clays are mineralogically similar – both containing a large fraction of kaolinite – with the major differences residing in the remaining impurities, usually quartz, montmorillonites, titania, and organic matter.

English china clays (also kaolinitic in nature) are obtained from primary and secondary deposits and differ slightly from the kaolins obtained from the United States in the types of mineral impurities. Micas are common (almost no montmorillonites are found), and the titania content is lower. The lower titania content in English clays corresponds to a high whiteness after firing, even though the iron contents are similar. A small amount of ionic substitution of iron in the titania lattice is responsible for the coloration after heat treatment.

As stated previously, kaolinite is a 1:1 sheet silicate composed of a $[Si_2O_5]^{-2}$ (tetrahedral) layer and an $[Al_2(OH)_4]^{+2}$ (octahedral) layer. Minerals such as talc, pyrophyllite, mica, bentonite, etc. are 2:1 sheet silicates in which the octahedral layer is sandwiched between two tetrahedral layers.[5] As demonstrated in Figure 1, the presence of 2:1 sheet silicate impurities in commercial clays is easily detected by powder X-ray diffraction as peaks at ~6° and ~8.8° 2θ (using $Cu_{K\alpha}$ radiation) for montmorillonite and mica, respectively. Similarly, it is routine to demonstrate that the amount of impurity 2:1 minerals is small (on the order of a few percent) in industrial kaolins and ball clays. To understand the colloidal nature of kaolinite, it is essential to recognize that the basal plane surfaces of a kaolinite particle are different – one will behave as does silica in water, and the other, as does alumina or, more accurately, aluminum hydrate.

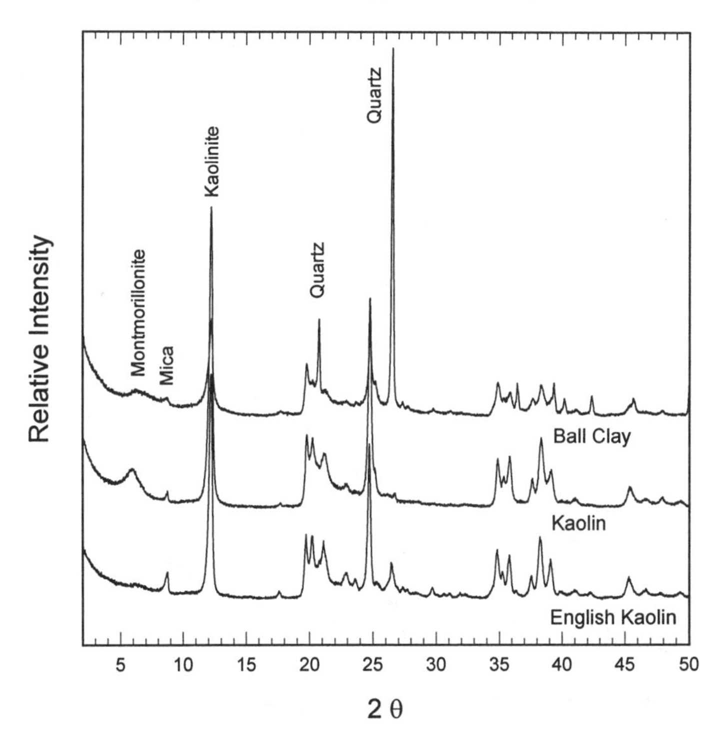

Figure 1. Powder x-ray diffraction patterns for a domestic ball clay and kaolin and an English kaolin. The diffraction peaks at ~6° and ~8.8° 2θ correspond to the interlayer spacings for mica and montmorillonite, respectively. The large reflection at ~12.3° 2θ represents the interlayer spacing for kaolinite. All three of the samples contain impurity quartz.

Direct evidence regarding the dual surface nature of kaolinite has been obtained via atomic force microscopy (AFM) studies conducted in conjunction with surface modeling.[6] These results confirm the presence of silanol and hydroxide surfaces on kaolinite particles. Indirect evidence occurs in the dispersant-clay interactions – the amount of poly(acrylic acid) (PAA) dispersant necessary to reach the minimum in a viscosity versus dispersant level curve is ~58% of the amount to coat the surface of an alumina powder (also obtaining a minimum in viscosity curve), as illustrated in Figure 2.[7]

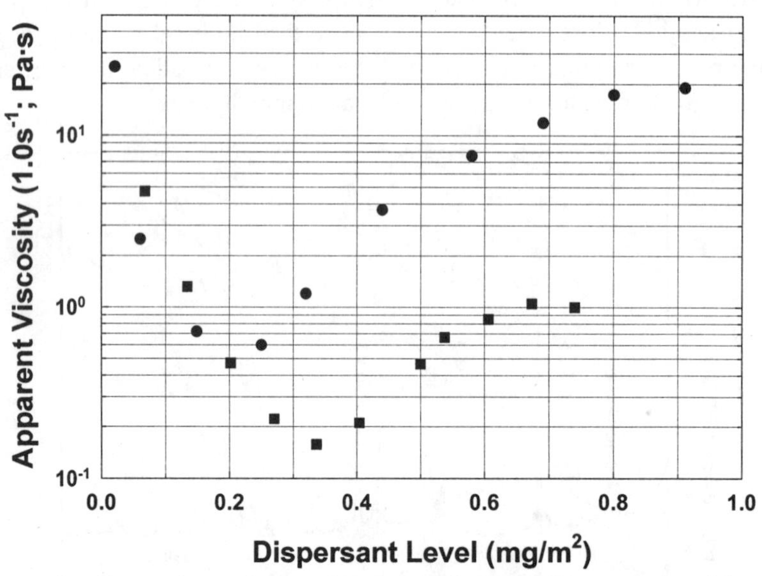

Figure 2. Comparison of viscosity versus Na-PAA dispersant level for 30 vol% alumina (ν) and kaolin (λ) suspensions at pH=8.5 (±0.2), corrected for the specific surface area of the powders (9.6 m^2/g and 21.4 m^2/g, respectively).[7] The data indicates that approximately 58% as much dispersant is necessary to reach the minimum in the viscosity curve for kaolin compared to that required by alumina.

THE COLLOIDAL NATURE OF KAOLINITE IN WATER

The prevailing perspective regarding the nature of kaolinite in an aqueous medium is that the basal plane surfaces are negatively charged and the edges are positively charged. This view is supported in the literature by studies demonstrating the agglomeration of negatively charged gold particles to the "positively-charged" edges of clay platelets as observed via transmission electron microscopy (TEM),[8] although the experimental conditions, specifically the pH at which the experiments were performed, is unclear. The lack of particles agglomerated to the faces of the kaolinite particles has been used to argue that the faces thus are negatively charged – a conclusion subsequently arrived at roughly forty years later, however, by other authors.[9]

Understanding several key concepts is essential to deciphering the nature of kaolinite in aqueous suspension. When an oxide particle is suspended in water, a series of acid-base reactions occurs that create charged sites on the particle surface. When the particle surface is in equilibrium with the aqueous suspension medium, the charge at the particle surface is net neutral (i.e., there are an equivalent number of positive and negative charges). The pH at which neutrality is achieved is called the point of zero charge (pzc). The pzc is unique for each oxide and is determined by the cation valence and the lattice bond strength.[1]

Colloidal theory introduces the concept of a double layer of ions surrounding a suspended particle to compensate for the potential at the particle surface. Close to the particle surface is a strongly held layer of ions, termed the Stern layer. The interface between the Stern layer and the diffuse layer is called the shear plane, and the potential at the shear plane is defined as the zeta-potential (ζ-potential). Similar to the pzc, the pH at which the ζ-potential changes sign is defined as the isoelectric point (iep). In the presence of an indifferent electrolyte (i.e., the anion and cation have similar affinities for the particle surface), the pzc and the iep are equivalent.[1,10-11]

In a kaolinite particle, one basal plane surface is composed of a silica-like surface while the other is composed of an alumina-like or aluminum hydroxide-like surface. (It is generally recognized that alumina forms an aluminum hydroxide surface in an aqueous medium.) The kaolinite particle edge is composed of a mixture of alumina and silica sites. It is herewith proposed that one basal surface should behave like silica while the other behaves like alumina in aqueous suspension. It is well documented that silica has an iep of 2.0-3.5 and aluminum hydroxide has an iep of 8.5-10.4. At suspension pH levels above the iep, the particle is net negatively charged; below the iep, the particle is net positively charged. Conservatively, then, the basal plane surfaces should be oppositely charged from pH = 3.5 to pH = 8.5. Therefore, at a neutral pH, the silica-like basal plane surface must be negatively charged, and the aluminum hydroxide-like basal plane surface must be positively charged. However, 2:1 sheet silicates possessing silica-like surfaces for both basal planes will have negatively charged basal-plane surfaces above pH = 3.5, with the edge charge changing as a function of pH. Obviously, the prevailing model of both basal plane surfaces as negatively charged cannot be rectified with the mineralogy of kaolinite.

The concept of a silica-like and an aluminum hydroxide-like surface is further supported by the amount of poly (methacrylic acid), or PMAA, dispersant required to disperse kaolin compared to α-Al_2O_3 (as presented in Figure 2). When corrected for the specific surface area of the powder (measured using N_2 adsorption), the amount of dispersant required to disperse kaolinite is approximately 58% of that required to disperse alumina based on suspension viscosity measurements. Other experimental results have indicated that PMAA has a weak affinity for silica surfaces.[12] Therefore, assuming that the clay booklets have been dispersed into individual particles, approximately one-half of the kaolinite particles – proposed to possess the alumina-like basal plane surface – are being coated by the PMAA. Again, the common explanation has been that the edge of kaolinite particles is being coated by the dispersant. This explanation, however, is inconsistent with the morphology of kaolinite particles, as will be demonstrated below.

THE MORPHOLOGY OF KAOLINITE

Kaolinite particles have a hexagonal habit, and the perfection of the platelets is dependent on the crystallinity of the particle. Figure 3 is a schematic illustration of an idealized kaolinite particle.

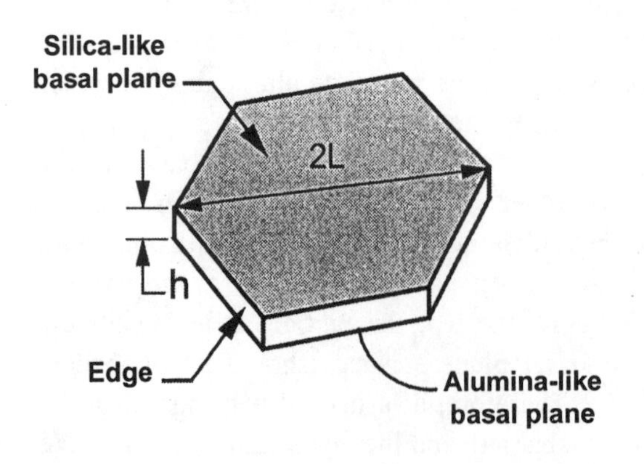

Figure 3. Schematic of an idealized kaolinite particle. The point-to-point dimension across the basal plane surface is defined as 2L; h is the platelet thickness.
In aqueous suspension, the silica-like basal plane surface charge should be negative and the alumina-like basal plane surface charge should be positive over a broad pH range. The charge on the edge is dependent on pH.

Well-crystallized kaolinite particles are clearly hexagonal with sharp corners and edges; poorly-crystallized kaolinites have hexagons with rounded corners and edges. If a perfect hexagonal parallelepiped is assumed, the basal plane to edge surface areas can be calculated and plotted as a function of the aspect ratio of the platelet thickness (h) to the point-to-point dimension (2L) across the basal plane surface. Thus, the total surface area (A_{tot}) of a hexagonal platelet is:

$$A_{tot} = A_{edge} + 2A_{basal} = 6hL + 2\left(\frac{3\sqrt{3}}{2}\left(L^2\right)\right) \quad (1)$$

To calculate the ratio of edge to basal plane surface areas, L is defined as 1, and the aspect ratio is defined as x=h/2L. Thus, the fraction of surface area due to the edge (f_{edge}) is:

$$f_{edge} = \frac{A_{edge}}{A_{tot}} = \frac{12x}{12x + 3\sqrt{3}} \quad (2)$$

As illustrated in Figure 4, f_{edge} is strongly dependent on the aspect ratio of the kaolinite particle. Similarly, the fraction of surface area of one edge and one basal plane surface also scales with aspect ratio.

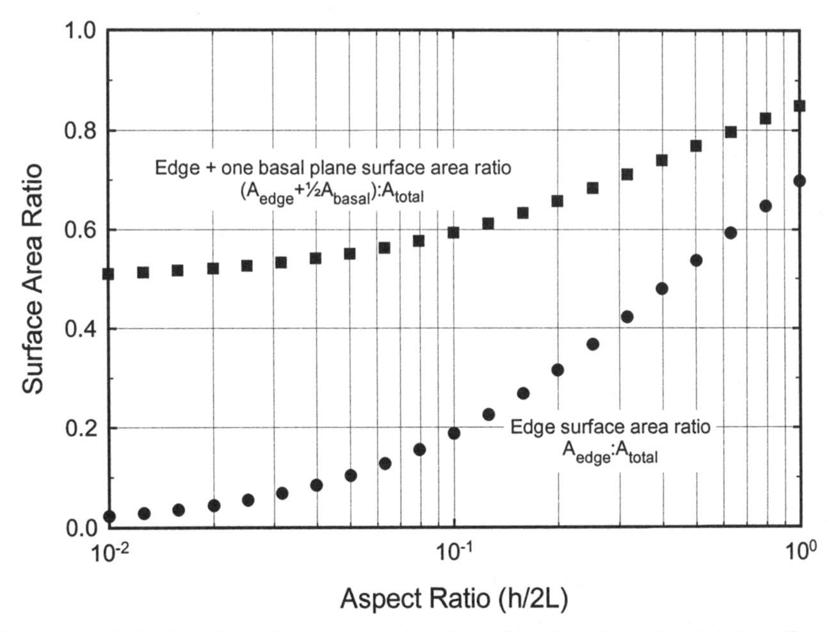

Figure 4. Calculated surface area ratios for edge (only) to the total surface area (λ) and the edge plus one basal plane surface to the total surface area (ν) as a function of aspect ratio (h/2L). These calculations assume a value for L of unity.

If the dispersant addition only adsorbs on the edge surfaces, significantly thicker kaolinite particles are necessary. For example, if a 50% dispersant level is required to disperse a kaolin sample compared to the that needed to disperse alumina, f_{edge} is equal to 0.5, producing a calculated edge thickness of 0.866L (or 0.433·2L), as schematically illustrated in Figure 5. Analysis of scanning electron microscopy images (Figure 6) of well-crystallized kaolinite particles indicates the average edge thickness (measured on 350 particles) to be 103 nm (±30nm) and the average point-to-point dimension (measured on 250 particles) to be 1.1 μm (±430nm). Using the average values for thickness and width, the estimated aspect ratio is 0.095. This produces a f_{edge} value of 18%, indicating that edge adsorption cannot account for the dispersant adsorption levels observed for kaolins.

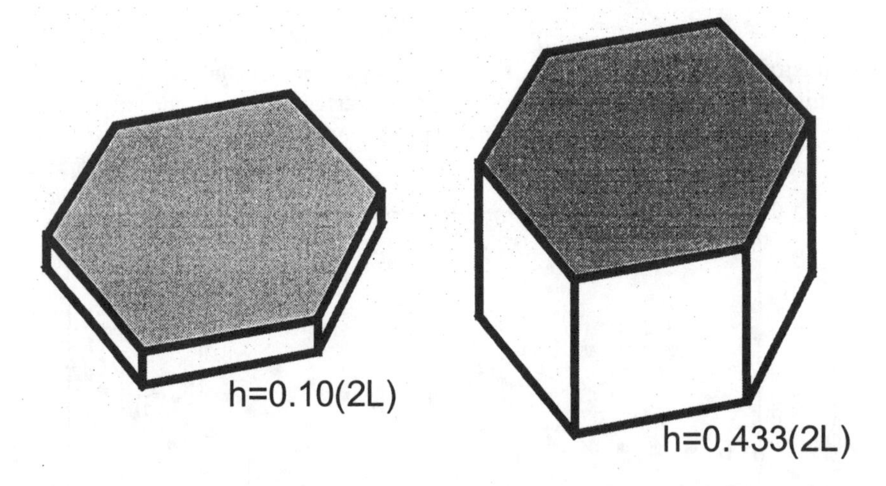

Figure 5. Schematic illustration of the aspect ratios of two kaolinite particles. The left drawing shows an average particle using a value similar to that obtained from the SEM images. The right drawing shows the aspect ratio necessary to allow the edge to account for one-half of the total particle surface area.

Clearly, dispersant must adsorb on a surface other than the edge. With the calculated aspect ratio of 0.095, the area ratio of one basal plane to the total surface area of a clay particle is 41%. However, the ratio of the edge area plus one basal plane surface accounts for 59% of the total surface area, which is perfectly consistent with the observed dispersant demands for kaolin compared to alumina (58%) based on viscosity measurements.

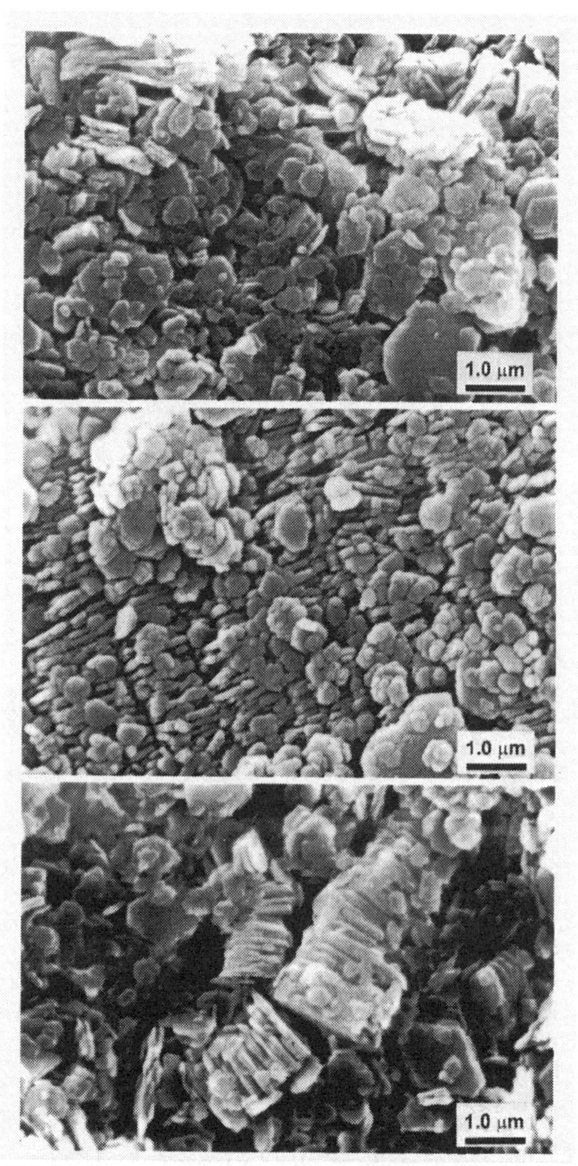

Figure 6. SEM photomicrographs of a well-crystallized Georgia kaolin (K-Ga1b) showing well-formed hexagonal platelets and sharply defined edges. Note the relative uniformity of thickness, the relatively broad distribution of platelet size, and the presence of "books." These samples were prepared by placing a drop of dilute aqueous suspension on a microscope mount, drying the sample, then sputter coating with Au-Pd.

SUMMARY AND CONCLUSIONS

Kaolinite, a 1:1 sheet silicate, is composed of a $[Si_2O_5]^{-2}$ tetrahedral layer and an $[Al_2(OH)_4]^{+2}$ octahedral layer, a structure requiring that the basal plane surfaces be oppositely charged over a broad range of pH. It is also observed that when corrected for specific surface area, one-half as much dispersant is necessary to disperse kaolins compared to alumina. Finally, calculations of the basal plane and edge surface area ratios (for idealized hexagonal platelets) indicate that significantly thicker kaolinite particles are necessary to justify adsorption of dispersants on the particle edges alone. However, adsorption on one of the two basal plane surfaces fits both the dispersant data and the mineralogy, supporting the proposal that there are two distinctly different basal plane surfaces on kaolinite particles.

REFERENCES

1. R. O. James, "Characterization of Colloids in Aqueous Systems," *Ceramic Powder Science; Advances in Ceramics, Vol. 21*, Eds., G. Messing, K. Mazdiyasni, J. McCauley, and R. Haber, Amer. Cer. Soc., Westerville, OH, pp. 401-403, 1987.
2. H. H. Murray and W. D. Keller, "Kaolins, Kaolins, and Kaolins," pp. 1-24 in *Kaolin Genesis and Utilization,* Edited by H. Murray, W. Bundy, and C. Harvey. Clay Minerals Society, CO, 1993.
3. T. Dombrowski, "Theories of Origin for the Georgia Kaolins: A Review"; see Ref. 2, pp. 75-98.
4. S. Powell, "Ball Clay Basics," Am. Ceram *Soc. Bull.,* **75** [6] 74-76 (1996).
5. W. D. Kingery, *Introduction to Ceramics;* pp. 78-79, 532-40. Wiley, New York, 1976.
6. M. Baba, H. Ishii, T. Okuno, H. Miyake, and S. Kakitani, "AFM Observation and Analysis for Atomic Arrangement on the Surface of Layer Silicates by a Novel Procedure of Image Processing," Poster Presentation at *The 11th International Clay Conference*, Ottawa, Canada, 1997.
7. W. M. Carty and U. Senapati, "Porcelain: Raw Materials, Processing, Phase Evolution, and Mechanical Behavior," *J. Amer. Ceramic Soc.*, 81 [1], 3-20 (1998).
8. P. A. Thiessen, "Wechselseitige Adsorption von Kolloiden," *Ztachr. Elektrochem,* **48** [12] 675-81 (1942).
9. H. van Olphen, *Clay Colloid Chemistry,* 2nd Ed., pp. 92-97, Krieger Pub., Malabar, FL, (1977).
10. R. J. Hunter, *Introduction to Modern Colloid Science,* Oxford Science, Oxford, U.K., 1993.
11. D. J. Shaw, *Introduction to Colloid and Interface Science,* 4th ed., Butterworth-Heinemann, Oxford, U.K., 1992.
12. G. G. Hong, H. Lee, B. R. Sundlof, and W. M. Carty, "Evaluation of Kaolin Surface by Rheological Behavior of Suspensions in the Al_2O_3-SiO_2 System," to be submitted to the *J. Amer. Cer. Soc.*

THE EFFECT OF SAMPLE PREPARATION ON THE PARTICLE SIZE DISTRIBUTION OF CERAMIC GRADE KAOLINS

Jeffrey C. Bruns
Albion Kaolin Company
1 Albion Road
Hephzibah, GA 30815

ABSTRACT

The particle size distributions of a pulverized and a slurry form, ceramic casting grade kaolin were evaluated using two different particle sizing instruments (a sedimentation method and a light scattering method) and three different sample preparation methods (ranging from low to high shear dispersion). Results from the experiment show significant differences between the instrumental methods, which are attributable to differences in the physics of the measurement. Sample preparation appears to impact the Sedigraph 5100 much more than the LA-910. However, differences can be seen in the data collected from both instruments as the amount of shear to which the kaolin is subjected increases. Overall this work illustrates the importance of understanding how particle size distribution data is acquired and how a sample is prepared before using the data to characterize a material such as kaolin.

INTRODUCTION

Particle size distribution (PSD) is a useful way to characterize fine particle materials, such as kaolin. There are two types of PSD instrumentation frequently used in industrial settings, x-ray sedimentation and laser diffraction. Both types of instruments have their advantages and disadvantages. However, they provide a means to rapidly determine relative particle size distributions that can be correlated to a particular product or process. The fact that the measurements are relative and not absolute should be carefully considered when PSD data is used, particularly when generated by different laboratories. This can introduce instrumental or sample preparation bias. Both techniques make the assumption that the sample consists of spheres, generating an 'equivalent spherical diameter' for materials, such as kaolin, which are not spherical. Particle shapes for kaolin range from single pseudo-hexagonal platelets of varying diameters, to stacks of platelets, all within the same sample. This presents a twofold problem when measuring a particle size distribution. First, care must be taken in the preparation to adequately disperse the kaolin, without delaminating stacks of platelets. Second, the behavior of a thin platelet when subjected to either a sedimentation or laser diffraction technique will give an equivalent spherical diameter different from the 'real' particle size as determined by techniques such as scanning electron microscopy.

No discussion of an instrumental technique would be complete without a description of the physical basis for that technique. Detailed descriptions can best be found in other references which cover both techniques described here.[1,2] A brief description of each is included to clarify points about the data collected.

The Sedigraph 5100, operates by a sedimentation technique and employs Stokes' law for determining particle diameters. It operates using the absorption of a x-ray beam coupled with a moving cell to allow the quick collection of particle size distribution data of a dispersed, low solids, suspension. The technique is often referred to as x-ray sedimentation, although that term is merely descriptive, since a finely collimated x-ray beam is used to measure the sample density across the cell. Advantages of sedimentation based instrumental techniques include speed (less than 30 minutes to collect a PSD from 50 to 0.25 microns) and reproducibility. The reproducibility quoted by the Sedigraph manufacturer is 1% relative. For kaolin the reproducibility is size dependent and varies considerably from the coarse to the fine end of the distribution. Disadvantages of the Sedigraph include the fact that for a plate shaped material, the particle size distribution acquired, is relative, not absolute and may be biased because of inconsistent settling velocities for particles buoyed by the sedimentation liquid. There are also issues of instrument to instrument variability, which is particularly evident in older Sedigraph units.

The Horiba LA-910 operates on quite a different principle than the Sedigraph and is one example of the many light scattering particle sizing techniques currently available commercially. Generally the technique is referred to as Laser Diffraction. However, this nomenclature is not completely accurate, since it is possible to measure particles smaller than the wavelength of the laser used. In the technique a dilute suspension of particles is rapidly circulated through a cell which has a laser (or finely collimated white light source) shining through it. As the particles pass between the light source and an array of detectors, they scatter light at angles which can be related to an equivalent spherical size. A computer rapidly collects and manipulates the raw data to generate a particle size distribution based on either the Mie or Fraunhof scattering theories. Advantages of laser diffraction techniques include speed (less than 60 seconds to acquire a particle size distribution in the range of 50 to 0.1 microns) and the ability to collect data either in suspension or powder form. An additional advantage is the reduction in errors associated with particle shape and buoyancy. Thin, plate shaped material will give an 'average' spherical distribution as it repeatedly passes through the sample cell. Therefore, the approximation to a sphere is somewhat more accurate. This does not, however, improve the measurement of the particle's true size. One final advantage, which is not significant for measuring kaolin, is that the technique can be used on organic systems which are generally x-ray transparent, making them unsuitable for analysis by Sedigraph. Disadvantages include the fact that the refractive index of the material must be accurately known and, if different size fractions have varied refractive indices, inaccuracy results. Representative sampling can be an issue, due to the very low solids suspension used in the technique. It is of minimal concern, however, since the measurement is the result of several tens of passes of the particles through the cell.

Both techniques, while imperfect, are used in industrial situations where a rapid means for acquiring a particle size distribution is important for process control and the determination of final product quality. The speed of both techniques relative to other more accurate particle sizing techniques such as scanning electron microscopy and sedimentation techniques employing hydrometers and centrifuges make them useful tools, when understood and used properly.

The experiment was designed to vary the amount of shear to which the kaolin suspension was subjected, while holding all other variables constant. The level of shear decreases in the order of Method 3 > Method 2 > Method 1. If there are effects attributable to the amount of shear, such as delamination of the kaolin stacks, they should be manifested as a finer particle size distribution. If there are effects which can be attributed to the dispersion, they should show up as differences in the behavior of Sample A and B, since the slurry was already pre-dispersed when subjected to the sample preparation procedure.

EXPERIMENTAL

All samples were acquired from standard production lots of kaolin destined for use in the ceramic industry. A sodium silicate dispersed, 71% solids kaolin slurry (Sample A) and a dry pulverized kaolin (Sample B) were selected for comparative purposes. Sample preparation was standardized to one of three methods listed in Table I. All dispersions were made at 7% solids using 0.05 % sodium hexametaphosphate (Calgon). The same Calgon solution was used for rinsing the Sedigraph 5100. Samples analyzed on the Sedigraph received an additional 1 minute of low shear mixing and 30 seconds of ultrasonic probe prior to loading into the instrument.

Table I: Sample Preparation Methods

Method 1	15 minutes low shear mixing on Mastertech autosampler. (or magnetic stir plate for the LA-910)
Method 2	10 minutes on high setting of Hamilton Beach mixer
Method 3	10 minutes on high setting of Waring Blender

Sample splits were taken of the materials from methods two and three and sent to an off site laboratory for analysis using a Horiba LA-910 laser diffraction particle size analyzer. Laser diffraction data for Method 1 was generated by submitting the original sample along with dispersion instructions to the off site lab. All Sedigraph runs were performed under standard instrument operating conditions, listed in Table II. Operating conditions for the Horiba LA-210 laser diffraction particle size analyzer are listed in Table III.

Table II: Sedigraph 5100 Operating Conditions

Analysis Type	High Speed
Liquid Viscosity	0.7378 centipoise
Analysis Temperature	34.0 degrees Celsius
Sample Density	2.600
Liquid Density	0.9945 g/cm^3
Reynolds Number	0.20

A total of 18 sample preparations and analysis runs were made of each method. For the LA-910 a minimum of one run was made, and a maximum of 10 sample preparations were performed, to allow calculation of standard deviations. All data presented is an average, with the exception of that for Method 1 on the LA-910.

Table III: Horiba LA-910 Operating Conditions

Relative Refractive Index	116-020*i*
Mixing Speed	2
Ultrasonicated	90 seconds

RESULTS AND DISCUSSION

As stated previously, the experiment was designed to vary the amount of shear to which the kaolin suspension was subjected, while holding all other variables constant. Table IV presents the Sedigraph 5100 data for each preparation method and clay, as well as the standard deviations for the measurement. These data are graphically represented in Figures I and II. It is obvious from the data that Method 1 (low shear stirring) generates a significantly coarser particle size distribution than either Method 2 or Method 3. This could be attributed to incomplete dispersion, however the same effect is seen for both Sample A and B, suggesting that delamination is playing a role. Particularly dramatic is the large increase in <0.5 micron particles when either Method 2 or 3 are applied, with increases of greater than 200% for the finest particle diameter measured. Also of note in the Sedigraph data is the fact that for all preparation methods, the standard deviation associated with the measurement is relatively constant, until you get to the finest fraction, 0.25 microns. This is an example of a limitation associated with the technique, which is not really well suited for measuring ultra fine particles.

Table IVa. Sedigraph 5100 Results for Sample A

Equivalent Spherical Diameter (microns)	Method 1 (% finer)	Method 2 (% finer)	Method 3 (% finer)
10	86.6 ± 0.3	90.8 ± 0.4	91.2 ± 0.6
5	75.3 ± 0.3	79.8 ± 0.5	80.1 ± 0.7
2	58.2 ± 0.2	63.4 ± 0.5	63.6 ± 0.5
1	46.3 ± 0.3	51.4 ± 0.4	51.6 ± 0.4
0.5	30.5 ± 0.3	36.4 ± 0.5	36.5 ± 0.5
0.25	15.3 ± 1.4	22.0 ± 1.5	22.2 + 1.3

Table IVb. Sedigraph 5100 Results for Sample B

Equivalent Spherical Diameter - microns	Method 1 (% finer)	Method 2 (% finer)	Method 3 (% finer)
10	88.6 ± 0.4	91.7 ± 0.6	91.9 ± 0.5
5	76.4 ± 0.4	80.4 ± 0.6	80.8 ± 0.6
2	55.9 ± 0.6	62.5 ± 0.5	62.7 ± 0.5
1	41.1 ± 0.6	48.8 ± 0.5	48.9 ± 0.4
0.5	22.5 ± 0.8	32.3 ± 0.5	32.5 ± 0.5
0.25	7.4 ± 1.1	16.1 ± 1.4	16.4 ± 1.5

Data for the LA-910 instrument shows some striking differences from the Sedigraph. Of particular note is that the particle sizes measured are significantly different, appearing much coarser than the Sedigraph measurement. This is due to differences in the measuring technique and how the data is reported (cumulative mass finer for Sedigraph and volume percent finer for the LA-910) and is a good illustration of how particle size, particularly for kaolin, is a relative

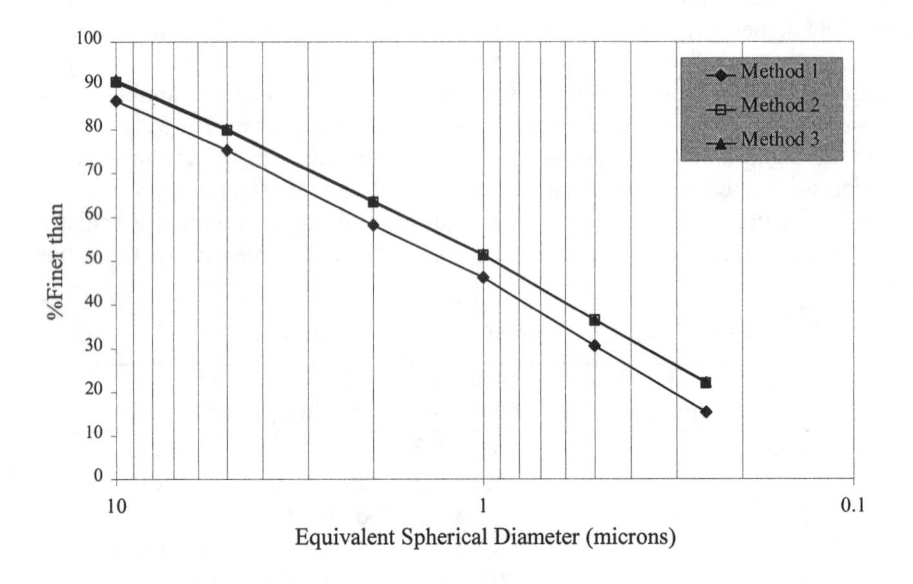

Figure 1: Sedigraph 5100 Data for Sample A

measurement. Immediately evident in the LA-910 data for Sample A is that Methods 2 and 3 appear to be slightly coarser than Method 1, the exact opposite of the Sedigraph data. Standard deviations for the measurement are much higher than for the Sedigraph and show the opposite behavior, in that the standard deviation for coarse fractions is higher than for fine fractions. This is once again attributable to differences in the techniques, with light scattering being a technique better suited for measuring fine particles. Virtually identical data were obtained for all three sample preparation methods (when you take into account standard deviations). This suggests that the light scattering technique may be insensitive to effects such as delamination of particles. This argument is consistent with the predicted behavior of platey particles rapidly passing in front of a detector while undergoing turbulent flow. Each pass in front of the detector might be viewed as a snapshot of the particle in one of its orientations, which taken together as a whole results in the particle sweeping out a pseudospherical volume.

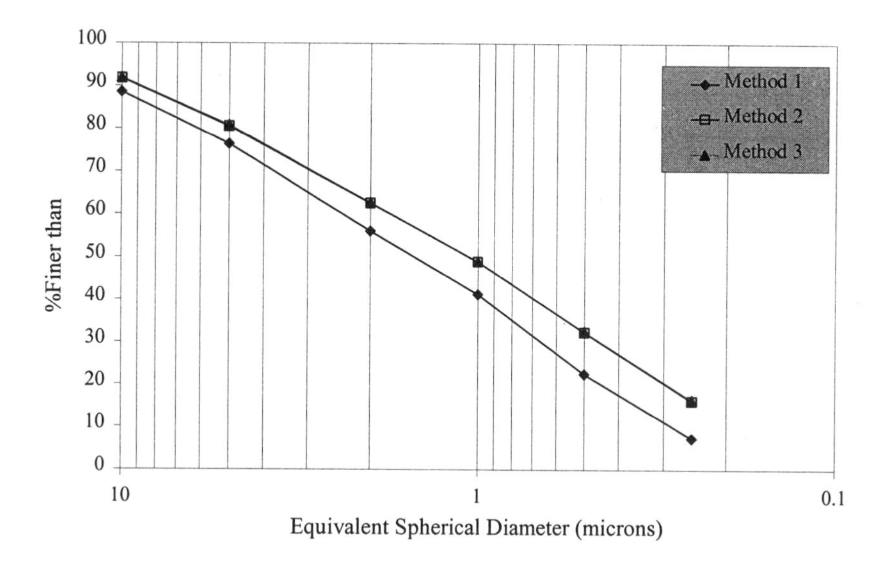

Figure 2. Sedigraph 5100 Data for Sample B

Table Va. Horiba LA-910 Results for Sample A

Equivalent Spherical Diameter (microns)	Method 1 (% finer)	Method 2 (% finer)	Method 3 (% finer)
5	58.0	55.3 ± 1.6	55.3 ± 1.5
2	36.7	33.9 ± 1.4	33.9 ± 1.4
1	24.7	22.4 ± 1.1	22.1 ± 0.9
0.5	15.6	14.1 ± 0.7	13.7 ± 0.7
0.2	2.0	1.9 ± 0.1	1.8 ± 0.1

Table Vb. Horiba LA-910 Results for Sample B

Equivalent Spherical Diameter – microns	Method 1 (% finer)	Method 2 (% finer)	Method 3 (% finer)
5	59.4	60.9 ± 2.0	60.8 ± 1.9
2	35.4	36.2 ± 1.3	36.3 ± 1.2
1	22.9	23.6 ± 0.7	23.5 ± 0.7
0.5	13.8	14.8 ± 0.3	14.6 ± 0.3
0.20	1.8	1.9 ± 0.1	1.8 ± 0.1

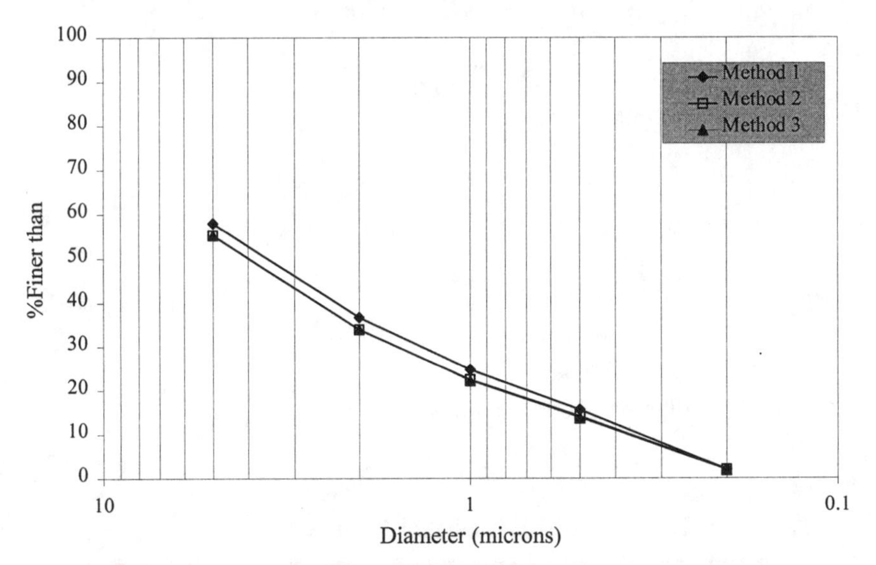

Figure 3. Horiba LA-910 Data for Sample A

CONCLUSIONS

It is very important when dealing with particle size distributions to have a complete understanding of the instrumental method employed, the material being analyzed and the sample preparation method. Even minor changes in sample preparation, can lead to significantly different particle size distribution, especially for techniques such as the Sedigraph. Comparisons of particle size distributions across different instrumental methods is difficult, since no two techniques will give the same answer. The best advice which can be given to anyone attempting to utilize particle size data, particularly for materials such as kaolin, is to make sure that as many parameters as possible are under control, including sample preparation and the technique used. It should also be understood that the particle size distribution for any material which is not spherically shaped, is relative. By having a good understanding of all the parameters involved, particle size distributions collected using instrumental methods provide a quick and easy way to characterize fine particle systems. As long as it is realized that the measurements are relative, not absolute, correlation can easily be established to industrial processes.

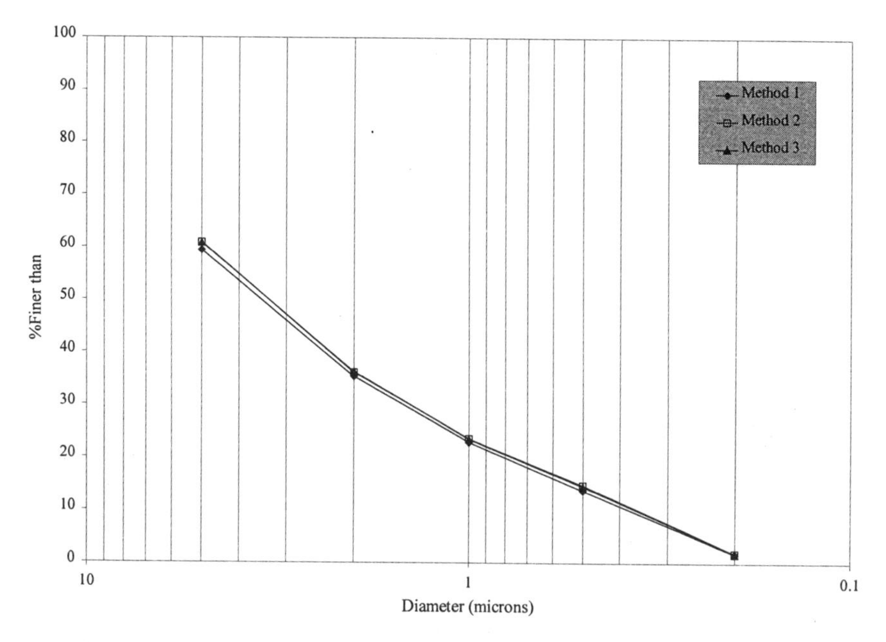

Figure 4. Horiba LA-910 Data for Sample B

REFERENCES

[1]P.A Webb and C. Orr, Analytical Methods in Fine Particle Technology, Micromertics Instrument Corporation, Norcross, GA, 1997.

[2]J.D. Stockham, Particle Size Analysis, Ann Arbor Science, 1977.

ACKNOWLEDGEMENTS

Thanks are extended to the Analytical Services Group of Thiele Kaolin Company for the Horiba LA-910 data.

KAOLIN PARTICLE SIZE DISTRIBUTION EFFECTS ON WHITEWARES-RELATED PERFORMANCE PROPERTIES

T. Adkins, J. Davis, C. Manning, M. Phillips, E. J. Sare
Dry Branch Kaolin Company
Route 1, Box 468-D
Dry Branch, GA 31020

On behalf of all my Dry Branch Kaolin colleagues, I'd like to first thank Professor Carty and the Whiteware Research Center at Alfred University for the opportunity to participate in this second Science of Whitewares Conference. For those of you not familiar with Dry Branch Kaolin, we are the kaolin division of DBK Minerals, which also owns Georgia Marble, the largest United States supplier of calcium carbonate products. Dry Branch Kaolin, formerly Georgia Kaolin, has been a long term supplier of both water-washed and air-floated kaolins to the whitewares industry for almost a century, including products such as our high strength 6-Tile, and high white firing HWF kaolin products.

With the excellent opening presentations already given in the conference, for once I won't spend any time distinguishing between kaolin and ball clay or other clays or processing. However, I would like to begin with one fundamental, but perhaps under-appreciated aspect of the use of kaolin in whitewares and other ceramic bodies, and I don't mean as an inexpensive alumina source. Rather, it's the role kaolin plays in overall particle packing. Consider a 10/30/30/30 calcined alumina, silica, nepheline syenite or feldspar, high plastic kaolin whitewares body. The relative particle size data as determined on a Sedigraph 5100 are summarized in Table I. Notice that the vast majority of the fine end of the overall distribution is the kaolin. Simply considering the overall proportion of components and the median particle size, the kaolin obviously plays a major role in obtaining the desired high strength, high density, and low relative porosity that is necessary in the final product for commercial success. With that in mind, we'd first like to present some relatively broad generalizations with respect to the relationship of particle size to various important physical properties, as a number of you suggested. These initial generalizations are for "ordinary", not "high plastic" kaolins. Thus, as the particle size increases there is a change in the overall particle shape from more plate-like to more block-like. In addition, as the particle size increases, the surface area, dispersant demand, dispersed state viscosity, and CEC, as determined by use of methylene blue, all decrease. Finally, as the particle size increases, the relative rate of dewatering increases markedly. The data summarized in the first two rows of Table II illustrate a number of these physical property dependencies. For these two non-"high plastic" kaolins, it can be seen that the dependence of the CEC on particle size is far less than that of the surface area; and even for this very fine kaolin, the CEC is still relatively low. In the case of the relative dewatering data, it was obtained using DBK's computerized Baroid filter technique as introduced at the 1994 fall ceramics meeting in Louisville, KY. As can be seen, even in the flocced state the fine particle size sample exhibits very slow dewatering characteristics.

With respect to mineral impurities, there are also a number of generalizations. Thus, as the particle size increases, the relative level of mica and quartz generally increases, while the relative level of iron and titanium-containing impurities generally decreases. Thus for high white-firing in a whiteware body, larger, higher overall chemical purity particle size products are generally used.

A number of whiteware body-related properties can also be generalized. In the unfired body both the water of plasticity and shrinkage generally decrease as the particle size increases. In contrast, the modulus of rupture increases. With respect to the fired body, the firing shrinkage and water absorption generally decrease, while the modulus of rupture and fired whiteness generally increase as the particle size increases.

In these generalizations, we've assumed "normal" or "traditional" kaolin particle size distributions. In the case of water-washed products there are a number of processing options available. If we deliberately modify these distributions using these options, particularly with respect to the fine particle end, major changes in this general behavior can be made. As an example consider the following two relative fine kaolins, both 90+% less than 2μ. As shown in Table III, these kaolins differ in particle size distribution primarily below 1.0μ. The effect of this type modification on surface area and methylene blue index are also summarized in this Table, and the relative change in surface area is extremely pronounced. The relative dewatering characteristics of these two materials are contrasted in Table IV. As can be seen, these nominally similar products exhibit very significantly different dewatering characteristics. Thus the dewatering rate of the lower fines kaolin is almost twice as fast as that of the normal distribution kaolin. In fact, the dewatering rate of this material begins to approach that typical of

Table I. Comparison of Whiteware's Casting Body Components: Particle Size Distribution

Sample	Density (g/cm³)	Sedigraph 5100			
		Median	% < 10μ	% < 5μ	% < 2μ
Calcined Alumina	4.02	3.6	96	77	12
Silica	2.65	10.4	49	27	11
Neph. Sy.	2.61	7.2	66	36	14
High Plastic	2.62	0.9	96	87	68

Table II. Comparative Physical Property Summary

Kaolin Type	Sedigraph 5100		BET (m²/g)	CEC Methylene Blue (meq/100g)	Relative Dewatering Rate (Baroid)[1] Filtrate Volume, ml.		
	% < 2μ	% < 1μ			10 min	50 min	100 min
Water-Washed Fractionated Coarse Particle	35	24	8.9	2.6	54	"B"[2]	
Water-Washed Fractionated Fine Particle	97	91	19.9	3.9	7	22	"B"[3]
Air-Floated High Plastic Intermediate Particle	66	50	22.5	10.8	12	30	43

[1] 55% solids, flocced to pH = 3.5
[2] "B" - Blowout @ 26 minutes, 70.4% cake solids.
[3] "B" - Blowout @ 99 minutes, 70.0% cake solids.

Table III. Fine Kaolin Physical Property Comparison

Kaolin Type	Sedigraph 5100			BET (m²/g)	CEC Methylene Blue (meq/100g)
	% < 2μ	% < 1μ	% < 0.5μ		
Water-Washed, Fractionated Fine Particle	91	80	62	17.6	3.4
Water-Washed, Fractionated "Low Fines"	91	76	42	10.9	2.9

Table IV. Dewatering Characteristics Summary - Fine Kaolins

Time, min.	Filtrate Volume, ml.	
	"Normal" Distribution	Low "Fines" Distribution
10	7	28
25	14	39
50	22	59
75	28	74
100	34	"B"
200	50	
300	63	

"B" - Blowout @ 99 minutes, 70.0% solids

Table V. Dewatering Characteristics Summary - Flocced "Coarse" Kaolins

Time, min.	Filtrate Volume, ml.	
	4.6μ Median Diameter	3.8μ Median Diameter
10	34	31
20	51	46
30	65	58
20	88	81
75	"B"[1]	"B"[2]

[1] "B" - Blowout @ 64 minutes, 73.0% solids.
[2] "B" - Blowout @ 72 minutes, 73.2% solids.

Table VI. Dewatering Characteristics Summary - Whiteware's Casting Body

Time, min.	Filtrate Volume, ml.	
	Lightly Flocced	Flocced
5	21	48
10	31	71
15	39	89
20	46	105
40	67	"B"
80	97	"B"
120	"B"[1]	"B"[2]

[1] "B" - Blowout @ 118 minutes, 78.4% solids.
[2] "B" - Blowout @ 22 minutes, 75.8% solids

significantly coarser, intermediate particle size kaolins. This result illustrates the large effect of fines on permeability. This is also illustrated by comparison of two coarse kaolins which have very similar distributions in the fine end. As shown in Table V, although these materials differ significantly in the coarse end of the distribution, they exhibit very similar overall dewatering behavior. In addition, properties such as surface area and CEC are essentially identical within experimental error.

To this point, we've deliberately excluded "high plastic" kaolins from discussion. These are naturally occurring smectite-modified kaolins found in some deposits in Georgia, not kaolins to which a smectite, such as montmorillonite, has been added artificially. In terms of physical properties, the following generalizations can be made. At comparable particle size, high plastic kaolins will have higher surface area, dispersant demand, viscosity, and CEC, while the overall casting rate will be reduced. Several of these are illustrated in Table II. As can be seen, for their relative particle size high plastic kaolins have proportionately very high methylene blue CEC and surface areas.

In terms of overall body properties, "high plastic" kaolins also generally exhibit similar to somewhat higher water of plasticity and drying shrinkage, and a higher modulus of rupture, than a comparable non-high plastic kaolin. In the case of fired body properties, the firing shrinkage and modulus of rupture are generally similar to higher than that of a comparable particle size non-high plastic kaolin. In contrast, the fired whiteness is generally lower to comparable.

Before leaving this section, we'd like to briefly "revisit" the 10/30/30/30 calcined alumina, nepheline syenite or feldspar, silica, high plastic kaolin body. The relative dewatering characteristics of this body at two different levels of flocculation are summarized in Table VI. As can be seen, the relative rates of dewatering of these flocced bodies differ by over 200%. Thus, the more difficult to characterize flocced state also plays a major role in the overall commercial utilization of kaolin.

We'd like to end by taking this opportunity to introduce a new surface characterization instrument, the Vertical Scanning Interferometer (VSI). A schematic is shown in Figure 1. We've been using our VSI to assist a number of our customers in a diverse number of markets. This type vertically scanning interference microscope enables three-dimensional, non-contact measurements of relatively large areas of surfaces to be made rapidly. Thus it differs from techniques such as the scanning electron microscopy. The use of this instrument

Figure 1. Interference Microscope

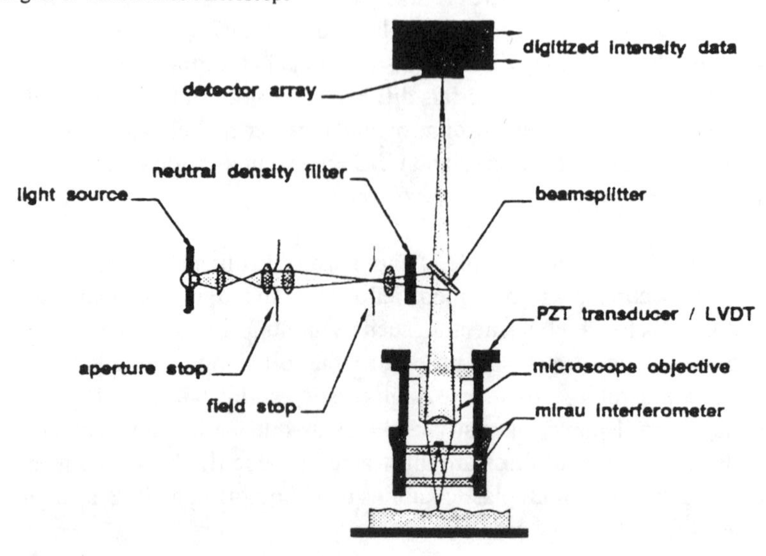

Figure 2. NIST Step-Height Standard

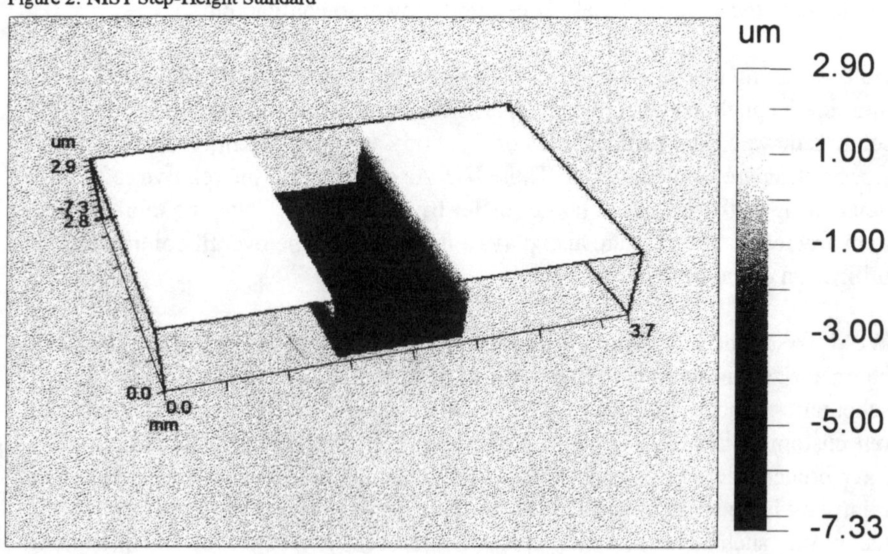

is illustrated in Figure 2 which shows the measurement of a NIST-certified ten-micron step-height calibration standard. As shown, a pseudo-three-dimensional representation is obtained over the area of measurement. In these Figures, red will always represent structure above the average surface, and blue represents the structure below. Our next illustration shows the internal surface of a porous ceramic catalyst substrate. As can be seen in Figure 3, this method not only three-dimensionally visualizes the surfaces, but can also be used to accurately measure the pore dimensions.

In addition to the relatively obvious use to characterize the exterior surfaces of whiteware products, the VSI can also be used to characterize the surfaces of the various mold surfaces used to produce the ware. Figures 4 and 5 show a representative portion of the surface of a gypsum casting mold both before and after use. In addition to the wear pattern, which can be seen by comparing these Figures, the relative size of some of the pores should be noted. Since the VSI is both a non-vacuum and non-contact instrument, it can be used to follow the effects of wetting on the overall surface and/or pores, for example, wear during glaze application. As a simple example of this type application, Figure 6 graphically summarizes the change in the overall surface roughness of a gypsum mold due to wetting. As can be seen, there was an almost 45% change in the overall roughness. From analysis of the corresponding contour surface topography it was found that these changes were associated with relatively pronounced increases in the pore dimensions.

In a final example of this technique, Figure 7 shows the surface topography of a body in which significant glaze-related problems were being experienced. These difficulties were shown to be associated with the relatively large depression areas which occurred in the body. As can be seen in Figure 8, the difficulty wasn't the relatively shallow depressions, per se, but rather the associated presence of relatively deep, narrow imperfections within these areas.

Figure 3. Catalyst Substrate Internal Surface #4

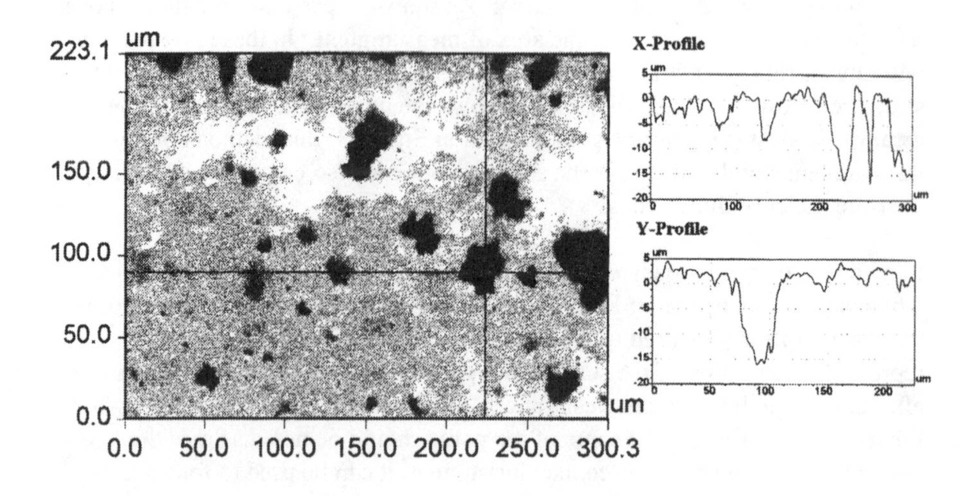

Figure 4. Mold A Unused

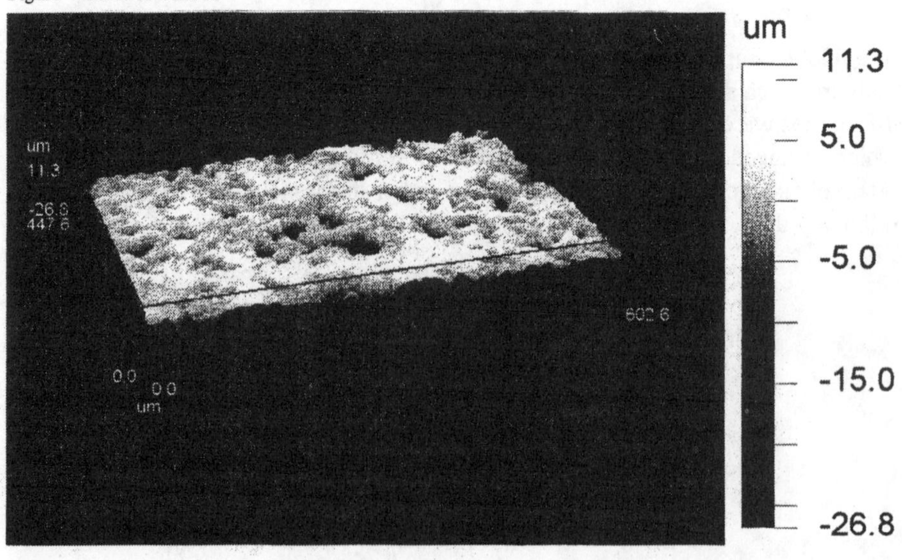

Figure 5. Mold A Used

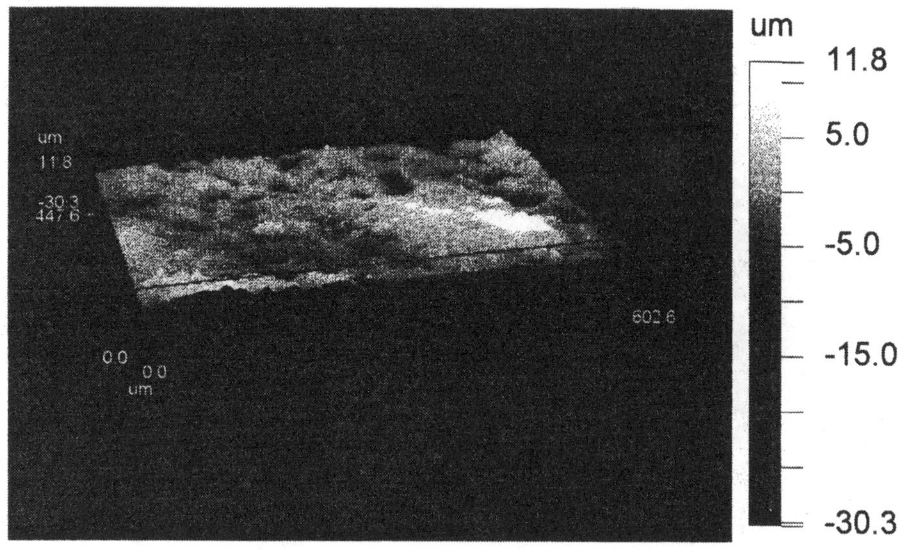

Figure 6. Roughness versus Time-After Wetting

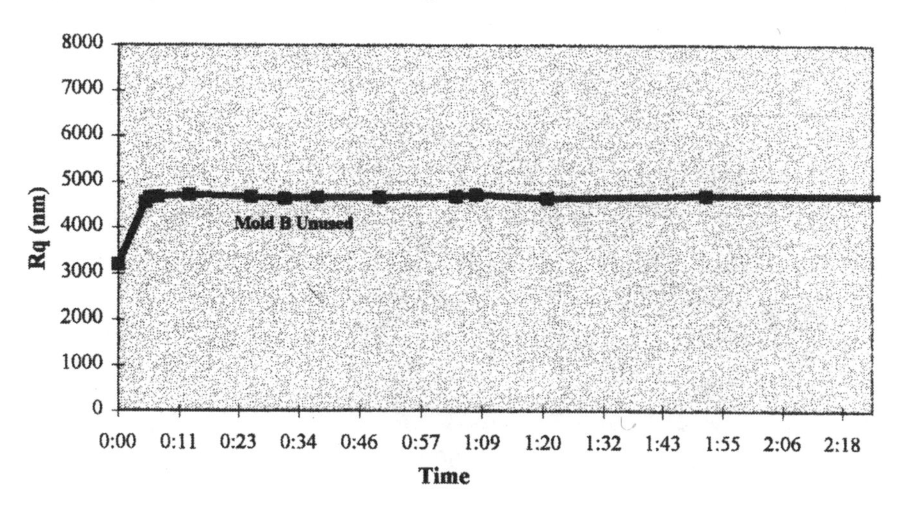

Figure 7. Base Whiteware Body-Glaze Upset

Figure 8. Base Whiteware Body-Glaze Upset

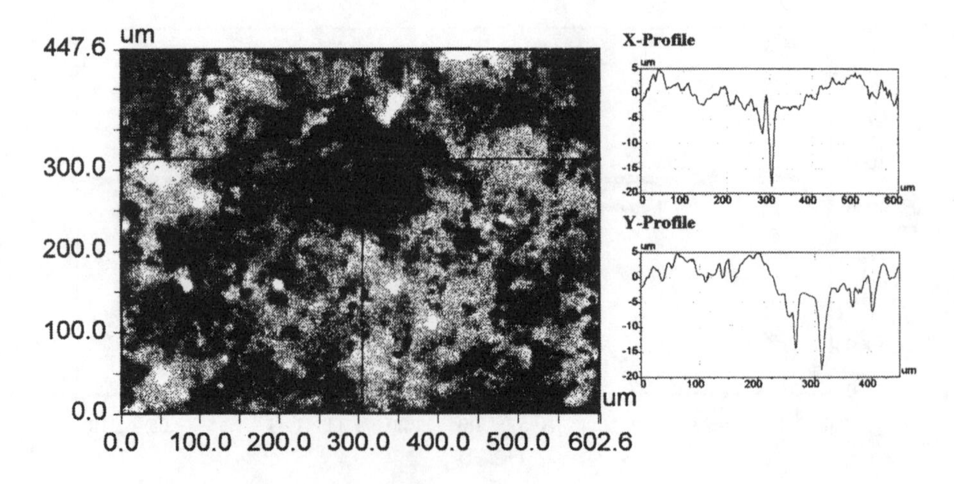

FTIR SPECTRAL ANALYSIS OF ORGANIC ADDITIVE/CERAMIC OR GLASS INTERACTIONS

R.A. Condrate, Sr. and D.H. Lee
NYS College of Ceramics
Alfred University
Alfred, NY 14802

ABSTRACT

Vibrational spectral techniques such as infrared (FTIR) spectroscopy provide useful structural information relating to the physical and chemical properties of coated materials such as either the nature of the structural species that are bonded on a ceramic or glass material, or what phases that are present in the investigated material. This paper will discuss the application of a particular infrared reflectance spectral technique, the so-called Diffuse Reflectance Infrared Fourier Transform (DRIFT) spectroscopic technique, which can be used to obtain structural information concerning the bonding of organic coatings on ceramic or silicate glass surfaces. The investigated organic coating materials include polyacrylates (PAA), polymethyl methacrylate (PMMA), oleic acid or stearic acid, while the substrates include alumina, barium titanate, bismuth cuprate superconductors, soda-lime-silicate glass or silica glass. The spectral data clearly indicates that such coatings bond to the substrate surface either by bonding their carboxylate groups to metal cations on the substrate surface or by hydrogen-bonding protonated carboxylate groups. The nature of the substrate surface and/or the conditions of sample treatment determine the nature and concentrations of the various bonding species that occur at the coating/substrate interface.

INTRODUCTION

Surface infrared spectroscopic techniques such as Diffuse Reflectance Infrared Fourier Transform (DRIFT) spectroscopy can generate useful structural information regarding how a particular organic coating bonds to a ceramic or glass surface. Such interface interactions directly relate to the physical and chemical properties of the coated materials. In this paper, the structural interface interactions between various organic coatings and either ceramic substrates such as alumina, barium titanate and bismuth cuprate superconductors or SiO_2-containing glass substrates such as soda-lime-silicate, sodium silicate and silica will be analyzed using DRIFT spectroscopy. The investigated organic coating materials such as oleic acid, stearic acid, polyacrylates (PAA) or polymethyl methacrylate (PMMA) possess carboxylate groups which can chemically interact with the substrate surface, and improve the bonding between the coating and the substrate.

EXPERIMENTAL DRIFT SPECTRAL PROCEDURES

Both transmission and DRIFT spectra were measured in the 400-4000 cm^{-1} region using a Fourier Transform Infra-red (FTIR) spectrometer (Nicolet model 60-SXR). Transmission spectra were measured either on KBr pellets of the various metal-oleates or on thin films of sodium polyacrylate that was dried on silicon windows. Each spectrum was scanned 32 times to generate a signal-averaged spectrum. DRIFT spectra were obtained for the uncoated or the coated powders, using a diffuse reflectance accessory. The investigated powder was placed in a micro sample holder without diluting with KBr. Each spectrum was scanned 2048 times to generate a signal-averaged DRIFT spectrum whose signal to noise ratio has been improved. The DRIFT spectra were obtained for various coated and uncoated powders using a DRIFT cell whose configuration is illustrated in Fig. 1. The cell is designed to generate a diffuse infrared spectrum for a powdered sample, using and integrating the spectral results with respect to all of the angles of reflectance with respect to the substrate surface that are possible for the cell.

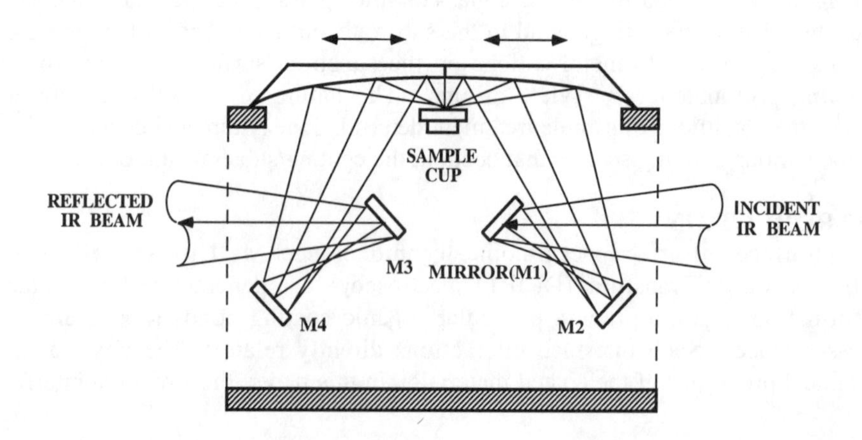

THE INVESTIGATION OF POLYACRYLATE DISPERSANTS ON ALUMINA

In many cases, useful structural information relating to the bonding at the dispersant/ceramic or glass powder interface can be obtained by looking first at the DRIFT spectra of the untreated substrate materials. Fig. 2 illustrates the DRIFT spectra for several different alumina samples with varying degrees of dispersion capability [1]. Sample A4 possessed the best dispersion properties, samples A1 and A2 possessed intermediate properties, and sample A3 possessed the worst dispersion properties. Bands in the hydroxyl-stretching region (3000-4000 cm^{-1}) clearly indicate that the alumina surface should contain bayerite- and/or gibbsite-type hydroxyl groups for better interaction with the dispersant. Also, the infrared bands in the carbonate region (1300-1700 cm^{-1}) indicate that formation of dawsonite [$NaAl(OH)_2CO_3$] on the alumina powder particles hinders the interaction of some powder particles with the dispersant. Further Drift spectral data indicate that the dawsonite can be removed, and the hydroxyl groups can be developed with sufficient washing [1].

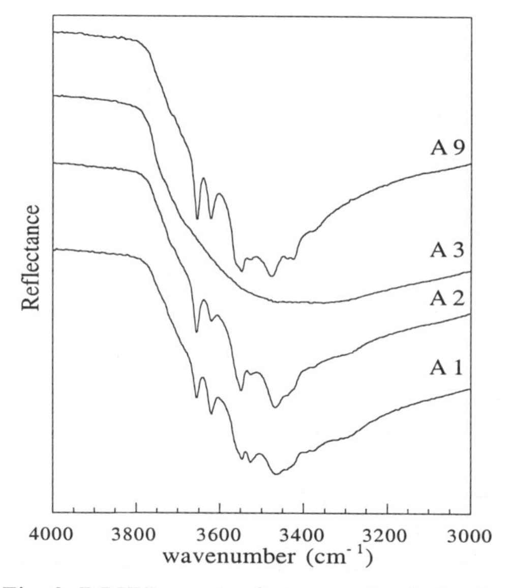

Fig. 2. DRIFT spectra for as-received alumina.

DRIFT spectroscopy can also indicate how polyacrylate polymer chains are bonded with alumina surfaces [2]. The nature of the bonding and the configuration of the polyacrylate species that are present on the substrate surface depend upon the conditions of treatment. For this polymer, The C=O stretching mode associated with the carboxylate group will shift its wavenumber location, depending upon how it bonds to the surface. If protonated carboxylate species form hydrogen bonds across the coiled polymer chains forming dimers, the C=O stretching mode occurs at ca.1710 cm^{-1}. If isolated protonated carboxylate groups are present or they are hydrogen-bonded side-ways along the chain with each other, the C=O stretching mode is located at ca. 1760-80 cm^{-1} and 1740 cm^{-1}, respectively. When the carboxylate groups bond to aluminum ions on the alumina surface, the band shifts to ca. 1600 cm^{-1}. In the case of ionized carboxylate groups (possibly bonded to sodium ions), the related infrared band is located at ca. 1570 cm^{-1}.

Fig. 3 illustrates the DRIFT spectra of dried polyacrylate-coated alumina powders that were prepared with variations during treatment in the pH of the starting solution and/or the sodium polyacrylate concentration. The infrared band in all of the observed spectra at ca. 1600 cm^{-1} clearly indicates that polyacrylate bonding with the alumina surface involves forming aluminum-carboxylate complexes. The spectral data also indicate that at higher pH and lower polyacrylate concentration, the concentrations of the side-ways hydrogen-bonded protonated carboxylate species and the isolated protonated carboxylate species is dominant with respect to the protonated carboxylate species. Conversely, at lower pH and higher polyacrylate concentrations, the hydrogen-bonded protonated carboxylate dimeric species is dominant. The structural model that best describes this variation in band intensities is a model in which the polyacrylate chains are stretched out at high pH and low polyacrylate concentration on the alumina surface, mainly forming metal-carboxylate bonds along the chain. While, at low pH and high polyacrylate concentration, the chains are more coiled up, allowing the formation of hydrogen-bonded dimeric units of the protonated carboxylate groups across the chain. Even in the latter case, the chains are strongly bonded to the alumina surface by the formation of aluminum-carboxylate bonds.

THE INVESTIGATION OF PROTECTANT ORGANIC COATINGS ON $BaTiO_3$

DRIFT spectroscopy also provides a useful tool to investigate the reaction mechanisms associated with surface degradation or surface protection on $BaTiO_3$ [3]. The following DRIFT spectral data clearly illustrate this possibility with $BaTiO_3$ powder as the substrate material. Surface degradation of this substrate material can easily occur either in a humid atmosphere containing carbon dioxide or in a water solution containing carbonate species. Fig. 4 illustrates the DRIFT spectra for $BaTiO_3$ powders (Ba/Ti=1.007) that were aged in humid air after calcination at 1200°C. The spectra clearly indicate the increased development of a $BaCO_3$ species with time as the sample was aged in humid air. Placing coating

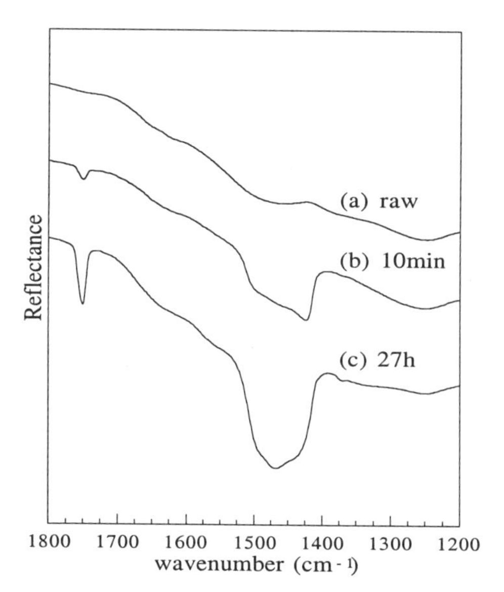

Fig. 4. DRIFT spectra for $BaTiO_3$ powders (Ba/Ti=1.007) aged in humid air after calcination at 1200°C.

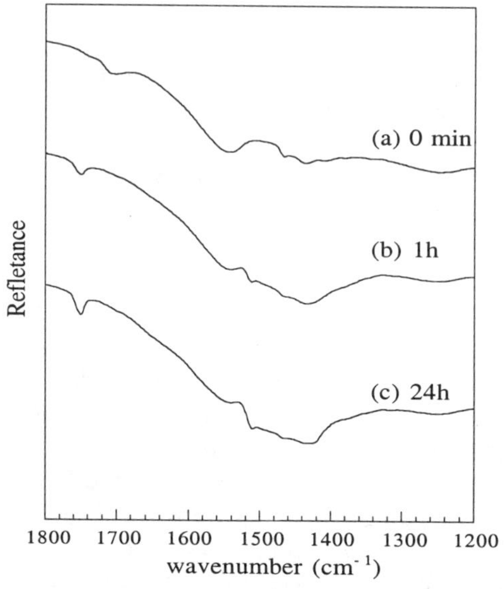

Fig. 5 DRIFT spectra for $BaTiO_3$ powders (Ba/Ti=1.007) aged in humid air after calcination at 1200°C followed by coating with oleic acid.

of oleic acid on the surface of $BaTiO_3$ powder can protect its surface from a carbonate attack. Similar spectral changes with respect to carbonate formation were noted for $BaTiO_3$ with the Ba/Ti molar ratio=0.992. However, its rate of carbonate species formation was slower than was observed for $BaTiO_3$ with the higher Ba/Ti molar ratio. Fig. 5 illustrates the DRIFT spectra for $BaTiO_3$ powders (Ba/Ti=1.007) that were aged in humid air after both calcination at 1200°C and oleic acid coating. One can observe that the relative concentrations of carbonate species on the surfaces of these coated samples are lower than on the previously illustrated uncoated samples.

THE INVESTIGATION OF LUBRICATING OR PROTECTANT ORGANIC COATINGS ON VARIOUS SiO_2-CONTAINING GLASSES

DRIFT spectroscopy can also produce structural information concerning organic coatings that are bonded on silicate-containing glasses for the lubrication or the protection of their surfaces [3,4]. Either oleic or stearic acid can be used to lubricate glass surfaces. Fig. 6 illustrates the DRIFT spectra that were obtained for freshly-ground soda-lime-silicate glass which was coated with oleic acid, and also, for synthesized metal oleates. The former spectrum is a DRIFT difference spectrum involving the subtraction of the spectrum of freshly-ground glass from that of the coated glass. Bands clearly indicate the presence of metal-carboxylate complexes involving the major metal ions found in the glass such as sodium, calcium, magnesium and aluminum ions. Also, the infrared band above 1700 cm^{-1} indicates the presence of physically-adsorbed protonated carboxylate groups. The relative intensities of these bands relate to the relative concentrations of the various metal-carboxylate complexes. Fig. 7 illustrates the DRIFT spectra of several different types of silicate glasses. The spectra for fused and fumed silica indicate that only physically-adsorbed protonated carboxylate groups occur on their surfaces. Also, the related band associated with the C=O stretching mode of the carboxylic acid group may be broader for fumed silica than that for fused silica due to the particle size differences between the two types of silicas.

Variations in the treatments of the uncoated or coated glasses can alter the structural organic species that are present on the glass surfaces. For instance, either changing the slurry solvent in which the coating material was dissolved or washing the glass before it was coated can cause changes with respect to the organic species on their surfaces. Fig. 8 illustrates the DRIFT spectra that were obtained for coated glass powders which were generated from slurry solutions involving benzene or ethyl alcohol as the slurry solvent. One may note that that the relative band intensity associated with the sodium-oleate species is more intense. This result is expected because benzene is non-polar and therefore, the related slurry solution will extract less sodium-containing species from the glass surface than one using ethyl alcohol. The effects of washing with water the glass

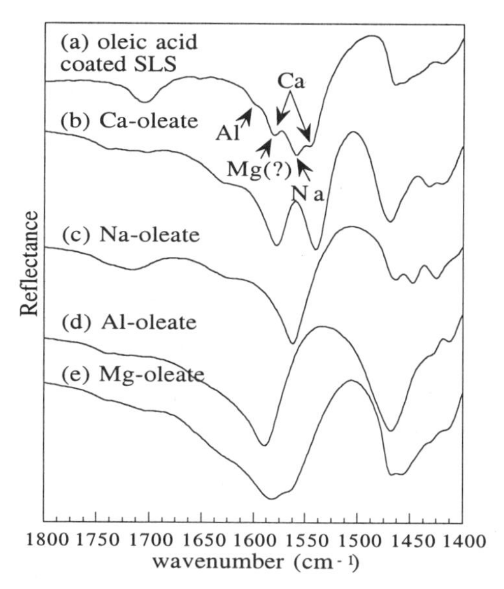

Fig. 6 DRIFT spectra for (a) the freshly-ground soda-lime-silicate glass coated with oleic acid and (b)-(e) synthesized metal-oleates.

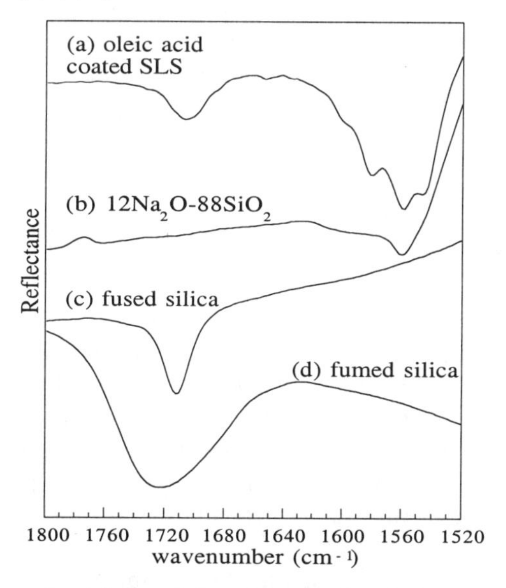

Fig. 7 DRIFT spectra for (a) the oleic acid-coated soda-lime-silicate glass, (b) the coated soda-silicate glass, (c) the coated fused silica glass, and (d) the coated fumed silica glass.

surface before coating can also be seen in Fig. 8. The spectral data clearly indicates that there is more sodium-oleate complex formation on the freshly-ground glasses than on the glasses that were washed in water. This is to be expected because sodium ions are highly soluble in water, and will be extracted from the glass surface. Therefore, the relative band intensities and the related concentrations for the other metal-carboxylate species will be higher with washing than that for the sodium-carboxylate species.

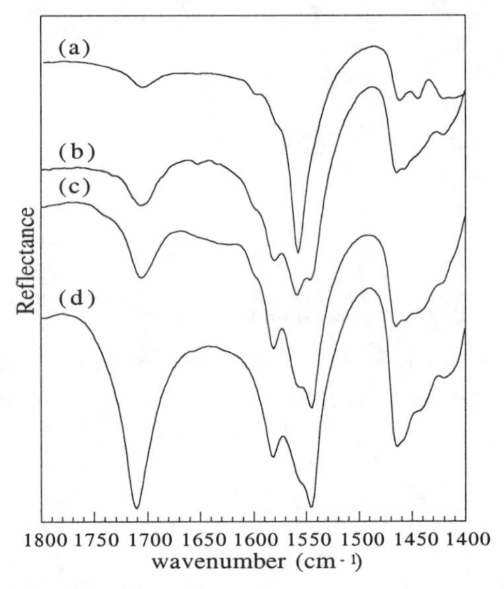

Fig. 8 DRIFT spectra for oleic acid coated on soda-lime-silicate glass from (a) benzene, (b) ethyl alcohol without agitation, (c) ethyl alcohol with agitation, and (d) ethyl alcohol after washing glass with water for 19h.

CONCLUSIONS

The following conclusions can be drawn on the basis of this investigation:

1.) DRIFT spectroscopy provides a useful and flexible tool for investigating the nature of adsorbed organic surface species on ceramic and glass powders.

2.) The surface species on alumina powders before and after coating with polyacrylate can be investigated by DRIFT spectroscopy, generating structural information concerning the nature of the bonding of the dispersant to the alumina surface.

3.) DRIFT spectroscopy can be used to determine the structural interactions that occur during both degradation and placing protective coatings on both $BaTiO_3$ and bismuth cuprate superconductors. Clearly, the spectral data indicate that the organic coating on the ceramic powder protects its surface from carbonate attack.

4.) Structural variations in the organic species that can occur on glass surfaces with variations in the treatment conditions can be determined from the DRIFT spectral data. For instance, the spectral data indicate that the relative concentrations of the various carboxylate complexes formed on the glass surface with the various metal cations in the glass composition varies with respect to the nature of the solvent.

REFERENCES

[1]D.H. Lee and R.A. Condrate, Sr., "An FTIR Spectral Investigation of the Structural Species Found on Alumina Surfaces," *Mater. Lett.* **23** [3] 241-246 (1995).

[2]D.H. Lee, R.A. Condrate, Sr. and J.S. Reed," Infrared Spectral Investigation of Polyacrylate Adsorption on Alumina," *J. Mater. Sci.* **31** [2] 471-478 (1996).

[3]D. H. Lee, "Thin Film Coatings on Glasses or Ceramics and Structural Determination by DRIFT Spectroscopy," Ph.D. Thesis, Alfred University (1998).

[4]D. H. Lee and R.A. Condrate, Sr., "Infrared Reflectance Spectral Characterization of Various Organic Coatings on Glasses," *J. Non-Cryst. Solids* **222** [3] 435-441 (1997).

MICROSTRUCTURAL EVOLUTION IN DENSE KAOLINITE, ILLITE AND SMECTITE CLAY BODIES

C. J. McConville,* W. E. Lee and J. H. Sharp
Department of Engineering Materials, University of Sheffield,
Sheffield, S1 3JD, United Kingdom

* Present address: New York State College of Ceramics at Alfred University,
2 Pine Street, Alfred, NY 14802, USA

ABSTRACT

Thermal transformations occurring in clay minerals from three different clay groups; kaolinites, illites and smectites were studied using X-ray diffraction and transmission electron microscopy techniques. After 3 hours at 1400°C, mullite, cristobalite and amorphous phases only remained in all three materials, but differences were observed in the types and sizes of crystals formed at intermediate temperatures. Crystal structure breakdown of the clay minerals took place at different temperatures; kaolinite below 600°C, illite above 900°C and smectite above 800°C. Spinel-type phases were formed after 3 hours at 1000°C in all materials, but the size, morphology and composition of the spinel crystals varied, with 5 nm single crystals in kaolinite, orientated 100 nm rods in illite, and randomly orientated 100 nm rods in smectite. Mullite crystals reached greater sizes in the illite (>10 μm needles) and smectite (>1 μm) materials than in the kaolinite (0.5 μm). Haematite crystals formed in the illite and smectite materials, and feldspars present as impurity in these clays, along with interlayer cations (potassium and calcium), resulted in large quantities of liquid phases being formed above 1000°C.

INTRODUCTION

Kaolinite ($Al_2Si_2O_5(OH)_4$) is one of the most widely used clay minerals. It is a major component of whiteware and fireclay bodies and is present in cements, bricks and roof tiles. Kaolinite is a two layer (1:1) silicate, with one layer of silica tetrahedra, and one layer of octahedral gibbsite. Illite ($K_{1.5}Al_4(Si_{6.5},Al_{1.5})O_{20}(OH)_4$) is a major component of brick and tile clays, as are the smectite group clays, which include the Ca-montmorillonites ($Ca_{0.33}(Al,Mg)_4Si_8O_{20}(OH)_4.xH_2O$). Both illites and smectites are 2:1 silicates, consisting of an octahedral layer surrounded by two sheets of silica tetrahedra. These materials also have interlayer cations, such as potassium and calcium, which charge-balance substitutions in the octahedral and tetrahedral layers, for example Mg for Al and Al for Si.

Previous research has established the phase transformation sequence observed when a typical kaolinite is fired.[1,2] At 500-600°C, dehydroxylation of the kaolinite takes place. This gives rise to metakaolin, which consists of the kaolinite lattice after most of the hydroxyl groups have been removed. As heating proceeds, the metakaolin is partially replaced by a spinel phase, and mullite begins to form. These transformations take place at 900-1000°C. At 1100-1200°C, cristobalite begins to crystallise, until eventually mullite and cristobalite are the only crystalline phases remaining in the body.

Grim and Bradley established the thermal transformation sequence of illite and montmorillonite.[3] The dehydroxylation of illite starts at about 350°C and is largely complete at 600°C. Its original lattice structure remains, in dehydrated form, until around 700°C, when it begins a gradual decomposition. At 850°C, a spinel-type phase begins to form from the octahedral layer. The tetrahedral silica layer forms a separate liquid (glass) phase, starting as low as 950°C. At 1100°C, a mullite phase develops, and remains until at least 1400°C. If heating is continued, the spinel phase dissolves in the liquid at about 1300°C. Any quartz present persists until 1050°C, when it too dissolves in the liquid phase. It does not recrystallise as cristobalite.

Montmorillonite dehydroxylates below 600°C, and the clay crystal structure breaks down by 800°C.[3] A spinel-type phase forms at about 850°C, and persists until between 1100°C and 1300°C, when it dissolves in the liquid resulting from reaction of the silica portion of the clay lattice and any alkalis present. Mullite formation begins around 1050°C, and increases with increased temperature, and cristobalite develops at higher temperatures.

Many researchers have examined the decomposition of individual kaolinite particles by transmission electron microscopy (TEM) by firing the material, and then depositing a suspension of particles on carbon-coated copper grids before TEM examination.[4] McConville et al.[5] have recently examined microstructural evolution on firing dense kaolinite bodies.

Although some TEM studies of the thermal breakdown of the 2:1 clay minerals such as illites and smectites have been made,[6] on the whole these materials have been neglected. The present study utilises X-ray diffraction (XRD) and TEM to determine the phase and morphological changes taking place in three different clay materials, allowing comparison of the microstructural features observed.

EXPERIMENTAL METHODS

Three different raw materials were used, one each from the clay mineral groups kaolinite, illite and smectite. The kaolinite material was a 99% pure English China Clays kaolinite (previously investigated by McConville et al.[5]), the illite material was Silver Hill clay, a model illite from Georgia, USA, with a potassium interlayer cation and kaolinite present as an impurity. The smectite was Berkbond clay, a commercially-used Ca-montmorillonite, which also contained illite as an impurity. The kaolinite was extruded into bars, and cut into specimens of dimensions: ≈ 8 mm x 15 mm x 15 mm. The Silver Hill and Berkbond clays were uniaxially pressed into pellets (≈ 10 mm x 10 mm) at a pressure of 250 MPa. All samples were fired in alumina crucibles to temperatures at 100°C intervals between 500°C and 1400°C. Heating and cooling rates for all samples were 3K/minute, and soak times were three hours. One pellet from each firing was crushed in an impact mortar, and ground to a powder (<75 μm) in an agate mortar. This powder was used for XRD analysis using the pressed cavity mount technique. A similar method was used for the unfired raw material. Standard TEM sample preparation techniques were used.

RESULTS

On heating, the dehydroxylation of kaolinite was largely complete by 600°C and the product, metakaolin, was amorphous to X-rays as indicated by the XRD pattern obtained after heating to 800°C shown in figure 1.

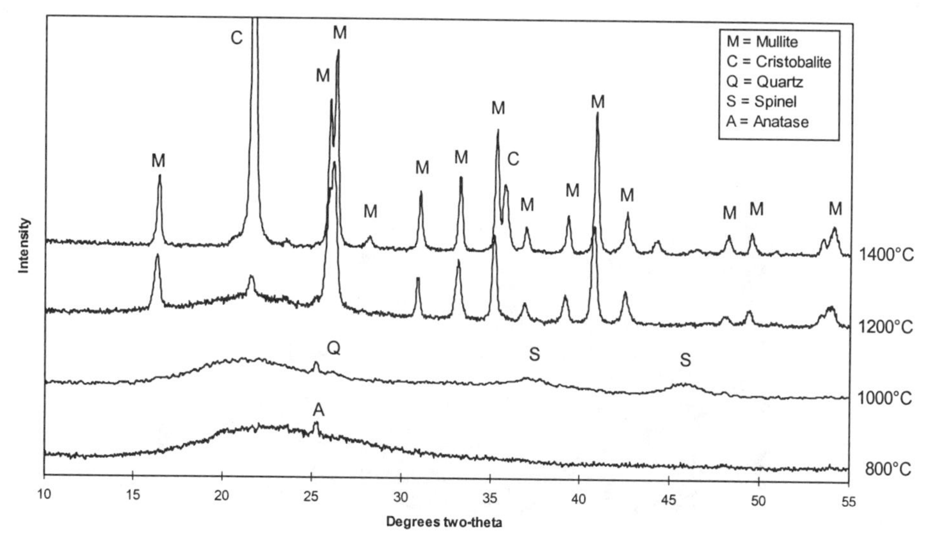

Figure 1. X-ray diffraction of dense kaolinite clay fired to the indicated temperatures.

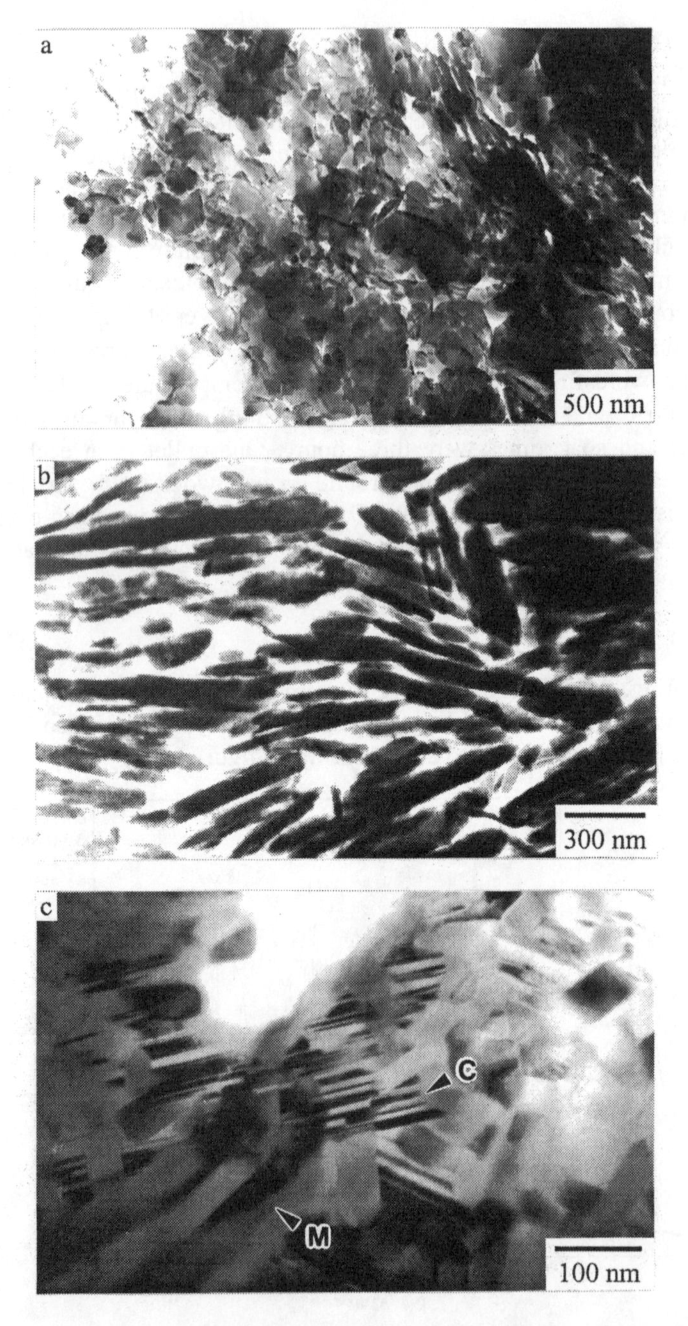

Figure 2a-c. Dense kaolinite clay fired 3h at (a) 800°C, (b) 1000°C and (c) 1300°C.

The metakaolin platelets retain their original pseudo-hexagonal kaolinite particle morphology, as in figure 2a. As heating continues, a spinel-type phase is formed from the aluminous sheet of the metakaolin. The spinel crystals, thought to be γ-Al_2O_3,[7] are approximately 5 nm in diameter, and surrounded by an amorphous, silica-rich matrix phase. Their size means that the spinel XRD reflections are very diffuse (figure 1) and they are seen as a 'texture' within the particles by TEM (figure 2b). Mullite is formed above 1000°C, and the spinel phase is lost by 1200°C (figure 1). Above 1200°C, the amorphous silica matrix crystallises around the previously-formed 100-200 nm mullite crystals to form cristobalite. Figure 2c shows cristobalite (marked 'C' - mullite is marked 'M') crystals containing planar defects. By 1400°C, prismatic mullite crystals have reached up to 0.5 μm in length.

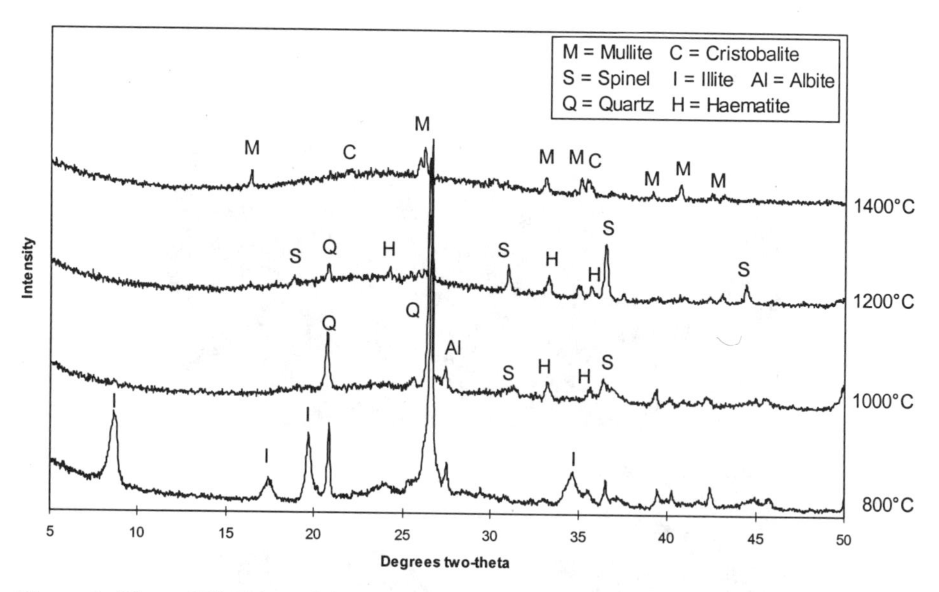

Figure 3. X-ray diffraction of dense Silver Hill clay fired to the indicated temperatures.

The illite component of Silver Hill clay is dehydroxylated to illite dehydroxylate on firing. This phase retains the illite structure, and the XRD peaks remain until above 800°C (figure 3). Figure 4a shows that the lath-like illite particle morphology has been retained after dehydroxylation. As heating continues, a spinel-type phase is formed from the aluminous sheet of the illite. The spinel crystals (S) are larger than those in kaolinite (100 nm - giving sharper XRD reflections (figure 3)), and appear to be orientated within the original clay mineral particles (figure 4b). Energy dispersive spectroscopy (EDS) suggests

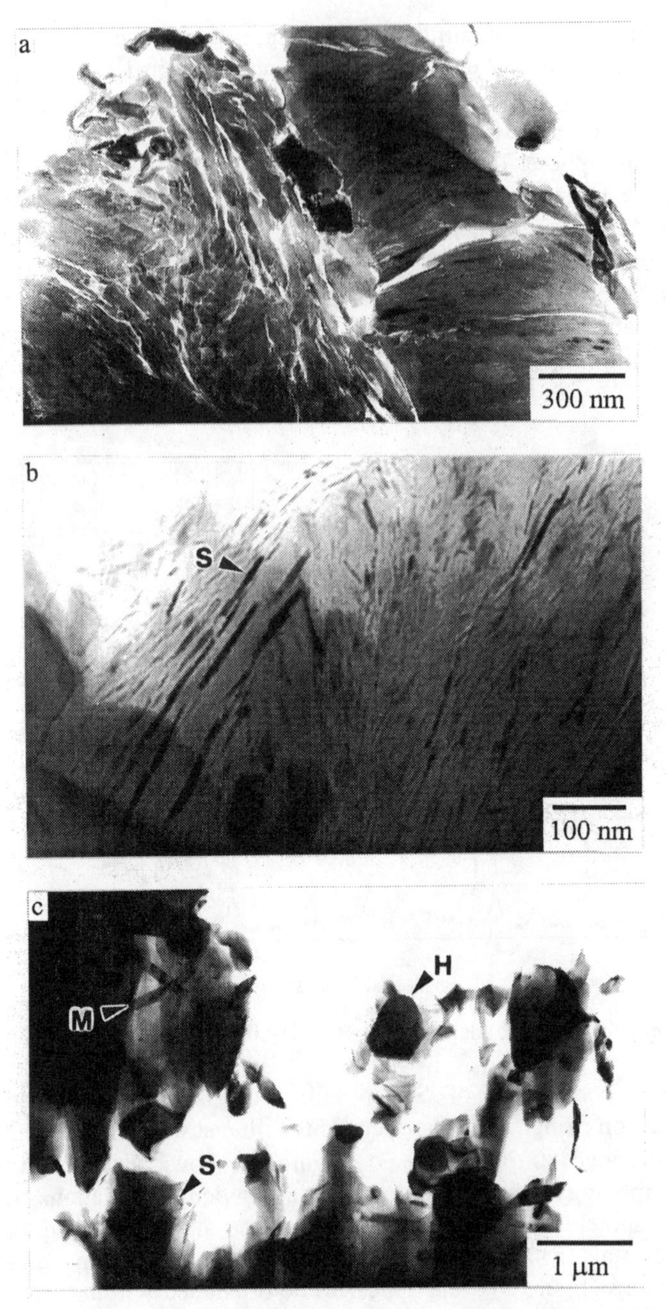

Figure 4a-c. Dense Silver Hill clay fired 3h at (a) 800°C, (b) 1100°C and (c) 1200°C.

they contain Mg as well as Al, arising from the alumina (octahedral) layer, and are surrounded by an amorphous, silica-rich matrix phase arising from the silica (tetrahedral) part of the clay lattice. Haematite (Fe_2O_3) and mullite crystals are observed by TEM at 1000°C, and by 1200°C by XRD. The sample consists mostly of glass above 1100°C, with just a few small peaks in the XRD patterns (figure 3). The feldspar albite, present in the raw material, melts above 1100°C and acts as a flux for the silica-rich glass phase, which dissolves the quartz by 1400°C. Mullite (M) (1 μm), haematite (H) (0.5 μm) and 0.25 μm spinel (S) crystals are observed at 1200°C (figure 4c), and spinel melts by 1300°C. The small amount of cristobalite detected by XRD (figure 3) might be formed from the kaolinite impurity present in the raw material. By 1400°C, mullite crystals have grown to >10 μm in length.

The Ca-montmorillonite (along with illite impurity) dehydroxylates around 500-600°C,[3] but the XRD reflections are present until structural breakdown above 900°C (figure 5).

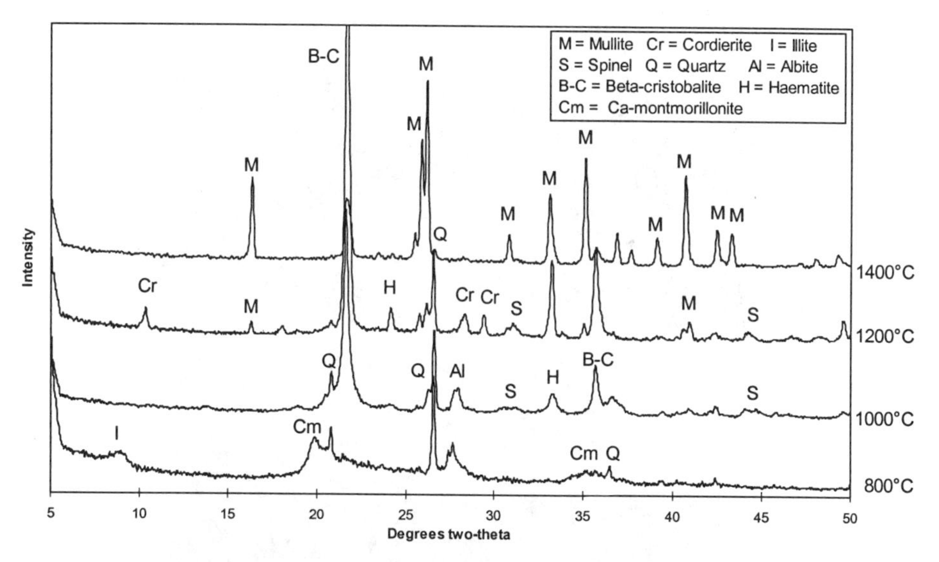

Figure 5. X-ray diffraction of dense Berkbond clay fired to the indicated temperatures.

The convoluted morphology of the clay mineral is still present at 800°C (figure 6a). Above 900°C, a spinel-type phase which EDS suggests contains Mg as well as Al, and of 100 nm size crystallises in the silica-rich liquid phase formed from the breakdown of the clay (figure 6b). Mullite and haematite are formed at 1000°C, and metastable β-cristobalite crystallises from the liquid. Figure 6c shows the variety of crystals formed within the silica-rich liquid matrix; rod

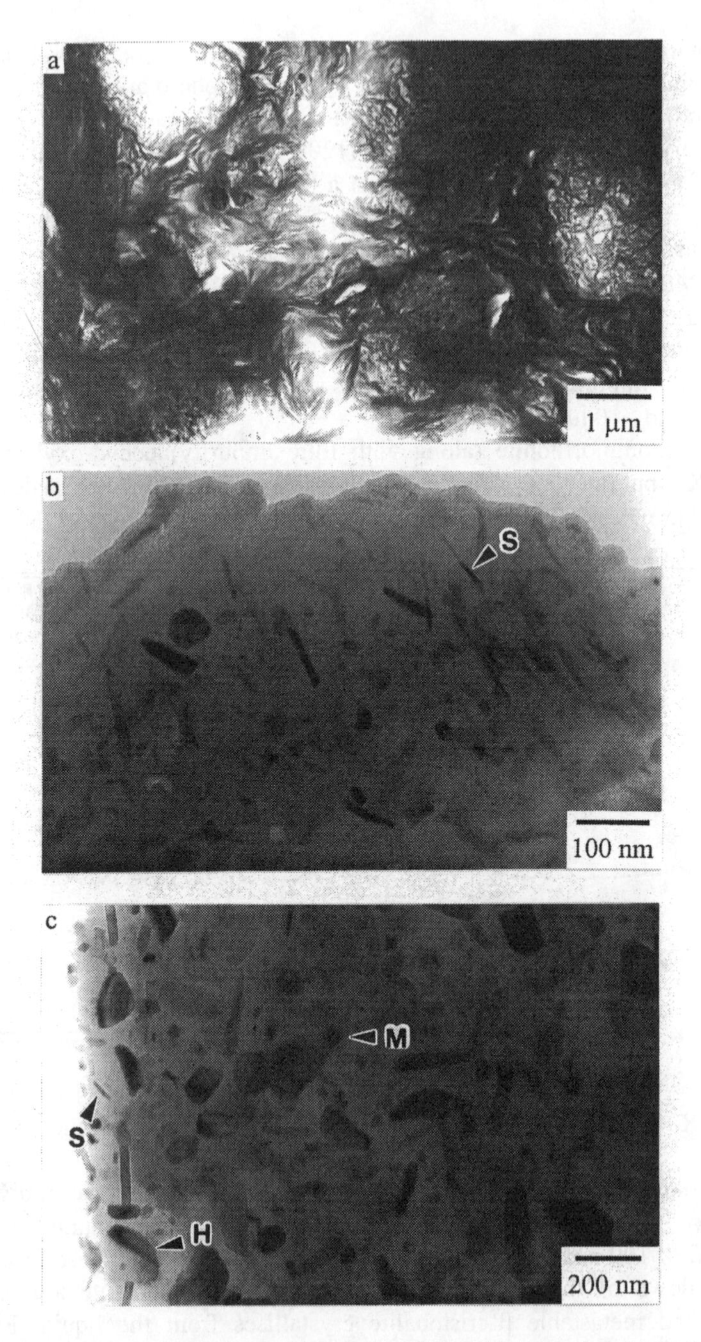

Figure 6a-c. Dense Berkbond clay fired 3h at (a) 800°C, (b) 1000°C and (c) 1100°C.

shaped spinel (S) and primary mullite (M), and 'coin' shaped haematite (H) crystals. The feldspar albite melts above 1100°C, and acts as a flux for the liquid phase, which dissolves all the accessory quartz by 1400°C. Cordierite formed in the 1200°C sample is probably an artefact of the cooling process (figure 5). Above 1200°C, haematite melts, mullite crystals reach sizes >1 μm at 1400°C, and EDS indicates that they contain Fe.

Figures 7-9 are schematic diagrams of the phases observed by TEM in each material after 3 hours at the indicated temperatures.

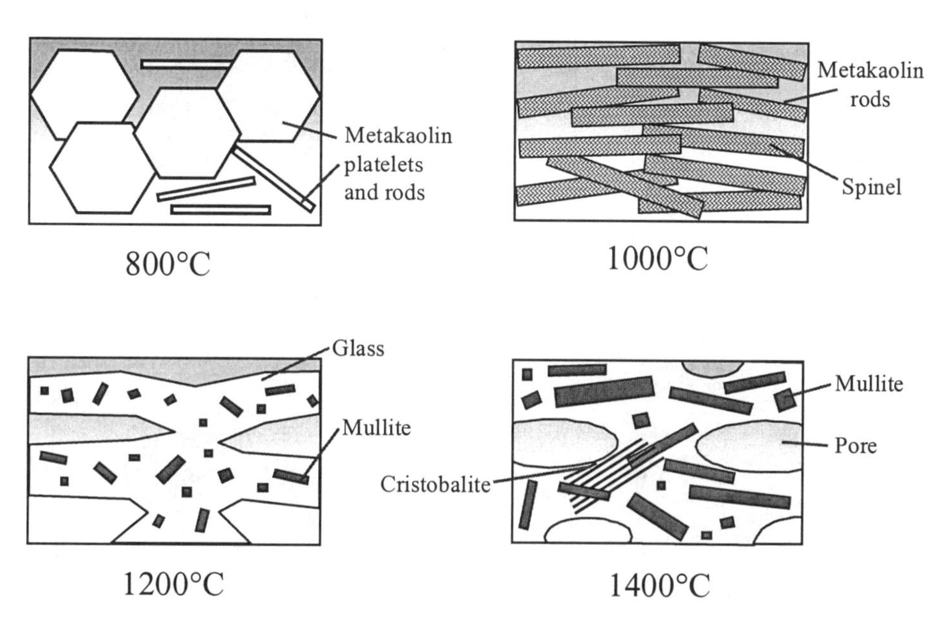

Figure 7. Schematic diagram of the thermal transformations observed in kaolinite clay.

DISCUSSION

The behaviour of the three clays on firing varied considerably. The formation of a spinel-type phase has been well documented in kaolinite[8] and the results of this study are consistent with previous findings.[5] The spinel-type phases in the 2:1 clays are considerably different. It is likely that they contain Mg, which was substituted into the octahedral layer, and the orientation of the spinel present in the Silver Hill material, which appears to be aligned inside the original illite laths, suggests that the spinel may form *in situ* from the octahedral layer in this material. The spinel-type phase in Berkbond clay, the smectite, does not show the same orientation, perhaps a result of the breakdown of the dehydroxylated clay mineral

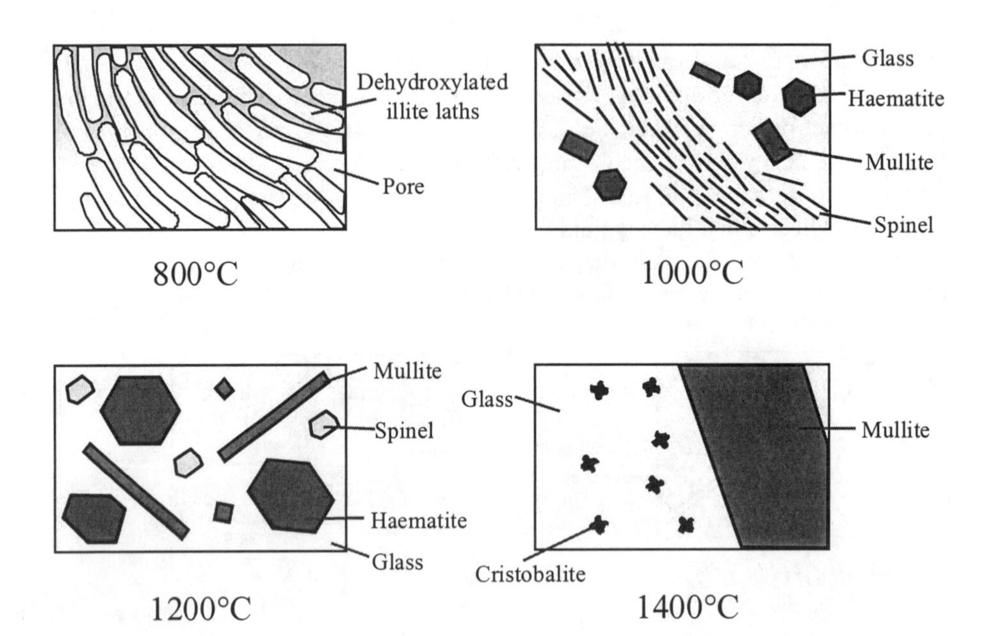

Figure 8. Schematic diagram of the thermal transformations observed in Silver Hill clay.

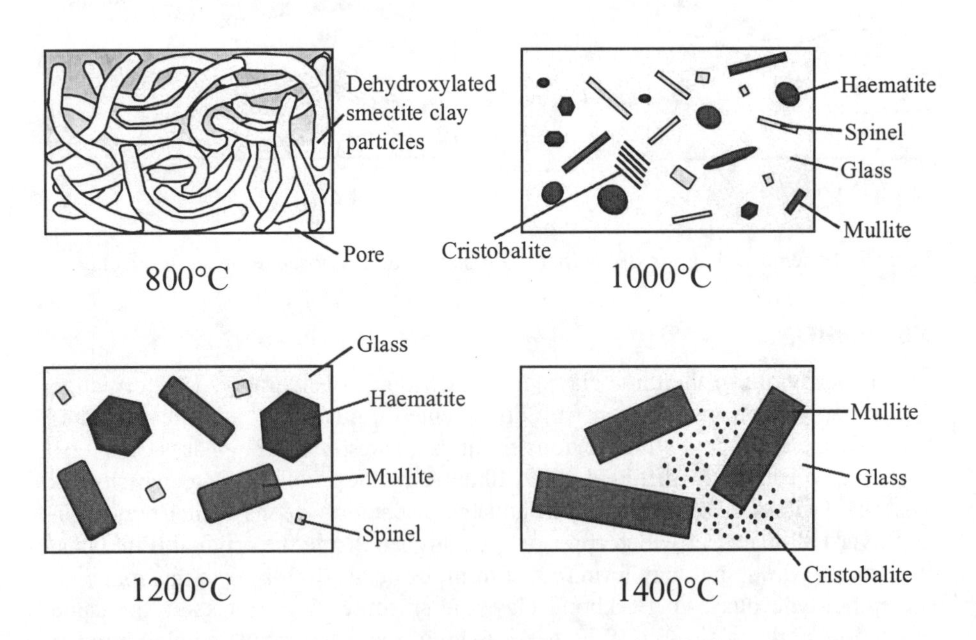

Figure 9. Schematic diagram of the thermal transformations observed in Berkbond clay.

structure which occurs at a lower temperature (≈850°C) than the Silver Hill (≈950°C). The spinel phase formation begins above 900°C, by which temperature the original 2:1 clay structure is lost in Berkbond, but not in Silver Hill.

Planar defects similar to those seen in the cristobalite crystals in kaolinite have been seen previously in siliceous materials. Carpenter and Wennemer[9] studied synthetic tridymites, which were found to contain cristobalite as mixed-layer phases with similar morphology to those in figure 2c. These were interpreted as (001) stacking disorders, possibly resulting from intimate intergrowths of cristobalite and tridymite. Cristobalite crystals with similar defects were also seen in the Berkbond (smectite) clay, along with secondary quartz, which crystallised on cooling, but little cristobalite was detected in the Silver Hill (illite) material. This was consistent with the findings of Grim and Bradley;[3] the cristobalite which did form may have arisen from kaolinite contamination in the raw mineral.

A far greater quantity of glass phase was observed in the 2:1 clays than in the kaolinite at temperatures over 1000°C. Cole and Segnit[10] reported that glass formation was greater in Fe-rich mica materials. Both the illite and smectite materials studied contained Fe, but also contained feldspars, which act as a flux, increasing the fluidity of the liquid (glass) phase. The interlayer cations, such as calcium and potassium may have a similar effect. In addition, a large amount of quartz impurity was present in both samples, which dissolved in the glass as temperature approached 1400°C, giving rise to more amorphous silica on cooling. It is speculated that the greater mobility of the fluid silica-rich liquids formed in the 2:1 clays allowed greater crystal growth to take place, as evidenced by the greater size of the spinel phase (5 nm in kaolinite, 100 nm x 10 nm in Silver Hill and Berkbond) and mullite crystals in these materials. Mullite crystals were only 0.5 μm in kaolinite after firing at 1400°C, but up to 10 μm in Silver Hill clay after firing to this temperature.

CONCLUSIONS

The three clays showed considerable differences in their firing behaviour, although all eventually formed mullite, cristobalite and glass on cooling from the higher temperatures. The intermediate phases showed great differences both in terms of the phases formed, and the size of crystals which developed; for example the spinel was far smaller (5 nm) in kaolinite than in either of the others (100 nm), and showed evidence of oriented growth in the Silver Hill clay. The spinel compositions were also different, with Mg present in that formed from the illite and smectite materials, whereas that from the kaolinite was γ-Al_2O_3. The properties and composition of the starting material appear to dictate the phases which form; for example the presence of iron in the illite and smectite materials gave rise to haematite, and the mullite in these materials at higher temperatures contained iron. It is likely that the presence of feldspars in the starting materials allowed more 'fluid' liquid to form from the silica portion of the clay lattices. This in turn allowed greater crystal growth.

ACKNOWLEDGEMENTS

C.J.M. would like to thank RBB Research & Development, Ltd. for supporting him in carrying out this research with a fully-funded Research Studentship.

Figures 2a-c were previously published in: McConville, C.J., Lee, W.E. and Sharp, J.H., 'Microstructural Evolution in Fired Kaolinite,' *British Ceramic Transactions,* **97**, [4], 162-168, 1998.

REFERENCES

[1]Brindley, G.W. and Nakahira, M., 'The Kaolinite-Mullite Reaction Series: I, A Survey of Outstanding Problems,' *J. Am. Ceram. Soc.*, **42**, [7], 311-314, (1959).

[2]Brindley, G.W. and Lemaitre, J., 'Thermal, Oxidation and Reduction Reactions of Clay Minerals,' in *'Chemistry of Clays and Clay Minerals,'* Mineralogical Society Monograph 6., Longman Scientific and Technical, Harlow, 1987.

[3]Grim, R.E. and Bradley, W.F., 'Investigation of the Effect of Heat on the Clay Minerals Illite and Montmorillonite;' *J. Am. Ceram. Soc.*, **23**, [8], 242-248, (1940).

[4]McConnell, J.D.C. and Fleet, S.G., 'Electron Optical Study of the Thermal Decomposition of Kaolinite,' *Clay Minerals*, **8**, 279-290, (1970).

[5]McConville, C.J., Lee, W.E. and Sharp, J.H., 'Comparison of Microstructural Evolution in Kaolinite Powders and Dense Clay Bodies,' *British Ceramic Proceedings*, **58**, 75-92, (1998).

[6]Clifford, J.F., 'High Temperature Reactions and Colour Development in Brick Clays,' Ph.D. thesis, University of Surrey, United Kingdom, 1984.

[7]MacKenzie, K.J.D. and Hartman, J.S., 'MAS NMR Evidence for the Presence of Silicon in the Alumina Spinel from Thermally Transformed Kaolinite,' *J. Am. Ceram. Soc.*, **79**, [11], 2980-82, (1996).

[8]Sonuparlak, B., Sarikaya, M. and Aksay, I.A., 'Spinel Phase Formation During the 980°C Exothermic Reaction in the Kaolinite-to-Mullite Reaction Series,' *J. Am. Ceram. Soc.*, **70**, [11], 837-42, (1987).

[9]Carpenter, M.A. and Wennemer, M., 'Characterisation of Synthetic Tridymites by Transmission Electron Microscopy,' *American Mineralogist,* **70**, 517-528, (1985).

[10]Cole, W.F. and Segnit, E.R., 'High-Temperature Phases Developed in some Kaolinite-Mica-Quartz Clays,' *Trans. Brit. Ceram. Soc.* **62**, [4], 375-395, (1963).

Processing

A CRITICAL REVIEW OF DISPERSANTS FOR WHITEWARE APPLICATIONS

William M. Carty, Katherine R. Rossington, and Udayan Senapati
NYS Center for Advanced Ceramic Technology—Whiteware Research Center
New York State College of Ceramics
Alfred University, 2 Pine Street, Alfred, NY 14802

ABSTRACT

Six common commercial dispersants: Na-polyacrylic acid (Na-PAA), Na-polymethacrylic acid, Na-silicate, Na-carbonate, and Na-hexametaphosphate, were evaluated in suspensions of kaolin and a porcelain body composition. The effectiveness of the dispersants, determined by apparent viscosity measurements, was normalized to the surface area of the suspended powders. The shift in pH associated with dispersant concentration was also monitored. Zeta-potential (ζ-potential) measurements of suspensions containing Na-PAA, Na-silicate, and Na-carbonate indicated that the anionic species are specifically adsorbed, proving that sodium is not responsible for dispersion in clay-based suspensions. An unanticipated result, however, indicated that the suspension rheology does not always correlate well with ζ-potential, suggesting that ζ-potential may not be a reliable means of predicting suspension stability.

INTRODUCTION

A comparison of dispersants used in industry or discussed in the literature is problematic due to several reasons. Industrially, dispersants are usually added on a need basis, and while the amounts added are usually noted, the conditions of addition are not easily quantified. It is also common to add dispersant a several times during a production schedule. Within the literature, although excellent work has been conducted in the past, the reported studies of dispersant effectiveness have been conducted under widely varying conditions, including variations in dispersant chemistry and ionic strength,[1-8] clay pre-treatment,[9-12] suspension composition,[1-16] and rheology or stability measurement technique.[1-16]

The colloidal nature of kaolinite adds additional problems to the interpretation of dispersant effects. To properly evaluate dispersants for whitewares, the particle surface characteristics must be understood. For kaolinite, a 1:1 sheet silicate, the particles possess one silica-like and one alumina-like basal-plane surface with a variable edge. The basal-plane surfaces will be oppositely charged over a broad pH range, but with the net surface charge usually negative. This is addressed in detail elsewhere.[17]

There is also a great deal of confusion regarding sodium as a dispersant. In the context of modern colloidal theory, Na^+, as a monovalent cation, is oppositely charged to the net-negatively charged particle, and thus must behave as a coagulant.[18] Work conducted prior to the application of colloidal theory to ceramics systems correctly deduced that monovalent was better than divalent, but their conclusion:[6] "*(a) the charge on the kaolinite particle controls the degree of deflocculation and is governed by the type of cation and (b) the stability of the system is controlled by the anion of the medium and is governed by the type of anion preferentially adsorbed*" (italics existing), can unfortunately be easily misinterpreted. The implication of this statement is that cations control the charge on the kaolinite particles. In the absence of specific adsorption, cations simply compress the double-layer leading to coagulation. It will be demonstrated clearly in this work that Na^+ cations are not functioning as dispersants.

Therefore, to provide an consistent evaluation platform, six sodium-based dispersants–Na-polyacrylic acid (Na-PAA), Na-polymethacrylic acid (Na-PMAA), Na-silicate, Na-carbonate (Na-ash), and Na-hexametaphosphate (SHMP) –were compared under uniform conditions, using a kaolin suspension and a porcelain batch suspension. Table I lists the chemical formulae for the dispersants and their common abbreviations. The suspension pH was monitored, but not adjusted with the addition of inorganic acids or bases. The efficiency of each dispersant was measured by the change in apparent viscosity as a function of the dispersant type and concentration, normalized to the surface area of the suspended powders. In addition, ζ-potential was measured using a five volume percent kaolin suspension with Na-PAA, Na-carbonate, and Na-silicate additions. Finally, a 1:1 mixture of Na-silicate and Na-ash was evaluated demonstrating a strong interaction.

Table I: Na dispersants evaluated in the suspensions with the corresponding chemical formulae and the abbreviation commonly used.

Dispersant	*Chemical Formula*	*Abbreviation*
Na-Polyacrylic Acid	$H\text{-}(NaC_3O_2H_3)_n\text{-}H$	Na-PAA
Na-Polymethacrylic Acid	$H\text{-}(NaC_4O_2H_5)_n\text{-}H$	Na-PMAA
Na-Hexametaphosphate	$(NaPO_3)_6$	SHMP
Na-Silicate (x=0.22)*	$xNa_2O \cdot (1-x)SiO_2$	Na-silicate
Na-Carbonate	$Na_2CO_3 \cdot 10H_2O$	Na-ash
Na-ash:Na-silicate blend	Equal parts by mass	1:1

*Ratio determined by ICP of the Na-Silicate solution, Acme Analytical Laboratories Ltd., Vancouver, British Columbia, Canada.

EXPERIMENTAL APPROACH

Suspension Compositions

The effectiveness of the dispersants was evaluated using two suspensions: one was composed only of kaolin (referred to in the text as *kaolin*); the other of a porcelain batch composition (referred to in the text as *batch*). The *batch* composition is listed in Table II. The two clays in the batch contribute 93.8% of the total *batch* surface area. Due to the higher viscosity of the *kaolin* suspensions compared to the *batch* suspensions it was necessary to prepare the *kaolin* suspensions at a lower solids loading (30 v/o) than that used for the *batch* suspensions (40 v/o) for reliable viscosity measurement. The difference in the suspension rheology is attributed to differences in particle packing.

Table II: The porcelain batch composition. The raw material percentages are shown on a dry weight basis (d.w.b.) and surface area by N_2-B.E.T.

Raw Material	*Wt. % (d.w.b.)*	*Specific Surface Area (m^2/g)*	*% of Batch Surface Area*
Kaolin	29.0	26.9	76.1
Ball Clay	7.0	25.9	17.7
Alumina	12.5	1.0	2.7
Quartz	29.5	0.9	1.2
Nepheline Syenite	22.0	1.1	2.3

Suspension Preparation and Dispersant Additions

All suspensions were prepared with distilled water using the typical industrial batching and mixing approach (i.e., clay slurried with water followed by the addition of the non-plastics, i.e., quartz, nepheline syenite, and alumina). To ensure homogeneity between individual batches, 13 liter suspensions were initially prepared at 35 and 45 v/o solids for the *kaolin* and *batch* suspensions, respectively. All suspensions were initially mixed using a high intensity mixer (SHAR, Inc., Fort Wayne, Indiana; 3 HP, 500 rpm, six inch impeller) without any dispersant addition.

To introduce the dispersants, the *kaolin* suspension samples were taken from the 13 liter batch and diluted to 30 v/o (specific gravity = 1.5) using distilled water and the appropriate dispersant mass to create the dispersant demand curve. The amount of dispersant added to the suspensions was based on the surface area of the suspension and dispersant mass. All dispersants were added as aqueous solutions and the concentration of dispersant in the solution was verified by chemical analysis and thermogravimetric analysis.

For the *batch* suspensions, dispersant concentrations ranging from 0.0 to 1.0 mg/m^2 were made by dividing the 13 liter batch into two suspensions. One suspension had no dispersant added to maintain a 0.0 mg/m^2 dispersant level; the other had dispersant added to create a 1.0 mg/m^2 "stock" suspension. Both were reduced to 40 v/o (specific gravity = 1.7) by the addition of the appropriate amount of distilled water. By blending these two endpoint suspensions in various ratios, the entire concentration range desired was achieved. For example, 10 ml of the suspension containing 1.0 mg/m^2 was blended with 90 ml of the 0.0 mg/m^2 suspension, producing a suspension with a dispersant level of 0.10 mg/m^2.

All samples were stored in sealed polypropylene bottles for two weeks prior to testing to eliminate aging as a variable.

For the 1:1 (Na-silicate + Na-ash) dispersed samples, two different experiments were conducted; the second set of experiments were initiated because of anomalous results from the initial results. The first set of suspensions were prepared by aging the Na-ash and Na-silicate separately and then blending (referred in the text as "aged 1:1"), and the second set of suspensions were prepared by mixing the 1:1 components with the suspension and then aging. With the "aged 1:1" suspensions, equal amounts of Na-ash treated suspension and Na-silicate treated suspension were blended for each level of dispersant, to produce 1:1 dispersant additions over the range of 0.0 – 1.0 mg/m^2. After mixing, the properties of these suspensions were measured without additional aging. These suspensions behaved like Na-ash at low concentrations and like Na-silicate at higher dispersant concentrations, as described below.

The second set of experiments were also conducted to evaluate the addition sequence, as well as the addition of a solution blend of Na-silicate and Na-ash. New *batch* suspensions were prepared and the dispersant additions followed the procedure outlined for the *kaolin* suspensions, and consisted of three groups as follows:

Group 1: Na-ash and Na-silicate were measured and blended to form a "premix" 1:1 solution. This solution was then added at the desired level to the non-dispersed suspension and mixed for two minutes.

Group 2: Na-ash solution was added to the non-dispersed suspension, mixed for two minutes, and then Na-silicate was added and the suspension and mixed for an additional two minutes.

Group 3: The order of addition was reversed from *Group 2.* The Na-silicate was added first followed by Na-ash, with the same mixing procedure as in *Group 2.*

Measurement Methods (rheology, pH, and ζ-potential)

The steady-state shear behavior of each suspension mixture was measured using a stress-controlled rheometer (SR-200 Dynamic Stress Rheometer, Rheometrics Scientific, Piscataway, NJ) with a 25 mm parallel plate geometry. Normally, a truncated cone-and-plate geometry would be desired, as this geometry provides a constant shear rate over the area of the plate, but the parallel plate geometry was chosen because of the potential for fixture damage due to the presence of (quartz and nepheline syenite) particles larger than the recommended 50 μm cone and plate separation distance. Viscosity was measured by a steady-state stress-sweep test from high to low stress, and the data were used to produce apparent viscosity versus shear rate curves. The apparent viscosity at a shear rate of 1.0 s^{-1} was calculated from a linear regression of log apparent viscosity versus log shear rate data.

The pH of the suspensions was measured using a standard combination electrode (Accumet 15, Fisher Scientific, Pittsburgh, PA).

ζ-Potential was measured via acoustophoretic mobility (Acoustosizer, Colloidal Dynamics, Warwick, Rhode Island). A five volume percent kaolin (EPK, Zemex Minerals, Inc.) suspension was used for ζ-potential measurements. The dispersant was added incrementally and the pH measured for each addition level.

RESULTS AND DISCUSSION

Effects of Dispersant Additions on Rheology

The effectiveness of the six dispersants evaluated is based on the apparent viscosity of the suspensions at a shear rate of 1.0 s^{-1}. Figure 1 shows dispersant effectiveness in the *batch* suspension; similar rheological trends were observed in the *kaolin* suspensions. The behavior of the six dispersants can be divided into three categories. The first dispersant category would be considered highly effective, which includes Na-PAA, Na-PMAA, Na-silicate, and SHMP; the apparent viscosity values all reduced by a factor of ~1000, over the evaluated dispersant concentration range. This change occurred over the dispersant concentration level of 0.0 to 0.20 mg/m^2. The similar behavior of Na-silicate and SHMP to Na-PAA and Na-PMAA, suggests that oligomization* occurs with Na-silicate and SHMP in an aqueous clay-based suspensions. Ineffective dispersant behavior is the second dispersant category, including only Na-ash, exhibiting apparent viscosity change of only a factor of ten. This small change occurred

* The correct terminology was discussed at length during the question and answer section of this presentation. According to the experts present, the term oligomerization should be used to describe the joining of a limited number of repeat groups, as apparently happens with Na-silicate and SHMP, thus causing the effects described.

over a large portion of the dispersant concentration range and was a more gradual change in viscosity compared to the highly effective dispersant group. The third dispersant category shows a combined effectiveness of the first and the second categories creating an anomaly; this group consists of the "aged 1:1" dispersant. Initially, the apparent viscosity behavior of the suspensions follows the path of the ineffective dispersant, Na-ash. But as the "aged 1:1" concentration increased, the suspension behavior trended towards the highly effective dispersants, as illustrated in Figure 2.

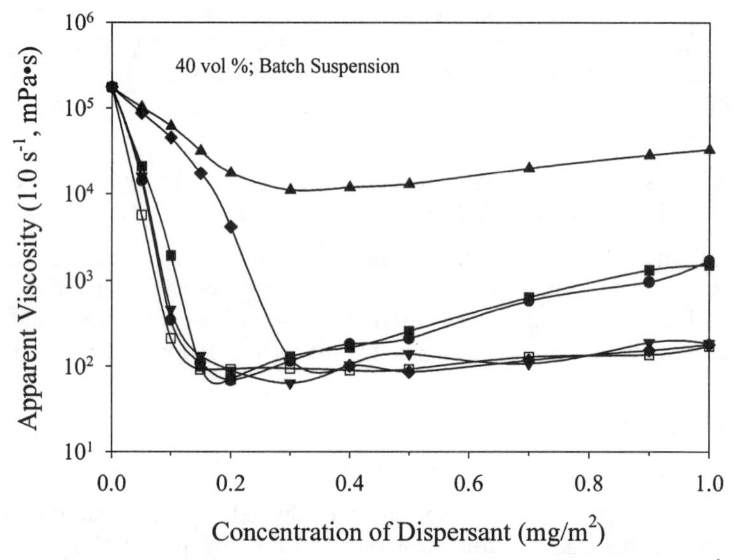

Figure 1. The apparent viscosity in the *batch* suspension at 1.0 s^{-1} as a function of dispersant concentration is shown for all six dispersants: Na-PAA (●), Na-PMAA (■), Na-silicate (▼), SHMP (□), Na-ash (▲), and 1:1 (♦). Similar results were measured in the *kaolin* suspensions.

Behavior of the 1:1 Blends

The rheology of the Na-ash and the Na-silicate suspensions is illustrated in Figure 2 along with the results for the "aged 1:1" suspension (Na-ash:Na-silicate). The apparent viscosity of the Na-ash suspension changed gradually, only decreasing by a factor of ten, over the dispersant concentration range tested. Above 0.30 mg/m^2, the apparent viscosity remained relatively unchanged, indicating that surface coverage was complete by 0.30 mg/m^2. The relatively small decrease in viscosity indicates that clay flocs remained in suspension. By comparison, Na-silicate caused a decrease in apparent viscosity by a factor of 1000, and the minimum occurred at the 0.2 mg/m^2 addition level. When aged suspensions containing the two dispersants were combined, the resulting suspensions were similar to the Na-ash case at low concentrations (below 0.15 mg/m^2) and the Na-silicate case at high concentrations (above 0.30 mg/m^2). The concentration range between 0.15 mg/m^2 and 0.30 mg/m^2 exhibited a transitional behavior.

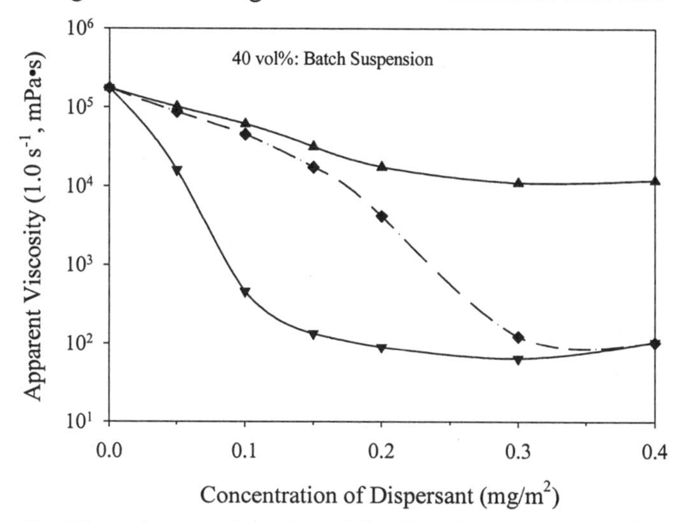

Figure 2. The apparent viscosity of the "aged 1:1" (♦) initially follows the rheological behavior of the Na-ash (▲). Above 0.15 mg/m^2 the "aged 1:1" behavior deviates from the Na-ash and eventually matches the behavior of the Na-silicate (▼) above 0.35 mg/m^2.

To explain the "aged 1:1" behavior of the suspensions, it is useful to consider a model system in which the suspension is composed of two clay particle "phases" within the suspension. The first phase consists particles coated with CO_3^{-2} ions from the Na-ash additions; the second phase coated with SiO_{3n}^{-2n} ions. (The "n" denoting a polyionic species, as suggested earlier.) Up to 0.15 mg/m^2 of the "aged 1:1" dispersant, the higher apparent viscosity of carbonate-coated phase dominates the suspension behavior, with the silicate-coated particles essentially not contributing to the rheology. At the 0.20 mg/m^2 addition level, excess silicate is now present, that can now displace CO_3^{-2} ions adsorbed on the clay particles, and the viscosity starts to deviate from the carbonate end-member, and as additional Na-silicate is added, it is added in excess, and the displacement of carbonate ions progresses until the suspension behaves like the Na-silicate end-member. If this hypothesis is correct, it indicates that complete surface coverage is obtained for SiO_{3n}^{-2n} ions at (or slightly below) the 0.20 mg/m^2 addition level.

The dispersant curves for *Group 1-3* were compared to "aged 1:1" behavior to assess the effects of addition sequence. As shown in Figure 3, the rheological behavior of the *Group 1-3* suspensions was essentially identical, and similar to the Na-silicate end member suspensions. The rheology of the *Group 1-3* suspensions is independent of the sequence of dispersant additions, implying that there is a strong preferential adsorption of the silicate anion over the carbonate anion. These results also indicate that the intermediate “aged 1:1” behavior was an artifact of the preparation technique. The "aged 1:1" samples were tested immediately following mixing of the aged suspensions, not allowing sufficient time for silicate anions to displace carbonate anions adsorbed on the particle surface. Assuming that silicate ions exhibit strong adsorption behavior, and that the minimum in the dispersion curve reasonably equates to surface saturation, there would be limited free silicate ions in solution below that necessary to reach the viscosity minimum, so limited displacement of carbonate ions would be possible. Once there were excess silicate anions in the Na-silicate suspension, carbonate anions are gradually displaced by silicate anions. The similarity of the *Group 1-3* dispersion curves indicates that the kinetics of the Na-silicate adsorption are faster then that of the Na-ash.

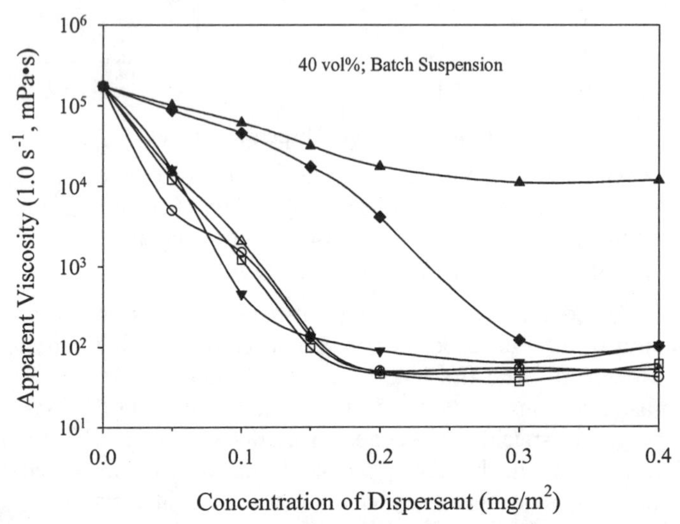

Figure 3. The addition sequence of the 1:1 dispersant components was shown to be unimportant with respect to the rheological behavior in the *batch* suspension. The plot shows the initial three dispersants evaluated: Na-ash (▲), Na-silicate (▼), and “aged 1:1” (♦), in addition to *Group 1* (△), *Group 2* (□) and *Group 3* (○).

pH Measurements

The addition of these six dispersants generated two distinct trends in the pH measurements as illustrated in Figure 4 for the *batch* suspensions. A pH of 7.4 was observed for the *batch* suspension without dispersant additions. The pH of suspensions prepared with Na-PAA, Na-PMAA, and SHMP remained relatively unchanged (7.3-7.8) over the entire dispersant range. Whereas suspensions prepared with Na-silicate, Na-ash, and the 1:1 blends exhibited a dramatic increase in pH with dispersant addition (10.2-10.9). (Similar behavior was observed in the *kaolin* suspensions, however the initial suspension pH was 5.4.) These results demonstrate that pH is not a critical factor in the suspension behavior with these samples, and is further supported by the ζ-potential data.

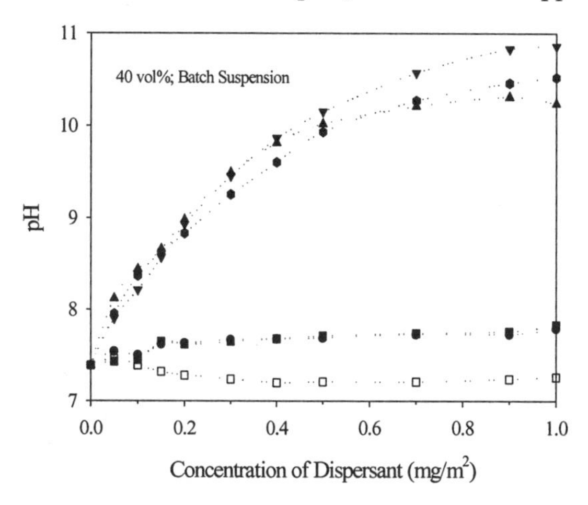

Figure 4. The change in the pH of the *batch* suspension as a function of dispersant concentration is shown for all six dispersants: Na-PAA (●), Na-PMAA (■), Na-silicate (▼), SHMP (□), Na-ash (▲), and "aged 1:1" (♦). The *kaolin* suspension exhibited similar trends but with an initial suspension pH of 5.2.

ζ-Potential Measurements

The ionic species responsible for dispersing a clay-based system has been heavily debated. By measuring the ζ-potential, the effect of dispersant additions may be directly related to the change of the particle surface charge thus identifying the ionic species responsible. The ζ-potential is calculated from the electrophoretic mobility measurements and is the net magnitude of the electrical potential at the shear plane. In the presence of an indifferent electrolyte, the ζ-potential scales with, and indicates the sign of, the potential at the particle surface.[19] The magnitude of the ζ-potential is generally accepted as a measure of the repulsive potential between suspended particles. For kaolin suspensions, it has been shown that a minimum ζ-potential of 25 mV is necessary for suspension stability.[16]

In these experiments, the dispersant solutions are definitely not indifferent electrolytes--the anionic polyelectrolytes specifically adsorb on the clay particle surfaces increasing the net negative potential. This also translates to an increase in the (negative) ζ-potential, as shown in Figure 5. In this example, Na-silicate was added at 0.05 mg/m^2 then the pH of the suspension adjusted with NaOH. Over the entire pH range evaluated, the Na-silicate treated suspension has a greater negative ζ-potential. From these results alone, it is clear that Na^+ cannot be responsible for dispersion in clay suspensions. This is further supported by another study demonstrating that Na^+ causes coagulation.[18]

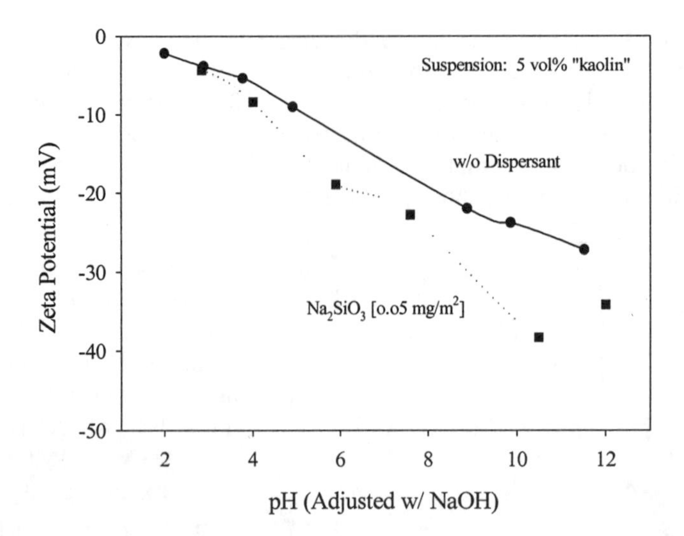

Figure 5. The change in the ζ-potential due to the change in pH is shown for two 5 v/o *kaolin* suspensions in a 10 mM NaCl background electrolyte. A more negative ζ-potential is observed over the evaluated pH range for the sample with 0.05 mg/m^2 Na-silicate addition.

The effect of Na-PAA, Na-ash, and Na-silicate dispersant additions on the ζ-potential of three kaolin suspensions is illustrated in Figure 6. In all three cases, the ζ-potential becomes constant above a certain concentration level, consistent with that necessary to obtain the minimum batch viscosity values. Inconsistent with theory and the rheology results, however, is the similar behavior of the Na-silicate and Na-ash suspensions. According to the ζ-potential values, these two suspensions should have similar rheology, not exhibit a difference by a factor of 100. By similar logic, the Na-PAA suspensions should have lower viscosity than the Na-silicate suspensions due to the greater ζ-potential magnitude in the Na-PAA suspensions. These results question the validity of the global application of ζ-potential measurements as an indication of suspension stability and dispersion. Obviously, this is an area for further study.

As the dispersant concentration increases, the pH increases. Figure 6 shows the data from Figure 5 plotted as a function of pH. Note that the pH was not adjusted, but changed with increased dispersant additions. The most negative ζ-potential of these two suspensions (-45 mV) occurs at a pH above 10.0; the most negative ζ-potential with Na-PAA occurs at a pH near 8.0. The ζ-potential in the Na-ash and Na-silicate suspensions was not effected by increasing pH above 9.0, further supporting the hypothesis that these suspensions are independent of pH within the pH range evaluated.

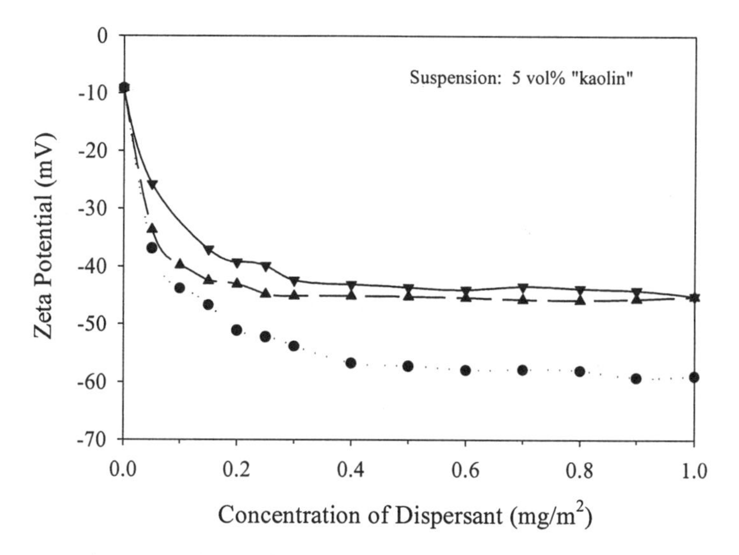

Figure 6. ζ-potential as a function of Na-PAA (●), Na-ash (▲) and Na-silicate (▼) concentration for a 5v/o *kaolin* suspension in 10 mM NaCl solution. The ζ-potential values of the Na-silicate and the Na-carbonate are nearly identical, while the Na-PAA is significantly greater. Note that these results do not correlate well with the rheology results in Figure 1.

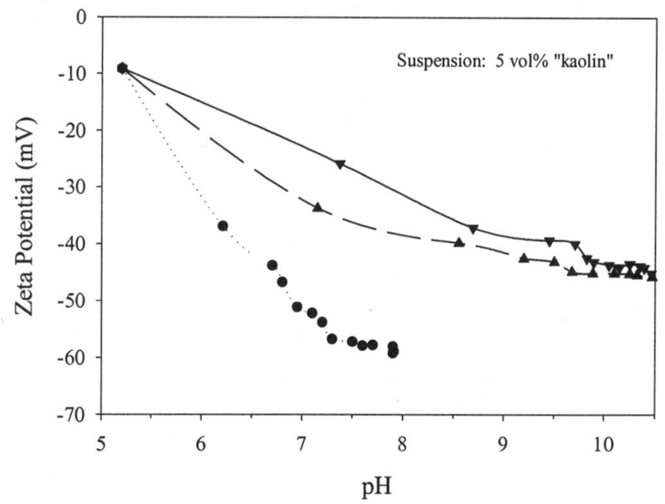

Figure 7. The ζ-potential as a function of pH for the data in Figure 6. The pH of these suspensions was not adjusted, but changes with dispersant level.

SUMMARY AND CONCLUSIONS

The effectiveness of these dispersants as measured by the apparent viscosity can be categorized as either highly effective (as with Na-PAA, Na-PMAA, SHMP, Na-silicate, and the 1:1 blends) or essentially ineffective (as with Na-ash). All of the highly effective dispersants performed similarly, although there were significant differences in suspension pH. The 1:1 dispersant was initially considered a combination of the highly effective and the ineffective dispersants, but subsequent investigations demonstrated that the intermediate results were an artifact of the suspension preparation technique.

These results demonstrate that dispersion is caused by the specific adsorption of the anionic polyelectrolytes, rather than Na^+, as indicated by the ζ-potential measurements. However, the ζ-potential results were also inconsistent with the rheology measurements, suggesting that the theory regarding ζ-potential, and its correlation with suspension stability and dispersion, may need to be addressed.

REFERENCES

1. L. Bergström, M. Sjöberg, and L. Järnström, "Concentrated Kaolinite Suspensions-Polymer Adsorption and Rheology," *Science of Whitewares*, Eds. V. Henkes, G. Onoda, and W. Carty, pp. 65-77, Amer. Ceramic Soc., Westerville, OH, 1996.
2. L. Järnström and P. Stenius, "Adsorption of Polyacrylate and Carboxy Methyl Cellulose on Kaolinite: Salt Effects and Competitive Adsorption," *Colloids Surf.*, **50**, 47-73, (1990).
3. B. Corradi, T. Manfredini, G. C. Pellacani, and P. Pozzi, "Deflocculation of Concentrated Aqueous Clay Suspensions with Sodium Polymethacrylates," *J. Am. Ceram. Soc.*, **77** [2], 509-513, (1994).
4. Rand, H. M. M. Diz, J. Li, and I. B. Inwang, "Deflocculation of Kaolinite Clay Suspensions by Sodium Silicate," pp. 231-236 in *Science of Ceramics* Vol. 14, Ed. D. Taylor, Butler and Tanner, London, 1988.
5. S. J. McDowell, "The Effect of Various Sodium Silicates and other Electrolytes on Clay Slips," *J. Am. Ceram. Soc.*, **10** [4], 225-237, (1927).
6. L. Johnson and F. H. Norton, "Fundamental Study of Clay: II, Mechanism of Deflocculation in the Clay-Water System," *J. Am. Ceram. Soc.*, **24** [6], 189-203, (1941).
7. B. Alince and T. G. M. van de Ven, "Stability of Clay Suspensions - Effect of pH and Polyethylenimine," *J. Colloid Interface Sci.*, **155**, 465-470, (1993).
8. S. Michaels, "Deflocculation of Kaolinite by Alkali Polyphosphates," *Ind. Eng. Chem.*, **50** [6], 951-958, (1958).

9. S. Michaels, "Rheological Properties of Aqueous Clay Systems," in *Ceramic Fabrication Processes*, pp. 29-31, Ed. W. D. Kingery. MIT Press, Cambridge, MA, 1958.

10. N. Celik, I. E. Melton, and B. Rand, "Rheological and Casting Behaviour of Kaolinitic Clay Suspensions," *Br. Ceram. Trans. J.*, **82** [4], 136-142, (1983).

11. W. G. Lawrence, "Theory of Ion Exchange and Development of Charge in Kaolinite-Water Systems," *J. Am. Ceram. Soc.*, **41** [4], 136-140, (1958).

12. Z. Zhou and W. D. Gunter, "The Nature of the Surface Charge of Kaolinite," *Clays and Clay Miner.*, **40** [3], 365-368, (1992).

13. T. E. R. Jones and S. A. Bullivant, "The Elastico-Viscous Properties of Deflocculated China Clay Suspensions," *J. Phys. D: Appl. Phys.*, **8**, 1244-1254, (1975).

14. H. Rasekh, K. W. Rose, and W. E. Worrall, "The Rheological Stabilisation of Clay-Water Suspensions: A Comparison of Various Methods," *Trans. Br. Ceram. Soc.*, **86** [4], 132-135, (1987).

15. W. E. Worrall and S. Tuliani, "Viscosity Changes During the Ageing of Clay-Water Suspensions," *Trans. Br. Ceram. Soc.*, **63** [4], 167-185, (1964).

16. R. J. Hunter and A. E. Alexander, "Surface Properties and Flow Behavior of Kaolinite Part I: Electrophoretic Mobility and Stability of Kaolinite Sols," *J. Colloid Sci.*, **18** [9], 820-832, (1963).

17. W. Carty, "The Colloidal Nature of Kaolinite," this volume.

18. K. R. Rossington and W. M. Carty, "The Effects of Ionic Concentration on the Viscosity of Clay-Based Suspensions," this volume.

19. R. J. Hunter, *Introduction to Modern Colloid Science;* pp.231-235, Oxford University Press, New York, 1994.

COMPARATIVE STUDY OF DIFFERENT SUSPENSION PREPARATION SYSTEMS FOR TILE MANUFACTURE

Enrique Sánchez, Vicente Sanz, Javier García-Ten, and Carlos Felíu.
Instituto de Tecnología Cerámica. Asociación de Investigación de las Industrias Cerámicas. Universitat Jaume I. Campus Universitario Riu Sec. 12006. Castellón. Spain.

ABSTRACT

A suspension preparation bench-scale, comparative study was carried out on the efficiency of the following equipment: disk disperser, concentric crown disperser and ball mill. The degree to which the solids constituents were incorporated and dispersed in the suspending liquid was monitored by measuring the solids particle-size distribution (direct method) and determining the suspension's flow curve and viscoelastic behaviour (indirect methods). A comparison was made of tile bodies formed from spray-dried powder samples made from the suspensions prepared with the different methods. The results confirmed the influence of the degree of solids dispersion on suspension behaviour and on the characteristics of the resulting body.

INTRODUCTION

Different systems are currently used to prepare suspensions for tile manufacture. The choice of a given system depends on the technical-economic efficiencies of the production facility involved, as well as targeted final product characteristics.

Before analysing the influence of dispersion in such suspensions, a series of concepts needs defining, which will be used in this study. *Aggregates* are groups of primary particles linked at their faces, exhibiting a significantly lower specific surface area than the sum of the surface areas of their constituent particles, which is why much energy is required for their breakup. *Agglomerates* are primary particle groups and/or aggregates that are linked at their edges and corners, in which the specific surface area of the group approaches the sum of the surface areas of the group's constituents. Given the difficulty of distinguishing aggregates from agglomerates in a suspension, we shall use the general term *particle groups* to cover both in this paper.

INFLUENCE OF DISPERSION IN SUSPENSIONS FOR TILE MANUFACTURE

The dispersion state of the solids comprised in a suspension for tile manufacture has a marked effect on behaviour during spray drying, as well as on the characteristics of both the unfired and fired body.

The size distribution of particles and existing particle groups directly impacts the viscosity of the suspension and consequently the characteristics of the spray-dried powder (size, shape, density and agglomerate hardness, flowability, etc.). On the other hand, having an incompletely dispersed system makes it keep on evolving to achieve greater dispersion. This gives rise to stability problems with regard to the viscosity of the suspension.

The presence of large particle groups can, as the literature reports, affect the microstructure and mechanical strength of a compact.[1] Microstructural changes can in turn affect green and fired body behaviour and properties. Finally, some researchers have reported glaze surface defects that could be due to the presence of incompletely dispersed particle agglomerates or aggregates.[2]

Owing to the manifest importance of the degree of solids dispersion in tile manufacturing, the present study has attempted to establish the most suitable laboratory methods for assessing the dispersion of the various suspension solid constituents. These methods were subsequently used to study how using different laboratory equipment affected dispersion in an industrial suspension.

EXPERIMENTAL

Materials

An industrial suspension typically used in Spain for manufacturing stoneware floor tile was used as a starting material. The suspension had been made up in a propeller mixer at low-speed stirring (250 rpm), and contained clay and previously dry ball-milled non-plastics. The solids contents consisted of a mixture of around 50 wt.% kaolinitic-illitic clay plus sodium-potassium feldspars. The specific surface area (BET) of the clay-feldspar mixture was 13 m^2/g. The suspension also contained 0.5 wt.% of a deflocculant blend comprising sodium metasilicate (MTS) and sodium tripolyphosphate (TPF) in a 3/1 weight ratio. The suspension solids content was 67 wt.%.

Equipment

The following laboratory apparatus was used in preparing the suspensions: crown disperser, disk disperser, ultrasonics dispersion system and ball mill.

The crown disperser basically consisted of two concentric crowns, involving a fixed stator and a rotor revolving at great speed ($\approx$ 2000 rpm). Particle agglomerate and aggregate breakup occurred mainly by impact when the material crossed the gap between the two crowns.

The disk disperser consisted of a 10-cm diameter disk impeller blade with serrations, set at right angles on a rotating shaft. The design yields relatively high shear in the region between the disk and the bottom of the container. Rotating speed was 1500 rpm. Shear was the prevailing deagglomeration mechanism, provided suspension viscosity was not excessively low.[3]

A planetary ball mill was also used, consisting of a 1000 mL porcelain container loaded with 200g alumina ball grinding media, whose diameter ranged from 1.5-2.0 mm. Finally, a laboratory ultrasonics dispersion system was also tested, simply for comparative purposes.

Experimental

Suspension preparation

Depending on the capacity of the system used, different volumes of starting suspension (hereinafter SS) were put into each system's dispersion or milling chamber. Running time was 15 min. in every case, except for the disk disperser (30 min.), given its lower power/suspension volume ratio. The suspensions were stored in closed containers after dispersion, pending characterization, referenced as follows: DDS (disk disperser), CDS (crown disperser), UDS (ultrasonics disperser) and BMS (ball mill).

Determination of the degree of dispersion

Direct particle-size analysis by laser diffraction (Malvern Mastersizer) was used. A small quantity of the suspension to be tested was placed in the measuring chamber with the ultrasonics generator switched off, in order to keep existing particle groups from possibly being broken up. For the same reason, pumping and suspension stirring rates were kept to a minimum.

Suspension viscosity and viscoelastic behaviour were determined as indirect methods of assessing degree of dispersion. Viscosity was determined on a rotational viscometer (Bohlin CS50), thermostatting at 25°C. After dispersion, the suspension was immediately transferred to the viscometer chamber to prevent ageing from affecting rheological behaviour. To keep existing particle groups from breaking up, suspension stirring prior to testing was held to a minimum. The resulting flow curve was used to determine viscosity at a shear rate of 0.01 s^{-1} ($\eta_{0.01}$).

The suspension's viscoelastic behaviour was measured by small amplitude oscillating tests in the same viscometer as before. The same precautions were also adopted relative to possible ageing and particle group breakup prior to testing. Test results were then used to calculate G' (storage modulus), representing the elastic behaviour of a suspension at rest.[4]

Characterization of unfired and fired test specimens

Certain suspensions were used to prepare spray-dried powder with a view to forming and characterizing green and fired test specimens. The suspensions were run through a pneumatic pilot spray nozzle (Niro Mobile) at a temperature of 250°C. The resulting pressing powder was used to form 0.7-cm thick cylindrical test specimens with a 4-cm diameter by uniaxial pressing in a laboratory hydraulic press. The compacts were formed at a powder moisture content of 5.5 wt.%, and pressing pressure of 25 MPa in every case (typical industrial conditions). The specimens were dried at 110°C in an electric laboratory oven, and cooled in a desiccator to room temperature (15 min.). Dry bulk density, ρ, was determined by measuring specimen dimensions and weight. Certain dry compacts were characterized microstructurally by mercury porosimetry (Micromeritics Poresizer 9310).

Dry compact strength was measured by three-point cross-bending testing (Instron 6027) at a constant loading rate of 1 mm/min. Prism-shaped specimens sized 1.5x8x0.7 cm were formed for this test by uniaxial pressing at the above conditions. The dry compacts were fired in an electric laboratory kiln at different peak firing temperatures ranging from 1120°C to 1200°C. The heating rate was 25°C/min., with a 6-min. dwell at peak temperature. Linear firing shrinkage (LS) and water absorption (E) according to standard EN-99 were determined for fired test specimens.

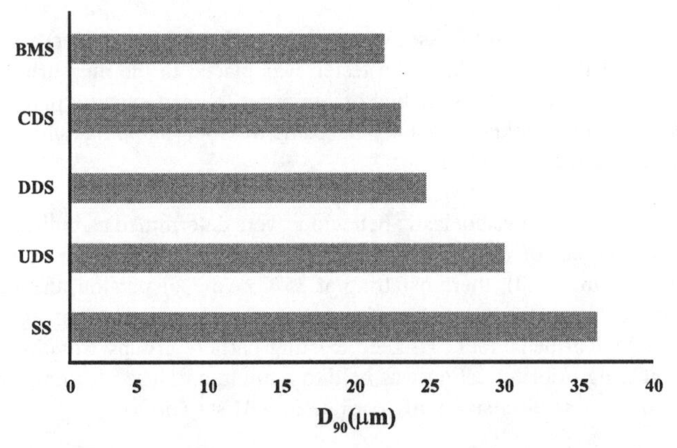

Figure 1. Values of characteristic particle diameter D_{90} in terms of the studied dispersion system.

RESULTS AND DISCUSSION

Evaluation of the degree of dispersion

On comparing the particle-size distribution curves found by laser diffraction for suspensions SS, DDS, CDS, UDS and BMS, the curve corresponding to SS was clearly observed to shift towards larger sizes. However, a comparison of the other suspensions was less straightforward. An analysis of the experimental data can be simplified by considering some parameters of the log-normal distribution curve to which these data can be fitted. The parameter used, owing to its greater sensitivity to the presence of particle groups in the suspension, was D_{90}, the size representing 90% of the cumulative curve screening. Figure 1 depicts the values of D_{90} in terms of the dispersion system used.

Figure 1 shows that the starting suspension (SS) exhibited the highest D_{90} values, indicating that the clay particle groups had not fully deagglomerated in the industrial low-speed stirring system. Comparing the efficiency of the studied dispersion systems reveals that:

- The application of energy in the form of ultrasonics reduced the number and/or size of the existing clay mineral particle agglomerates and/or aggregates less efficiently than the other studied systems. The decrease in D_{90} compared to the SS suspension was about half that exhibited by the other dispersion systems. This may have been due to the low power/suspension volume ratio of the laboratory equipment employed.
- The disk disperser efficiently reduced the number of clay particle groups, though twice the time was needed compared with the crown disperser or ball mill.
- The crown disperser produced the lowest number of clay particle agglomerates and/or aggregates. The resulting D_{90} value was comparable (slightly higher) than the value attained by ball milling. However ball milling, besides effectively breaking up clay mineral particle groups, also reduces non-plastics particle size. The crown disperser thus acted as a good disperser of clayey materials, while also having a milling effect, owing to the impact mechanism comminuting non-plastics grains that could not be broken up in water.

Figure 2 depicts the viscosity obtained at low shear rates ($\eta_{0.01}$) and the storage modulus (G') for the foregoing suspensions. It can be observed that $\eta_{0.01}$ and G' were minimal for the SS sample.

According to the literature, the viscosity of a concentrated suspension at rest or at sufficiently low deformation rates rises with the number of interparticle bonds or contacts.[5] The literature describes a similar dependence for a suspension's elasticity.[4] Thus, when a suspension contains inadequately deagglomerated particle groups, the possible number of surface contacts drops, leading to a drop in $\eta_{0.01}$ and G'.

The results obtained for $\eta_{0.01}$ and G' (Figure 2) confirmed the characteristic diameter (D_{90}) findings (Figure 1). The lowest viscosity and storage modulus values found for the SS indicated the greater presence of clay particle groups in this suspension than in any of the suspensions obtained with the tested dispersion systems.

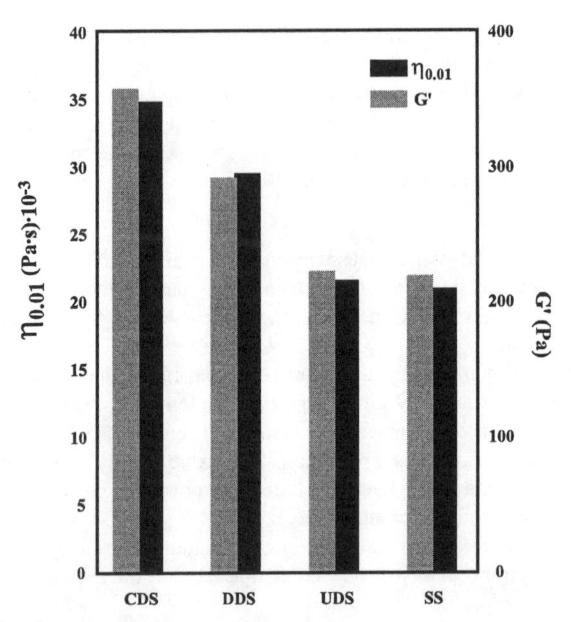

Figure 2. Viscosity values at low shear rates ($\eta_{0.01}$) and elastic modulus (G') for the studied dispersion systems.

It can be observed on comparing Figures 1 and 2 that as D_{90} dropped (i.e., the degree of dispersion rose), $\eta_{0.01}$ and G' increased, so that an inverse correlation could be established between the sequence of characteristic size variation of the existing particle groups in the suspension and the values of the rheological parameters. To verify this relation, Figure 3 plots the variation of $\eta_{0.01}$ and G' with characteristic particle diameter D_{90}. Both rheological parameters varied quite similarly on potentially reducing with D_{90}.

The good match found between the data obtained by the determination of particle-size distribution and indirectly, by the assessment of rheological behaviour from the flow curve ($\eta_{0.01}$) as well as from the material's viscoelastic characterization (G'), validate either of these indirect methods as a suitable procedure for evaluating the degree of dispersion in a ceramic suspension. This last conclusion is particularly relevant, since while laser diffraction or viscoelasticity techniques require costly scientific instruments and highly qualified operators, viscosity is determined with instruments that are easy to use and can be readily and inexpensively incorporated into industrial process control.

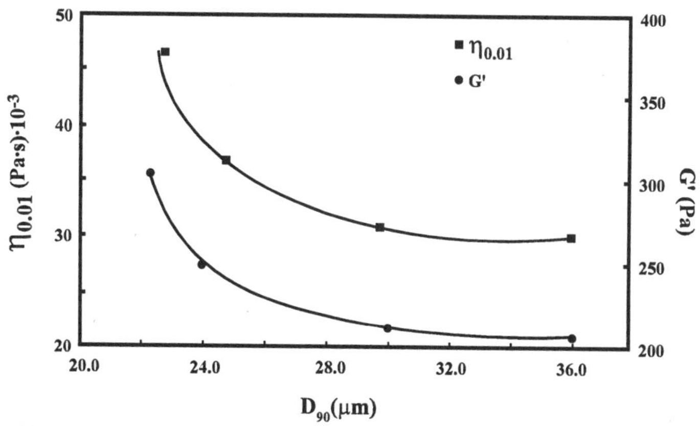

Figure 3. Relation between viscosity at low shear rates ($\eta_{0.01}$), elastic modulus (G') and characteristic particle diameter D_{90} for the studied suspensions.

Effect of clay particle group dispersion on body behaviour and properties

Spray-dried powder samples were prepared from two of the tested suspensions: SS and DDS. The reason for choosing the disk disperser suspension was the good degree of dispersion found and the system's resemblance to equipment currently used in industry. The spray-dried powder samples produced from these suspensions were respectively designated SSP and DDP.

Compacts were formed from these spray-dried powders and several green and fired properties were determined. Table I shows the specimen dry bulk density, dry mechanical strength and Dp_{90} pore diameter, representing the pore size below which 90% cumulative pore volume was found. Though no differences were established in bulk density or pore diameter, this was however the case with dry mechanical strength. The dry mechanical strength of the specimens formed from DDP spray-dried powder was about 17% higher than that exhibited by the SSP specimen.

Table I. Bulk density (ρ), mechanical strength (σ) and pore diameter (Dp_{90}) of dry compacts formed from spray-dried powders SSP and DDP

Sample	ρ (g/cm^3)	σ (MPa)	Dp90 (μm)
SSP	1.97	3.0	0.02
DDP	1.97	3.5	0.02

This difference was likely due to the mechanical strength (yield point) of the existing particle groups. Thus, the high mechanical strength of these groups, as well as their relatively small size compared with spray-dried powder granules hindered their deformation and/or breakup during pressing, so that they remained virtually unaffected inside the compact. The presence of non-deformed particle agglomerates and/or aggregates in a compact reduces interparticle contact area and interagglomerate bonding, both of which lead to reduced mechanical strength.

These data match the findings on using spray-dried powder with different moisture contents. Amoros et al.[6] have shown that even at minimum microstructural differences, a compact's mechanical strength was highly sensitive to the presence of agglomerates, always exhibiting lower mechanical strength in compacts with drier, hence less deformable agglomerates.

Figure 4 presents the vitrification diagram (variation of linear shrinkage LS and water absorption E of the body versus peak firing temperature) for SSP and DDP samples. The variation of both parameters with temperature was quite similar for both specimens. LS rose with firing temperature as a result of the parallel drop in apparent porosity (assessed as water absorption). The porosity reducing mechanism in this type of product is liquid-phase sintering.

The only significant difference between both specimens was the shift towards lower temperatures (about 8°C) of specimen DDP LS and E curves, as a result of the slightly more fluxing behaviour of this spray-dried powder compared to that of SSP. As expected, the presence of non-dispersed clay mineral particle groups produced a drop in the solid's effective specific surface area and therefore reduced reactivity during specimen heat-treatment in firing.

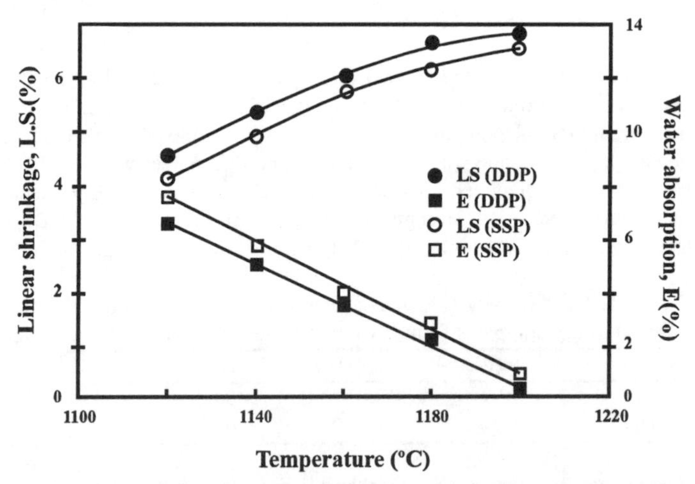

Figure 4. Variation of linear shrinkage (LS) and water absorption (E) versus peak firing temperature for SSP and DDP samples.

CONCLUSIONS

- The suspension prepared on an industrial scale by a dispersion system with low-speed stirring contained clay mineral particle groups that had not been completely dispersed.
- These particle groups were identified and quantified by laser diffraction based on the determination of a particle-size distribution characteristic diameter. This technique therefore allows monitoring a suspension's degree of dispersion.
- It was moreover shown that viscosity at low shear rates ($\eta_{0.01}$) and elastic modulus G' were highly sensitive to the clay mineral particle group content and size in a suspension, so that they can be used as indirect methods for evaluating a clay suspension's degree of dispersion.
- Using the foregoing techniques, the efficiency was determined for certain dispersion systems in breaking up the clay particle groups contained in the starting suspension. The best degree of dispersion was obtained with the disk disperser and with the concentric crown disperser (the latter in half the operating time of the former). Both systems broke up clay particle groups, though the crown disperser also had a certain milling effect, reducing non-plastics particle size, judging by the similarity between the characteristics of this suspension and the one produced in a ball mill.
- It was shown that the mechanical strength of the dry compact was extremely sensitive to the presence of non-dispersed clay particle groups: dry mechanical strength decreased as clay particle group content rose. Moreover, on using an inadequately dispersed suspension, a spray-dried powder was obtained whose optimum firing range shifted to slightly higher temperatures as a result of lower reactivity.

REFERENCES

[1]P.F. Messer, "Uniformity in Processing", *Trans. J. Br. Ceram. Soc.*, **82** [5], 156-162 (1983).

[2]J.L. Amorós; V. Beltrán; A. Blasco; J.E. Enrique; A. Escardino, and F. Negre, "Defects in Ceramic Tiles" *(In Spanish)*. pp. 70-87 AICE - Instituto de Tecnología Cerámica, Castellón (Spain), 1991.

[3]T.C. Patton, *Paint Flow and Pigment Dispersion: a Rheological Approach to Coating and Ink Technology*, 2nd ed. Wiley, New York, 1979.

[4]V. Sanz; E. Sánchez; S. Giménez, and V. Beltrán. "Evolution of the Rheological Behaviour of Concentrated Clay Suspensions", *7th Mediterranean Congress of Chemical Engineering*. Barcelona: Fira, 1996.

[5]J. Marco; E. Sánchez, R. Gimeno, F. Lucas, M.Rodríguez, P. Negre, C. Felíu, C.; and E. Bou, "Rheological Behaviour of Glaze Suspensions. Influence of Frit Solubility, pH, Water Hardness and Additives", Cer. Acta, **8** [6], 35-51, (1996).

[6]J.L: Amorós, C. Felíu, E. Sánchez, and F. Ginés, "Influence of Spray-Dried Granule Moisture Content on Dry Mechanical Strength of Porcelain Tile Bodies", this volume.

UNDERSTANDING WHITEWARES SUSPENSIONS USING ZETA POTENTIAL

P. R. Jackson, B. R. Heywood*, J. Michael and J.F.Birtles

Ceram Research
Queens Road
Penkhull
STOKE-ON-TRENT
Staffordshire
ST4 7LQ

* Keele University
Keele
Staffordshire
ST5 5BG

OBJECTIVES

The objectives of this paper are four-fold:

- To discuss, in simple terms, the Guoy-Chapman double layer theory (in the context of its application to ceramic particles dispersed in water) as a pre-requisite for explaining particle-particle interaction as suggested by the Derjagin- Landau-Verwey-Overbeck (DVLO) theory.

- To describe the AcoustoSizer®: A technique capable of simultaneously generating zeta potential and particle size information for high solids suspensions.

- To present AcoustoSizer® data for mono ball clay systems.

- To share some ideas for the application of zeta potential to Whiteware production problems involving multi-component suspensions.

THE INTERACTION OF CERAMIC PARTICLES IN AQUEOUS SUSPENSION

By regarding ceramic powders as spheres that assume a negative charge when dispersed in water, we can consider a double layer model as pictured in Figure 1. For materials like silica, the negative charge arises from the surface silanol (Si-OH) groups de-protonating in water to form Si-O$^-$ species. This negative surface charge will attract counter-ions (positive in this instance) that are held closely to the surface. Beyond this Stern layer is a more diffuse layer, containing an excess of positive ions; this layer if further delineated by its interface with the bulk medium (where an equal number of positive and negative ions exists). A plot of energy vs distance from the surface shows that there is a sharp fall off in potential energy in the Stern layer.

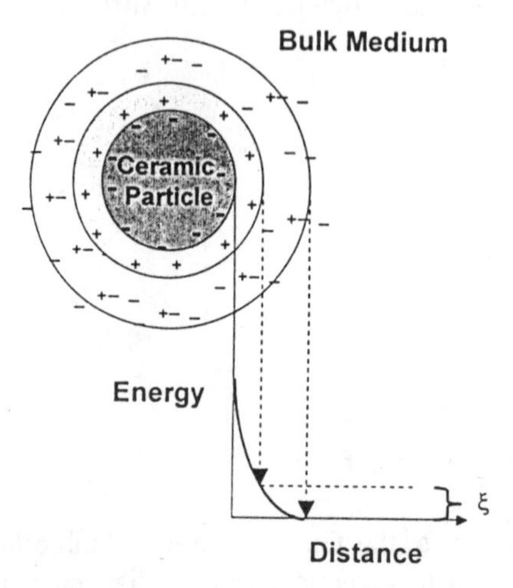

Figure 1. Double layer model and associated zeta potential ξ

There is therefore still a net negative charge at the edge of the Stern layer, but the positive ions contained within go a long way to cancelling the particle charge. There is more of an exponential decay in energy throughout the diffuse layer. Applying a voltage across the suspension causes the particle and stern layer to move as one, relative to the diffuse layer and bulk medium.

Zeta potential can be defined as the electrical potential at the surface between two phases, i.e., the potential at the boundary between the particle with its most closely associated ions and the surrounding medium. Zeta potential depends greatly on the size of the double layer: The larger the layer, the more particles want to remain apart (deflocculated); smaller double layers encourage flocculation. Factors governing the size of double layer (apart from particle surface charge) include ionic strength of the solvent, concentration of electrolyte and charge of electrolyte: For example, an M^{2+} counter-ion can more effectively cancel a negative surface charge than an alkali M^{+} ion. This results in the surface charge being satisfied at a shorter distance from the surface. Particle flocculation will therefore be encouraged.

There is much controversy concerning the double layer model. Workers like Guven [1] suggest more complex models in which the particle first induces a layer of water molecules and dehydrated counter-ions. Beyond this is a layer of hydrated cations. Certainly hydrated ions have a role to play, since degree of deflocculation is mediated by the alkali counter-ion series (Li+, Na+, K+) is dictated by the radius of the hydrated, not dehydrated ion.[2] Electropositivity of counter-ions is another factor considered important. [2]

The DVLO theory describes the interaction of two particles as depending on repulsive and attractive forces. The former arise from interaction of double layers (and Born repulsion at very short distance), whilst the latter are claimed to be due to Van der Waals (VDW) forces. Flocculation and deflocculation are determined by the size of the repulsive force, which as indicated above, relates to double layer size. The DVLO theory is however contended: For example, McBride [3] asks whether the Langmuir / Levine theory for particle – particle interaction is not more pertinent. Essentially this theory questions whether VDW forces are relevant in clay – water suspensions over the distances claimed. Instead, an electrostatic attraction is proposed, where the counter-ions in the two double layers re-arrange into a single counter-ion cloud which attracts both particles; repulsion then arises from osmosis in the counter-ion cloud.

For the purposes of this paper, double layer and DVLO theory will be accepted.

MEASUREMENT OF ZETA POTENTIAL

Zeta potential has the capability of augmenting rheological data and so improving our understanding of ceramic suspension behaviour.

Measurement of the phenomenon called electrophoresis is often used to calculate zeta potential. Essentially a voltage is applied across two electrodes immersed in the suspension. The negatively charged particle migrates to the positive electrode; viscous forces tend to oppose this movement but an equilibrium velocity is quickly established. Counter-ions move in the opposite direction. The particle velocity can be determined by for example detection of scattered light. If parameters such as the di-electric constant of the medium, particle size and suspension viscosity are known, zeta potential can be calculated. [4]

The main draw-back with light scattering techniques is the requirement for very dilute suspensions that are far-removed from the reality of whiteware body, glaze slops at 60 wt.%+.

The AcoustoSizer® offered by Colloidal Dynamics allows simultaneous measurement of zeta potential and particle size in high solids suspensions. Figure 2 is a simplified diagram showing the principle of the technique. The suspension is stirred in the sample cell which contains pH and conductivity probes. Ceramic rods heat and cool to maintain temperature to ± 0.1°C. A short pulse (featuring several cycles) of alternating voltage is applied across the electrodes that are embedded in opposing walls of the cell. Ceramic particles in suspension are attracted towards the positive electrode. Given the alternating voltage, the particles actually migrate between the two electrodes.

In reality the mobility of the particles is in-commensurate with their chemical environment (Stern layer, diffuse layer, counter ions etc). The resultant lag generates a sound wave which is detected by the pressure transducer which in turn is integral within the glass rods. This is recorded as a sinusoidal response, termed the Electro-kinetic Sonic Amplitude (ESA). The ESA signal is highly complex, containing sound responses for different particle size media.

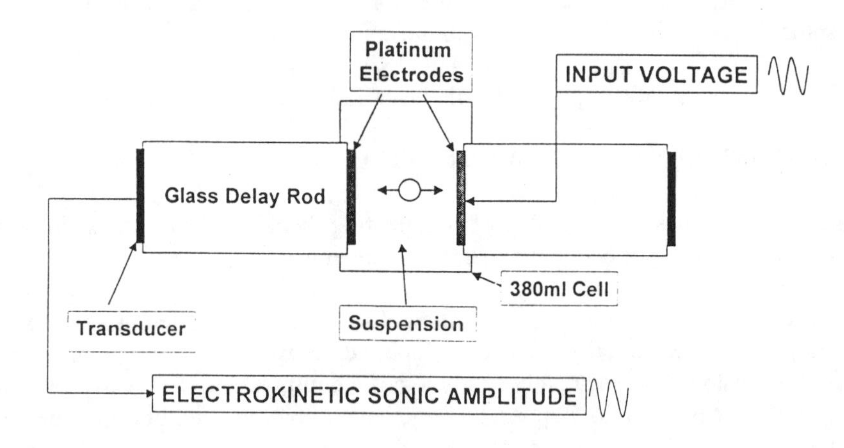

Figure 2. Simplified schematic of AcoustoSizer ® measurement cell

Figure 3 superimposes the sinusoidal wave-forms for the applied alternating voltage and the particle velocity it creates. By comparing the relative amplitudes of the two signals and the time delay (phase lag) zeta potential and particle size can be calculated. Each zeta potential / particle size measurement actually involves the application of not one but thirteen voltage pulses, each of increasing frequency; this takes about a minute. As the frequency of alternating voltage increases, an increasing percentage of particles show reduced velocity and increased phase lag; this is because the particles experience increased inertial forces, with inertial force being proportional to particle size.

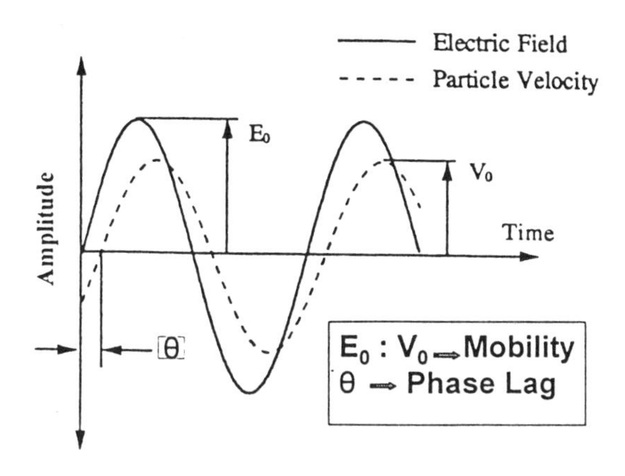

Figure 3. Sinusoidal wave forms for applied voltage and particle velocity

The AcoustoSizer® has titration capabilities and it is common to investigate the zeta potential of a suspension over a pH range or possibly as a function of additive (e.g. surfactant) concentration.

The interested reader is referred to papers written by the inventors for a more detailed description of the technique.[5]

ACOUSTOSIZER® DATA FOR BALL CLAY SUSPENSIONS

Ceram Research are working with Keele University in the UK on a Post-graduate Training Partnership (PTP) initiative. This is a scheme devised by the

Department of Trade & Industry (DTI) and Engineering & Physical Science Research Council (EPSRC) in England. Working in the overall field of particulate processing, the aim of the scheme is to generate PhD students with improved industrial awareness and contacts, thus rendering them more employable. One student within this scheme, Jo-Anne Michael is using the AcoustoSizer® to characterise mono ball clay suspensions. Figures 4 and 5 show zeta potential and particle size vs pH data for "as supplied" clays.

The simple model representing ceramic particles as perfect spheres with even charge distributed over the whole surface area is inappropriate for clay materials.

There is a need to recognise the platelet shape and the negative charge on the face (due to isomorphous substitution in the silica tetrahedra and alumina octahedra) relative to the positive edge charge (arising from broken bonds). Software is available to re-calculate AcoustoSizer® data to take account of these aspects.

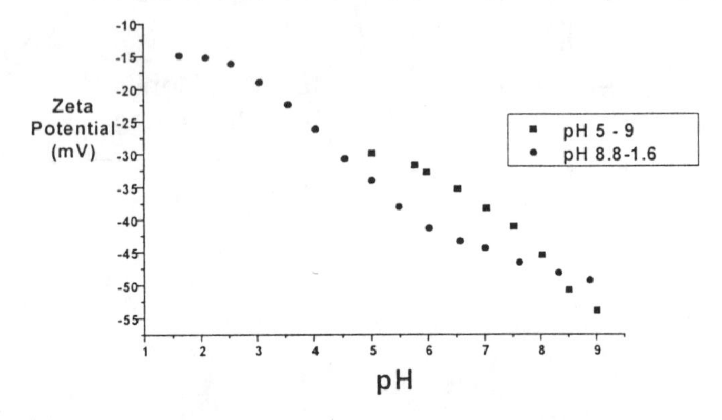

Figure 4. Zeta potential vs. pH for untreated ball clay

What should be noted from Figure 4 is the fact that the titrations from acid to base and base to acid do not coincide. It is postulated that when adding base, there is a reluctance to break down a clay particle interaction that is enhanced by the active organic component. A corresponding plot for the same clay stripped of organics shows that there is much better correlation between zeta potential data for acid and base titrations.

The particle size vs pH plot for "as supplied" clay in Figure 5 may at first sight appear misleading: The bars indicating the range of particles sizes present at each pH show that the largest agglomerates occur at high pH. The single particle size seen at low pH is associated with an AcoustoSizer® response from equal-sized entities within a house of cards structure. As base is added, the structure dissociates, to leave a mixture of small and larger structures. The mean particle size is fairly constant over the entire pH range; the mean does become closer to the minimum particle size with increased pH, suggesting increased generation of primary platelets.

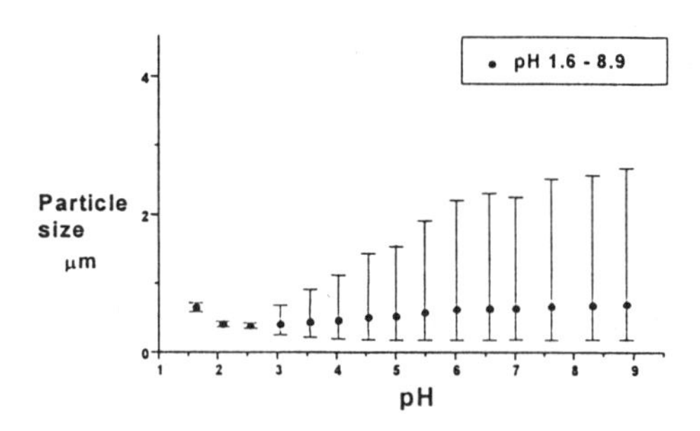

Figure 5. Particle size distribution vs. pH for untreated ball clay

The corresponding particle size plot for treated ball clay shows similar trends but with the overall mean particle size lower, suggesting organic adhesion to the clay particle raises the effective particle size. Also the largest particles (generated once again at high pH) are not as large; this is probably due to the absence of organic media to encourage particle linking.

Planned work will use (a) Techniques like Electron Microscopy, FTIR-microscopy (b) Simplified models for humic acid (featuring for example alkyl chains with carboxylic acid and amine groups) to help explain the AcoustoSizer® data.

Work is being extended to monitor the re-addition of commercial humic acid. Early data shows that at high pH, additions of humic acid (up to 400 mg added to a 2 wt% clay suspension) increase the negative zeta potential in ball clay suspensions by around 10%. Additions at low pH have less effect on zeta potential. Such differences can be at least partly explained from soil science papers written by workers like Kretzschmar.[6] It is postulated that humic acid adopts a coiled structure due to COOH groups at low pH allowing increased interaction with the positive clay edges. At high pH uncoiled humic acid units can interact with clay particles at a number of points resulting in less overall adsorption but (due to increased levels of COO^- groups) a more negative surface charge.

INDUSTRIAL APPLICATIONS

Looking at mono-systems represents one very necessary end of the spectrum. Once each mono system has been characterised, studies can move onto 2- and 3-component systems and a determination of whether or not suspension behaviour is based on additive rules.

Whilst acknowledging that a complete understanding is currently impossible, Ceram are applying the AcoustoSizer® technique to the analysis of real-life multi-component systems. Work is currently at an empirical level, looking for discrepancies in zeta potential for "good" vs "bad" suspensions. The following indicates the industrial problems being tackled:

Casting Slip

As sanitaryware producers increase production in low labour cost regions such as the Pacific rim and Eastern Europe, there is pressure to use cheaper local raw materials in body formulations. Unfortunately, setting up suspensions with novel materials to existing rheological conditions can give poor casting results.

Martin Stentiford et al. at Watts Blake & Bearne [7] warns of these dangers and shows that "fluid" and "gel" regions (calculated from viscosity vs time, density and deflocculation information) can be established for a given body. Setting up a body suspension close to the fluid / gel transition point is reported to ensure good casting. Ceram are starting work to assess whether there is a role for zeta potential as a reliable QC guide. Different measurements for "good" and "bad" bodies set up to a fixed rheology have been recorded. The change is not necessarily a change in zeta potential, but can be differences in pH, electrolyte level all of which can influence particle interaction and so casting rates, cast properties.

Unleaded Glazes

La Course [8] reported at the first Science of Whitewares conference that (unleaded) glaze frits leach appreciable amounts of Na, K etc. to generate a high suspension pH and a more de-flocculated state. This in turn can lead to alkali attack of the depleted frit particle surface layer and the formation of $Si(OH)_4$ which can start to reverse pH before an equilibrium is established. The large initial pH rise vs time and subsequent cycling to an equilibrium pH have been monitored using the AcoustoSizer®. There have been several instances in the UK of unleaded bone china glaze rheology drifting over a six, seven hour shift towards a higher fluidity. During automatic spraying, with biscuit ware travelling through the spray booth on rotating spindles, there is a gradual decrease in the quality of deposited glaze; ripples tend to arise, especially where little or no pre-heat has been used. Such ripples are not completely fired out in the glost piece. Ceram researchers are trying several ways to combat this, including the use of zeolite additions.

Further glaze work aims to use zeta potential to characterise re-claimed glazes, especially where filtration techniques (such as the "Xtract", "Renovex" systems) have been used to de-water re-claim suspensions.

There is a danger that cations leached from suspended frit particles will be lost in the re-claim glaze. Not only will this affect the rheology of virgin / re-claim glaze mixtures, there is a danger of raising the glaze viscosity during firing.

Consistent Plastic Bodies

Tableware producers using ball clay in their bodies often report variable pug roll plasticity, despite careful control of component and body suspension rheology. Whether plasticity is measured using "potters thumb" or more sophisticated techniques based on capillary rheometry, the fact is that waste pug is generated. Ceram researchers are about to begin a study where body suspension zeta potential is correlated with pug roll plasticity after filter-pressing and de-air pugging.

It is postulated that using additives to ensure a constant, ideal zeta potential (and so water / hydrated counter-ion environment around particles) will deliver a plastic body with consistent shear properties.

Other potential industrial applications include the appraisal of novel surfactants, quality control of water-based inks and studies to improve the efficiency of spray-dried granulate production. The list is by no means exhaustive and the need for better suspension control at all stages of whitewares production should encourage the development of a cheaper, more practical zeta probe, based on ESA. Ceram Research has developed a device for automatic measurement and adjustment of suspension rheology. This concept could possibly be extended to zeta potential, using acid / base or surfactant addition to maintain a constant suspension zeta potential prior to processing.

ACKNOWLEDGEMENTS

The valuable input of Watts Blake Bearne and Co PLC. as sponsors of Joanne Michael is acknowledged.

REFERENCES

[1]N. Guven, "Molecular aspects of clay-water interactions" *Clay – Water Interface and its Rheological Implications,* Ed. N. Guven and R.M. Pollastro, Clay Minerals Society (Boulder, Colorado, 1992) ISBN -1-881208-04-4

[2]W.E.Worrall, "Properties of clay-water systems" *Clay and Ceramic Raw Materials,* Applied Science Publishers 1975. ISBN 0 85334 631 3

[3]M.B. McBride, "A critique of diffuse double layer models applied to colloid & surface chemistry" *Clays and clay minerals,* **45** [4] 598-608 (1997)

[4]P. McFadyen, "Determining zeta potential" Brookhaven Instrument Ltd. Reprint from *Ceramic Industries International,* June 1993

[5]D.W. Cannon, "New Developments in electroacoustic methods and instrumentation" *Workshop proceedings NIST,* 40-66 (1993). Ed. G.Malghan

[6]R.Kretzschmar, D.Hesterberg & H.Sticher, "Effects of humic acid on surface charge and flocculation of Kaolinite" *Soil Sci. Soc. Am. J.* **61** Jan-Feb (1997)

[7]M.J. Stentiford, N.L. Whetton, J.M. Woodfine "The evaluation of sanitaryware body composition". *Fourth Euro ceramics,* **10** 15-20 (1995). Ed. P Gambara, G. Pasquali, M. Biadigo.

[8]"Glaze problems from a glass science perspective" by W.C. Lacourse & W. Mason. Proceedings of the Science of Whitewares Conference, Alfred University, July 16-20, 1995. Ed. V.E.Henkes, G.Y. Onoda, W.M. Carty.

INTERACTION BETWEEN BORATES AND CLAY SUSPENSIONS

Mark D. Noirot
U.S. Borax, Inc.
26877 Tourney Rd.
Valencia, CA 91355

INTRODUCTION

Borates have been used in ceramic glazes for centuries and are well known for their glass forming abilities and influence on glaze properties. There is interest in using borates in ceramic bodies to impart new properties and enhance the sintering process. Work by Hong and Messing[1] indicates that borate addition can enhance mullite formation in aluminosilicate gel systems. They found that B_2O_3 doping decreased the mullite transformation temperature and lowered the activation energy for mullite nucleation and growth processes. This evidence suggests that more traditional whitewares formulations may benefit from the properties of borates.

However, there are complications incorporating borates into clay suspensions. Many borates are soluble to the extent where "salt effects" on clay suspensions are expected to cause flocculation and unsuitable flow properties. Most glazes that use boron compounds incorporate them as preformed glass frit to avoid this problem. Interestingly, the effects of soluble borates on colloidal systems are distinct from simple inorganic salts and in some circumstances borates appear to behave as mild deflocculants.[2] This is an initial investigation of the interaction of borates with kaolinitic and illitic clay particles in aqueous systems through isothermal adsorption and rheological studies.

EXPERIMENTAL METHODS

Isothermal Adsorption Studies

Each sample was prepared by charging clay suspension to a centrifuge tube, adding a small volume of borax or boric acid solution, and a complementary volume of water so that all samples attained the same final solids density. For example, a 30.0 g sample of 10 vol % kaolin slurry was charged to a centrifuge tube then 3.00 ml borax solution (1200 ppm B) and 2.00 ml of water (HPLC grade) were added to make up one sample. Samples were thoroughly mixed using a vortex mixer, then conditioned in a constant temperature bath at 30 °C for 7 days. Then samples were centrifuged, 21000 rpm, for 2 hours and clear centrate was decanted into clean plastic beakers. Solution was taken up by plastic syringe and forced through 0.45 μm Nylon filters. The filtrate was analyzed for boron content by ICP emission spectroscopy. Results are shown in Figures 1 and 2. Borates selected for this study range from commercially available bulk quantities, to experimental compounds. Details are presented in Table I.

Table I. Borate Properties and Information[3]

Commercial / generic name	*Chemical Formula*	*Formula Weight*	*Comment*
-- boric acid	$B(OH)_3$	61.83	commercial refined product solubility 5.46 % wt at 25°C
Neobor® borax 5 mol	$Na_2O{\cdot}2(B_2O_3){\cdot}5(H_2O)$ or $Na_2B_4O_7{\cdot}5H_2O$	291.35	commercial refined product solubility 4.43 % wt at 25°C
Firebrake ZB® zinc borate	$2ZnO{\cdot}3(B_2O_3){\cdot}3.5(H_2O)$	434.66	commercial refined product solubility << 1 % wt at 25°C
-- calcium-hexaborate Nobleite (synthetic)	$CaO{\cdot}3(B_2O_3){\cdot}4(H_2O)$	336.86	experimental synthetic compound solubility ~ 1 % wt at 25°C
Hydroboracite	$CaOMgO{\cdot}3(B_2O_3){\cdot}6(H_2O)$	413.3	beneficiated mineral ore (Argentina) solubility < 1 % wt at 25°C

Clay Suspension Preparation

Two clay systems were chosen based on commercial interest and availability. Peerless # 2 Kaolin (R. T. Vanderbilt Company) is representative of the kaolins found near Bath, Georgia, USA. The specific surface area measured 14.1 m^2/g (multi point BET; Micromeritics, Gemini), density reported 2.62 g/cm^3. This kaolin was used to prepare clay suspensions with no other mineral component. The second clay chosen was a red illitic clay mixture (Villar) used in tile bodies in Spain. Villar samples were provided by the Instituto de Tecnología Cerámica (ITC), Spain, -density 2.6 g/cm^3, -specific surface area 43 m^2/g. Complete characterization of the sample was not carried out. Darvan 811, (sodium poly-acrylic acid, NaPAA) was used where dispersant is indicated.

Adsorption Suspensions: Low solids content suspensions, about 10 vol %, were prepared by charging 450 g high purity water (18.2 M ohm resistance) to a vessel, then adding 131.0 g kaolin. Solids were blended under low shear conditions followed by dispersion with ultrasonication. Suspensions were conditioned in a constant temperature bath at 30 °C overnight before further formulation. Identical procedures and quantities were used to prepare illitic clay suspensions.

Rheological Studies: Measurements were made using a Paar Physica DSR4000 instrument and software with parallel plate (50 and 75 mm diameter) sample configuration. Plate separation used was 1 mm for all kaolin samples and 1.5 or 2 mm for illitic samples. Villar contains a coarse fraction which requires greater plate separation. Apparent viscosity measurements were using this sequence:

shearing at 500 s^{-1} for 30 s, no shear for 30 s, shear ramp (upward) from 0.1 to 500 s^{-1}, shear ramp (downward) from 500 to 0.01 s^{-1}, all measurements conducted under steady shear conditions, at a temperature of 25 °C. The apparent viscosity at a selected shear rate of 1.0 s^{-1} was determined by regression analysis of a selected linear portion of each decreasing shear flow curve.

RESULTS AND DISCUSSION

Adsorption

The quantity of borate bound to the surfaces of kaolin and illite clays was different as might be expected, Figures 1 and 2. Kaolin appears to have little or no detectable bound boron from borax or boric acid across the low concentration range studied, while the illite clearly shows a minor interaction with both. The quantity of boron bound to illite was more than 10 times less than the amount of NaPAA dispersant needed to reach the minima in viscosity. Very little borate is strongly bound to the surface of illite and essentially no detectable amount to the surface of kaolin in the concentration range ~0 to 250 ppm B (0 to 23 mmol/l).

This minor amount of surface adsorption is not likely to cause the much greater magnitude effects on rheology found at higher boron concentrations. There are more plausible explanations involving different solution species. Performing isothermal adsorption studies at much higher boron concentrations is difficult

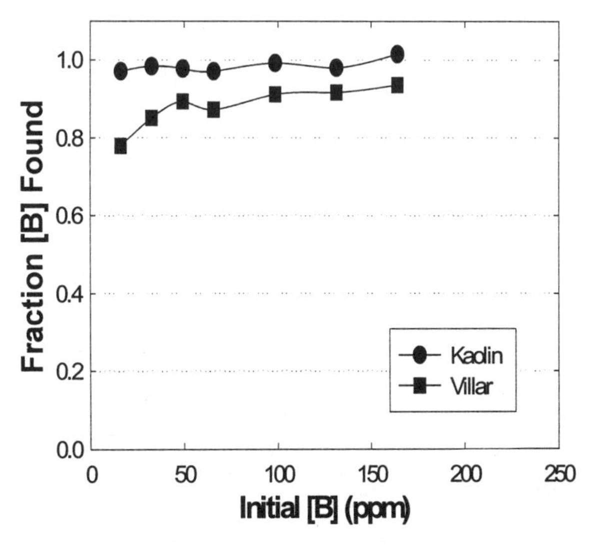

Figure 1. Boric Acid Adsorption Isotherms. Initial concentration of boric acid ($B(OH)_3$, expressed as ppm B in solution) versus the fraction of boron found after conditioning at 30°C for 1 week.

since the amount of boron in solution compared to the fraction on the surface is very large (even if the total surface loading becomes more significant). Analytical methods have trouble resolving small differences in large boron concentrations.

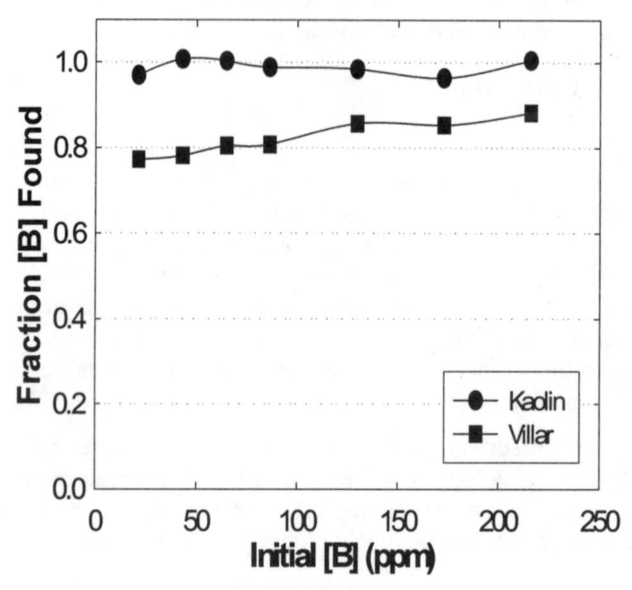

Figure 2. Borax Adsorption Isotherms. Initial concentration of borax ($Na_2O \cdot 2(B_2O_3) \cdot 5H_2O$, expressed as ppm B in solution) versus the fraction of boron found after conditioning at 30°C for 1 week.

Kaolin Rheology

Additions of borates to relatively high slurry density kaolin (29 vol %) and illite (35 vol %) clays had profound effects on the rheology in most cases (Figure 3). The flowcurves obtained were non-Newtonian and thixotropic. Generally, addition of borates to dispersed systems (NaPAA levels > 0.1 mg/m^2) increased viscosity to a maximum value near that of the non-dispersed clay. There were several cases where interesting differences were found. There is difficulty in presenting the information uniformly because the solubility limits of the borates are exceeded in several series. This makes it impossible to properly represent concentrations in the solution phase (without chemical analysis) and difficult to represent as a fraction of total solids. Presentation is further complicated by the incongruent dissolution of the low solubility metal borate compounds. For instance, zinc borate dissolves very slowly, to small extent, and does not result in simple stoichiometric concentrations of zinc and boron in solution.[4] The zinc forms $Zn(OH)_2$ at the pH of the kaolin slurries (about 6.8). Both synthetic

Nobleite and Hydroboracite behave similarly; they dissolve slowly and to fairly small extent. The Ca^{2+} and Mg^{2+}, on the other hand, likely remain as more soluble species. With these limitations in mind, addition of the low solubility borates to kaolin and illite are shown in Figure 4. Borate expressed as moles Boron added per gram of dry clay.

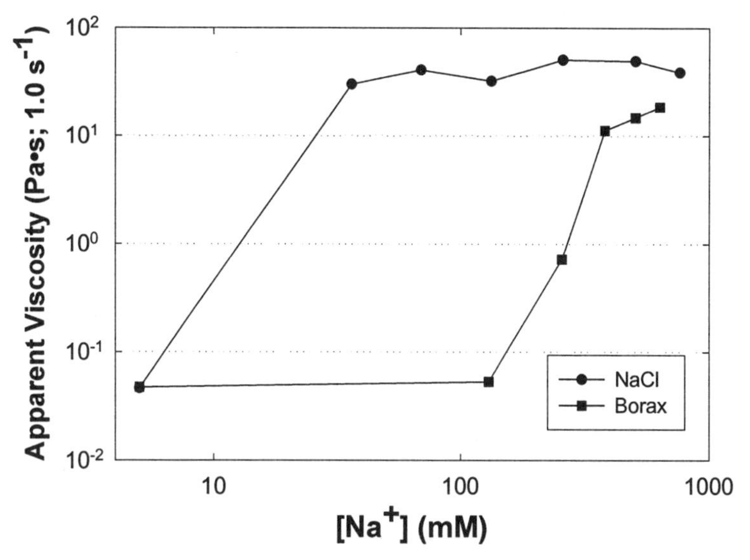

Figure 3. Apparent Viscosity, Kaolin with Sodium Chloride and Borax. Kaolin, 29 vol %, 0.10 mg/m^2 NaPAA, at 1 s^{-1} shear rate. About 5 mmol/l Na^+ contributed by NaPAA to each sample.

Illite Rheology

Several more detailed studies were carried out with the red illitic clay from Spain. Three initial states of dispersion were investigated with borax and boric acid additions. $CaCl_2$ and NaCl additions were made for comparison. The results are plotted in Figure 5. Addition of Ca^{2+} as $CaCl_2$ quickly flocculated the system at very low concentrations. Addition of Na^+ as NaCl gave similar but less rapid increase in viscosity. Once again, the borate significantly altered the effect expected for a given Na^+ concentration by offsetting the concentration where flocculation begins. Separate work by Moreno and Sans agrees with the nature of these findings.[5]

In fact, borax has a distinct dispersing effect obvious in the middle of the addition range. Each of the partially dispersed illitic suspensions shows a reduced viscosity at 300 mmol/l Na^+ (600 mmol/l B), Figures 5 and 6. Boric acid alone does not appear to disperse the clay. Sodium cations are needed for borates to have a dispersing interaction with the surface. In the case of borax, a favorable change in pH might account for some of the observed decrease in viscosity, however, the actual changes in pH are very subtle here and cannot account for this change alone. Boron solution chemistry provides a more likely explanation.

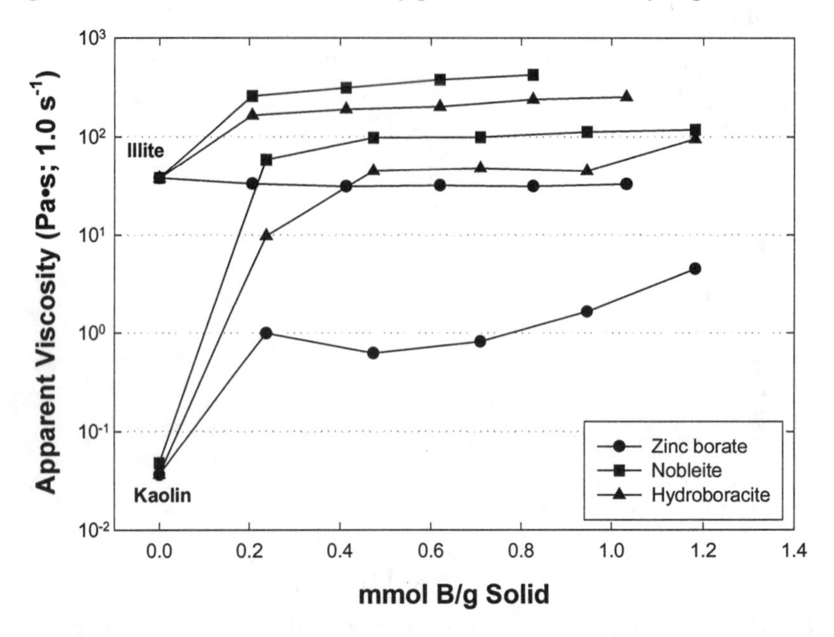

Figure 4. Apparent Viscosity, Illite and Kaolin with Low Solubility Borates. Illite, 35 vol %, 0.20 mg/m^2 NaPAA. Kaolin, 29 vol %, 0.10 mg/m^2 NaPAA, at 1 s^{-1} shear rate. Illite suspensions, 3 upper curves, Kaolin suspensions, 3 lower curves.

Soluble boron species tend to oligomerize at high borate concentrations. The equilibria and number of species involved are fairly complex but have been studied by NMR experiments.[6] Several equilibria have been quantitatively characterized by Maya using Raman spectroscopy[7] and are expressed below. Condensation reactions and hydroxide exchange equilibria are important. The free water generated by these reactions is one of the driving forces for oligomerization from concentrated solutions.

$$B(OH)_3 + OH^- = B(OH)_4^-$$
$$2\,B(OH)_3 + B(OH)_4^- = B_3O_3(OH)_4^- + 3H_2O$$
$$2\,B(OH)_3 + 2B(OH)_4^- = B_4O_5(OH)_4^{2-} + 5H_2O$$
$$4\,B(OH)_3 + B(OH)_4^- = B_5O_6(OH)_4^- + 6H_2O$$

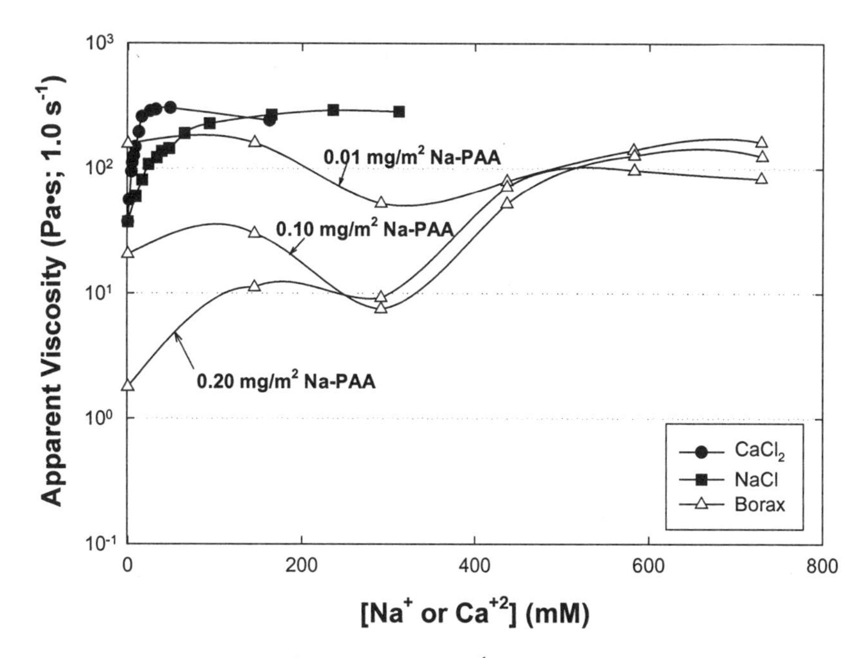

Figure 5. Apparent Viscosity (at 1.0 s^{-1}), Illitic Suspensions with Borax, NaCl, and $CaCl_2$, plotted versus cation concentration. Suspensions with $CaCl_2$ and NaCl have 0.10 mg/m^2 NaPAA; borax suspensions with NaPAA as indicated. Note that the relative minima suggested by the borax curves at [Na^+] = 300 mol/l ([B] = 600 mmol/l) could be anywhere in the range 200-400 mmol/l.

Borate trimers, tetramers, and pentamers all demonstrate acid base chemistries which can form polyanionic species. These are the most likely species to interact with the clay surfaces in much the same way as sodium silicate (a common dispersant). These polyanions can localize "excess" negative charge near the clay surface increasing the interparticle repulsive forces. Experiments using boric acid combined with NaCl, NH_4Cl, and the alkali metal series of chlorides should help demonstrate how the cation is involved and essential for interaction at the surface of the clay. Recent NMR studies by Kim and Kirkpatrick have already shown differences in adsorption of Na and Cs onto illite.[8] Many of the mechanistic questions could be studied by ^{11}B NMR techniques.[9]

Silicate oligomers are more kinetically stable and well defined than similar borate compounds. Polyacrylic acid dispersants are also quite stable. However, borate oligomers, in solution, undergo rapid hydroxide exchange, and rapid kinetic rearrangements. This probably explains why the magnitude of the dispersing effect is lower and only observed at high concentrations of borate in solution. As sodium concentrations increase above 300 mmol/l, the effect of Na^+ cation suppressing the double layer becomes more significant and eventually flocculates the system. Without any borate, this effect is observed around 100 mmol/l Na^+. Illite with borax additions offsets flocculation to nearly 400 mmol/l Na^+.

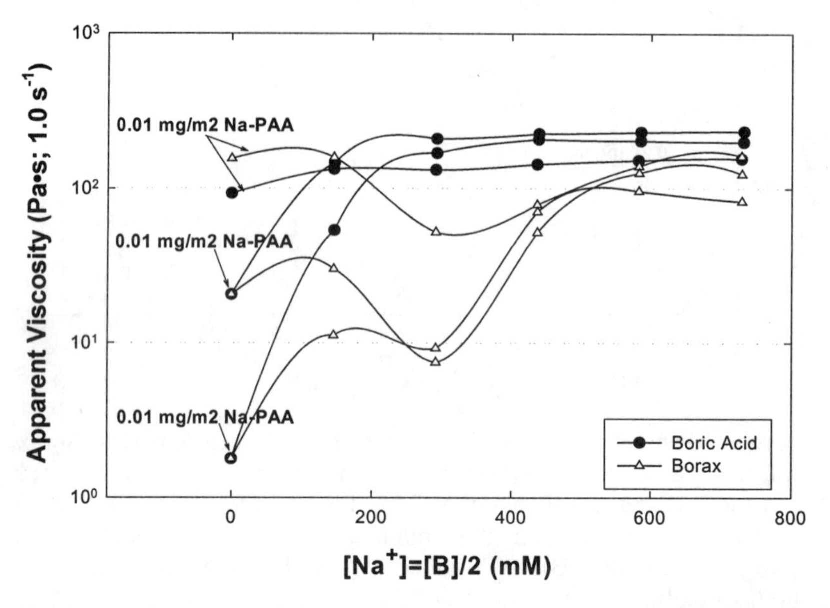

Figure 6. Apparent Viscosity, Illitic Suspensions with Borax & Boric Acid. Three loadings of NaPAA are noted, shear rate 1 s^{-1}. Boric Acid expressed as equivalent [B].

Nobleite, Hydroboracite, and zinc borate effects upon illite are shown in comparison to effects upon kaolin in Figure 4. Surprisingly, the addition of zinc borate to illite showed no significant change. The suspension retained its original flow character even at the highest borate additions. The other illite systems increased viscosity about 10 fold. Commercial processing of these slurries would be difficult. Attempts were made to correct or reduce the increased viscosity by adding more sodium polyacrylic acid dispersant. Results varied and are shown in Figure 7.

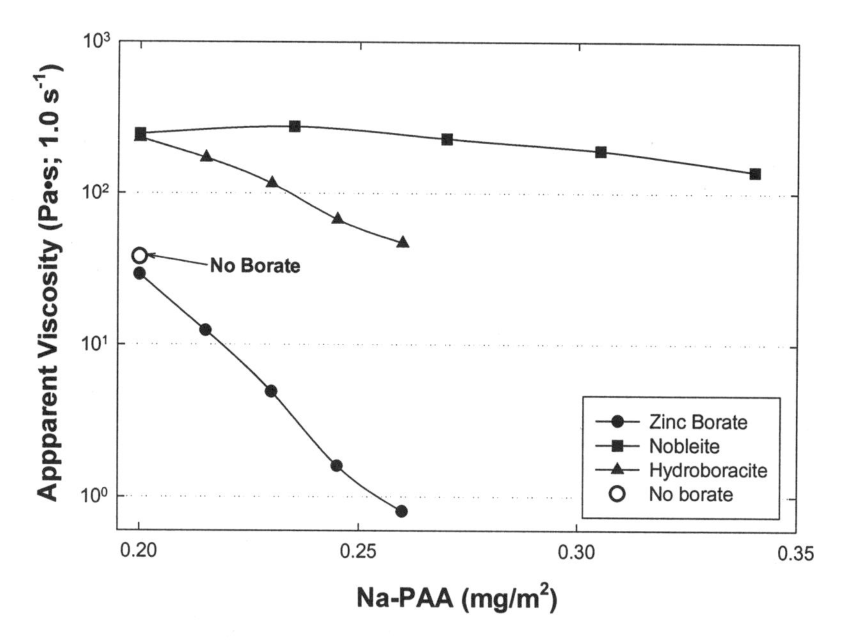

Figure 7. Addition of NaPAA to Illite-Borate systems. Open circle represents illite system with no borate added: Villar, 35 vol %, 0.20 mg/m^2 NaPAA. Borate concentration in all other samples corresponds to 1.2 mmol B / g clay.

CONCLUSIONS

Borax was found to have a dispersing interaction with illite at about 600 mmol/l B and 300 mmol/l Na^+. Boric acid alone did not show this effect. It is likely that anionic oligomers of boron are the effective dispersing agent for clays, similar to sodium silicate. Borax effectively inhibits flocculation of the illite systems with relatively high Na^+ concentrations but does flocculate the system at the higher concentrations examined.

Soluble borates can be added to the kaolin and illitic systems with acceptable effects below critical concentrations. Both the solution concentration of M^+ and borate are important factors. It was hoped that low solubility Hydroboracite and synthetic Nobleite would not flocculate these suspensions. Unfortunately, the effects were profound at all additions studied. Flocculation is most likely due to the small amounts of Ca^{2+} and Mg^{2+} that enter solution as a fraction of the borate dissolves. The influence of the cation is not offset by borates since very little boron enters solution. Essentially no oligomers are formed at very low boron concentrations. The zinc borate demonstrated uniquely lower impact on kaolin rheology and essentially no impact on illite rheology. This is likely due to overall lower solubility of zinc borate and formation of zinc hydroxide as trace zinc enters solution. Comparison of the isothermal adsorption studies, carried out at very low boron concentrations, with the rheological studies, carried out over a wide range of high concentrations, demonstrated that borates can have significant effects on clay surfaces at high concentrations where anionic oligomers are likely involved. At lower concentrations, where discrete boron monomers predominate in solution, the borate interaction with clay surfaces is very minor.

REFERENCES

1. S. H. Hong and G. L. Messing, "Mullite Transformation Kinetics in P_2O_5-, TiO_2-, and B_2O_3-Doped Aluminosilicate Gels," *Journal of the American Ceramic Society*, **80** [6] 1551-59 (1997).
2. J. H. Howles and W. M. Carty, "Incorporation of Boron Compounds into a Whiteware Body," NYS College of Ceramics at Alfred University, July 1997, unpublished results.
3. All borate compounds tested were provided by US Borax Inc., 26877 Tourney Road, Valencia, CA, 91355.
4. Zinc borate ($2ZnO{\cdot}3B_2O_3{\cdot}3.5H_2O$) dissolution in pure water results in a steady state concentration of $[Zn^{2+}] < 0.4$ mmol/l. Zn^{2+} in solution is expected to have significant impact on the double layer, but the concentration achieved is quite low near neutral pH.
5. A. Moreno and V. Sans, Instituto de Tecnología Cerámica, Castellon, Spain. Personal communications.
6. C. G. Salentine, "High Field ^{11}B NMR of Alkali Borates. Aqueous Polyborate Equilibria", *Inorganic Chemistry*, **22** [26], 3920-8 (1983).
7. L. Maya, *Inorganic Chemistry*, **15**, 2179, (1976).
8. Y. Kim and R J. Kirkpatrick, "NMR T_1 Relaxation Study of ^{123}Cs and ^{23}Na Adsorbed on illite," *American Mineralogist*, **83**, 661-5 (1998).
9. H. D. Smith, Jr. and R. J. Wiersema; "Boron-11 Nuclear Magnetic Resonance Study of Polyborate Ions in Solution," *Inorganic Chemistry*, **11** [5], 1152-4 (1972).

THE EFFECTS OF IONIC CONCENTRATION ON THE VISCOSITY OF CLAY-BASED SUSPENSIONS

Katherine R. Rossington and William M. Carty
NYS Center for Advanced Ceramic Technology—Whiteware Research Center
New York State College of Ceramics
Alfred University, 2 Pine Street, Alfred, NY 14802

ABSTRACT

The influence of anions (Cl^{-1} or SO_4^{-2}) and cations (Na^{+1}, Ca^{+2}, or Mg^{+2}) on the rheology of a typical whiteware suspension was evaluated as a function of concentration and dispersion level. Using a typical whiteware composition consisting of clay, nepheline syenite, quartz, and alumina, a suspension was prepared and dispersed using sodium polyacrylic acid (Na-PAA). The effect of salt additions (NaCl, $CaCl_2$, $MgCl_2$, Na_2SO_4, $CaSO_4$, and $MgSO_4$) to the whiteware suspension on the rheology and pH was measured. The results indicate the viscosity of the suspension is dependent on the cation level (Na^{+1}, Ca^{+2}, or Mg^{+2}) and is essentially independent of the anion level (Cl^{-1} or SO_4^{-2}). Zeta-potential measurements support the assertion that cations are responsible for the change in viscosity. Finally, the experimental results strongly suggest that double-layer compression, rather than specific cation adsorption, is the mechanism responsible for the change in suspension rheology.

INTRODUCTION

The question of which ionic species, the cation or the anion, is responsible for coagulation of a clay-based system continues to be disputed. For this discussion, coagulation is defined as the reversal of actions taken to disperse a suspension, usually by the addition of ions that compress the double-layer surrounding a colloidal particle. Flocculation is the condition that results from taking no action to improve suspension stability.

Increasing either the ion concentration (n_i^0) or charge (z_i) of the counter-ion compresses the double layer eventually leading to coagulation. An approximation of the double layer thickness (L, in nm) is calculated as the inverse of the Debye-Hückel parameter (κ), where ε is permittivity, k is the Boltzman constant, T is absolute temperature, and e is the proton charge:[1,2]

$$L = \frac{1}{\kappa} = \left[\frac{\varepsilon k T}{e^2 \sum n_i^0 z_i^2} \right]^{1/2} \tag{1}$$

Converting the ionic concentration (n_i^0) to mol•l^{-1} (denoted c_i), and assigning a temperature of 25∘C, Equation 1 simplifies to:[2]

$$\kappa = 3.288\left[\frac{1}{2}\sum\left(c_i z_i^2\right)\right]^{\frac{1}{2}}\left(nm^{-1}\right). \tag{2}$$

The Debye-Hückel approximation of the double layer thickness is predicated on the use of symmetrical electrolytes, i.e., 1:1 or 2:2 salts such as NaCl or $MgSO_4$, and therefore is not applicable to asymmetrical electrolytes, i.e., 1:2 or 2:1 salts such as Na_2SO_4 or $CaCl_2$.[1,2]

In addition to compression of the double-layer, coagulation can be caused by specific adsorption of ions from the suspension medium. Specific adsorption requires near surface interaction between the ions from the bulk solution and the ions on the surface of the particle. Two indications of specific adsorption are: a shift of the point of zero charge with increased ion concentration, and reversal of the sign of the ζ-potential at sufficiently high ionic concentrations.[2]

These coagulation mechanisms, compression of the double-layer and specific adsorption, are dependent on the magnitude of the particle charge and ionic strength of the suspension medium. Kaolinitic clays dominate the behavior of a typical whiteware suspension because clays have, by far, the greatest surface area in the batch. In addition, the charge on the particle surface can be controlled by the addition of dispersant, the effectiveness of which is dependent on the affinity of the dispersant for the clay surfaces.[3] Therefore, at a constant dispersant level, the surface charge can be assumed to be constant and the onset of coagulation can then be determined through suspension rheology measurements by changing the ionic concentration. It is then straightforward to determine whether cations (positively charged ions) or anions (negatively charged ions) are responsible for coagulation of a whiteware suspension, as well as the relative contributions of ionic charge (i.e., monovalent or divalent). Finally, the changes in rheology can be linked to the double-layer thickness approximations and to ζ-potential measurements to identify whether compression of the double-layer or specific adsorption is the responsible coagulation mechanism.

EXPERIMENTAL PROCEDURE

Batch Composition

The effects of ionic concentration and dispersion level were evaluated using suspensions with a typical whiteware composition; the batch composition is listed on a dry weight basis (d.w.b.) in Table I. The percentage of clay listed includes both kaolin and ball clay, which combined account for 93.8% of the total batch specific surface area.

Table I: The batch composition used for all viscosity measurements.

Raw Material	Wt. % (d.w.b.)	Specific Surface Area (m^2/g)	% of Batch Surface Area
Kaolin	29.0	26.9	76.1
Ball Clay	7.0	25.9	17.7
Alumina	12.5	1.0	2.7
Quartz	29.5	0.9	1.2
Nepheline Syenite	22.0	1.1	2.3

Batching and Ionic Strength Adjustment

Three thirteen-liter stock suspensions at 35 volume % (a specific gravity of 1.6, or 59 w/o) were prepared using a high-intensity mixer (3 HP Mixer, SHAR, Inc., Fort Wayne, Indiana; 500 rpm; six inch impeller) at dispersant levels of 0.00, 0.02, and 0.05 mg/m^2. Individual 200 ml suspensions were apportioned from the stock suspensions. The ionic concentration of the 200 ml suspensions was adjusted by adding saturated salt solutions diluted to the appropriate level with distilled water (as described below). These additions reduced the solids loading of all suspensions to 30 volume % (1.5 specific gravity or 54 weight %). After the ionic strength of each individual suspension was adjusted, the containers of suspension were sealed for a minimum of two weeks prior to viscosity measurements.

Verification of Ionic Strength and Raw Material Dissolution

All salt additions were introduced into the suspensions as saturated solutions to insure a reliable and consistent salt concentration. This addition route also eliminated two potential error sources: the hydration of the salt surface due to the hydroscopic nature of the salts (thus introducing a weighing error), and the formation of localized salt crystal-suspended particle clusters due to locally high ionic strengths. The concentration of each saturated salt solution was determined by the solubility limit of each salt as listed in Table II. The low solubility of $CaSO_4$ limited the $CaSO_4$ concentrations in the suspensions.

The ionic contribution from the raw materials was also evaluated by the dissolution of the individual materials. Suspensions of ball clay, kaolin, and nepheline syenite were prepared using distilled water at 10 volume % solids loading. The supernatant from each suspension was collected for analysis (Inductive Coupled Plasma, ICP, ACME Analytical Laboratories, Vancouver, B.C.) after aging for one week and centrifuging. A comparison of the ionic contribution from these raw materials and the distilled water used throughout this study are shown in Figure 1; the ionic contribution from the water is insignificant compared to the contribution from raw material dissolution.

Table II: Solubility limits of the salts used to adjust ionic strength.[4]

Chemical Formula*	Reported Solubility (g/100cc)	Reported Solubility (mM)	Measured via ICP (mM)**
NaCl	35.7	6.16×10^3 [Na^+]	6.09×10^3 [Na^+]
½ $Na_2SO_4 \cdot 10H_2O$	5.5	3.41×10^2 [Na^+]	2.90×10^3 [Na^+]
$MgCl_2 \cdot 6H_2O$	167	8.21×10^3 [Mg^+]	5.23×10^3 [Mg^+]
$MgSO_4 \cdot 7H_2O$	71.0	2.88×10^3 [Mg^+]	2.86×10^3 [Mg^+]
$CaCl_2 \cdot 2H_2O$	97.7	6.71×10^3 [Ca^+]	5.88×10^3 [Ca^+]
$CaSO_4 \cdot 2H_2O$	0.241	1.40×10^1 [Ca^+]	1.42×10^1 [Ca^+]

* Reagent grade chemicals, Fisher Scientific, Pittsburgh, Pennsylvania.
** Acme Analytical Laboratories Ltd., Vancouver, British Columbia, Canada

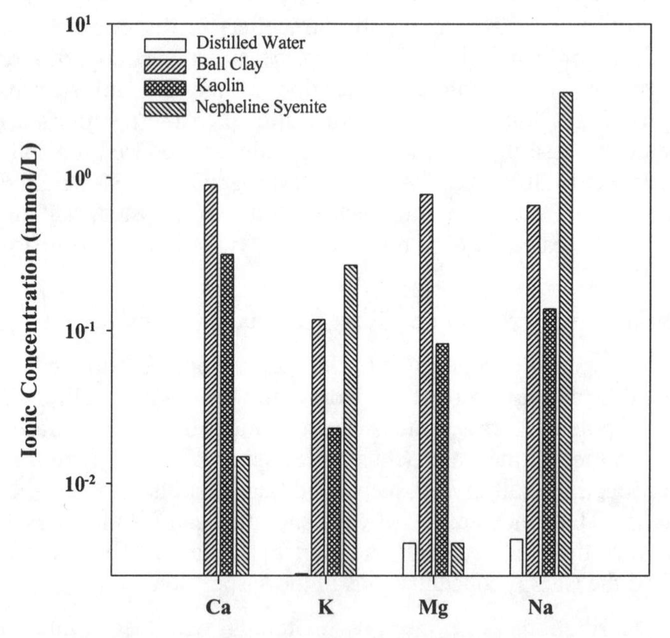

Figure 1. The ionic contribution from the distilled water is negligible compared to the dissolution from the raw materials.

ζ-Potential Measurement

ζ-Potential was measured via acoustophoretic mobility (Acoustosizer, Colloidal Dynamics, Warwick, Rhode Island). A five volume percent kaolin (EPK, Zemex Minerals, Inc.) suspension was used for ζ-potential measurements. The ionic strength was adjusted as described above.

Rheology Measurement

Steady-state shear behavior was measured using a stress-controlled rheometer (SR-200 Dynamic Stress Rheometer, Rheometrics Scientific, Piscataway, New Jersey); a cone and plate geometry was used for all measurements. Using a steady state, stress-sweep test, from high to low stress, log shear rate versus log viscosity data was generated for all suspensions. The apparent viscosity values reported for each suspension were extrapolated from a linear regression of the log shear rate versus log viscosity, to 1.0 s^{-1}. Similar apparent viscosity trends were measured with an industrial-type viscometer (LVTD, Brookfield Engineering Laboratories, Inc., Stoughton, Massachusetts).

RESULTS AND DISCUSSION

Salt Additions

The effects of the salt additions on apparent viscosity were similar for all three dispersion levels. For purposes of this discussion, most examples of salt additions presented in this section are at a dispersion level of 0.02 mg/m^2.

Figure 2 illustrates the change in apparent viscosity with increasing ionic strength using $CaCl_2$ as an example. Three distinct regions are identified. In Region I (low salt concentrations), the viscosity is independent of the ionic strength. Region II shows a dramatic change in the apparent viscosity with a small increase in salt concentration, and encompasses the critical coagulation concentration (CCC) level. At high salt concentrations, denoted as Region III, viscosity again becomes independent of ionic strength. The third region was not obtained with $CaSO_4$ additions due to the previously mentioned solubility limitations.

The effects of the chloride salts on the apparent viscosity are shown in Figure 3; again, the three regions previously discussed are clearly evident. The plateau of Region III appears to be similar for all three salts. The CCC shows one behavior for the two divalent cations (Ca^{+2} and Mg^{+2}) and one for the monovalent cation (Na^{+}). The monovalent salt cation concentration necessary to reach the CCC is eight to ten times that of the divalent cation concentration. Similar behavior is observed with the sulfate salt additions, as illustrated in Figure 4.

The comparison of the divalent sulfate anion versus the monovalent chloride anion, normalized to the cation level, is shown for the three cations, Na^{+}, Ca^{+2}, and Mg^{+2}, in Figures 5, 6, and 7, respectively. The apparent viscosity behavior of the divalent anion and monovalent anion salts is similar for the Na and Ca salts. A slight deviation of this trend was noted with the Mg salts; $MgSO_4$ coagulates the suspension at a lower concentration than $MgCl_2$, when normalized to the the cation concentration. These results indicate minor differences in the affinity of the ionic species for the particle surfaces. It is clear, however, that specific adsorption is not a strong contributor to coagulation.

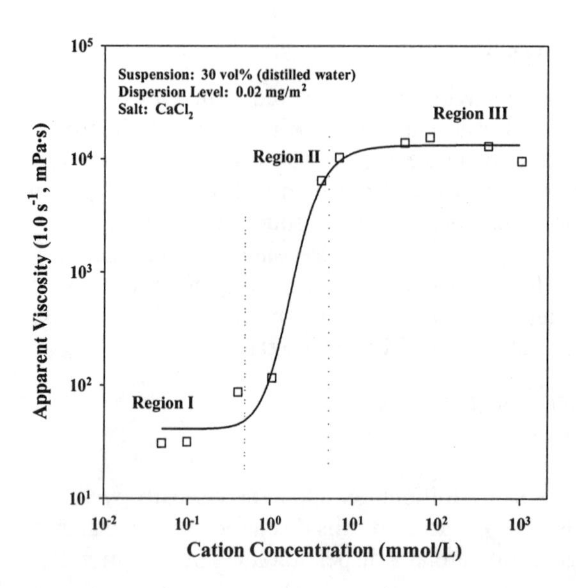

Figure 2. The effects of $CaCl_2$ additions on the apparent viscosity of the suspension. Similar trends were measured on the other two dispersion levels.

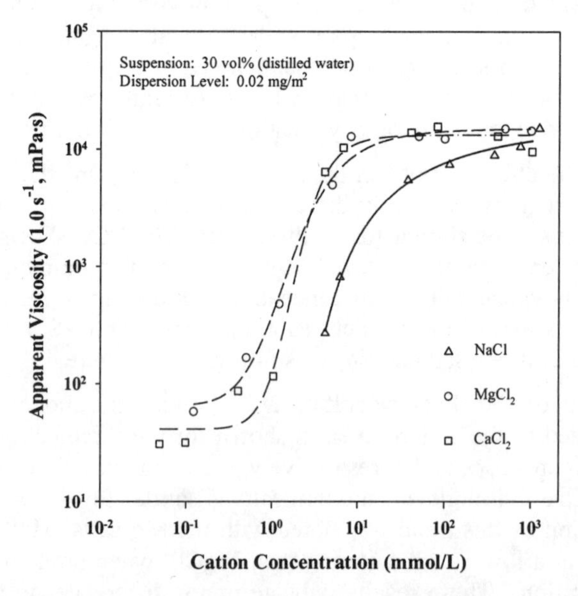

Figure 3. In the presence of Cl^-, the amount of Na^+ required to reach the CCC is 8 times more than Mg^{+2} and 10 times more than Ca^{+2}.

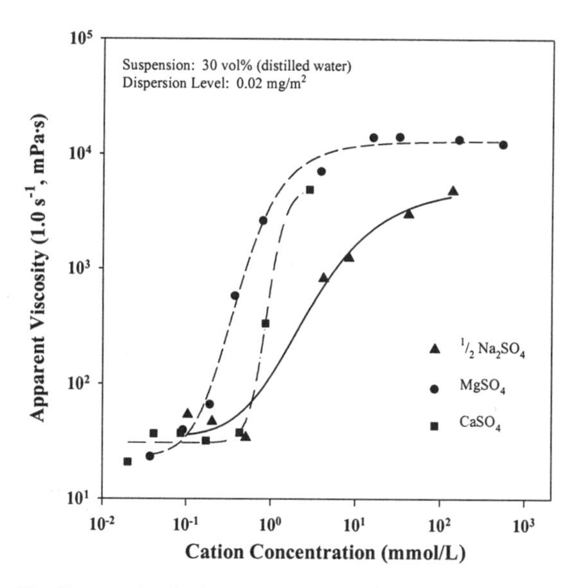

Figure 4. Similar to the behavior of the Cl^- salts (shown in Figure 3), in the presence of SO_4^{-2}, the amount of Na^+ required to reach the CCC was approximately 10 times more than Mg^{+2} and 5 times more than Ca^{+2}.

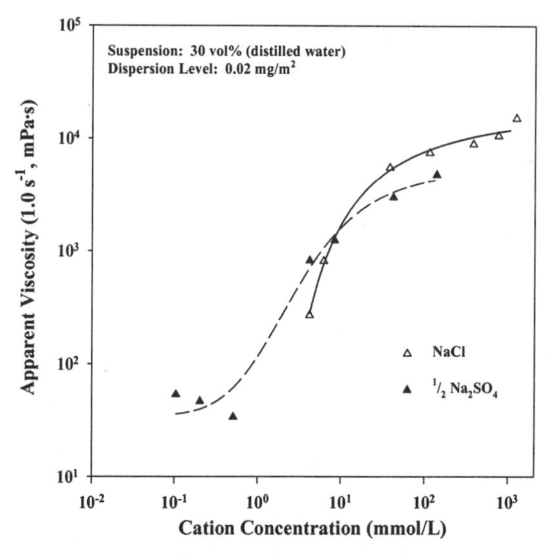

Figure 5. The curves for Na^+ associated with both Cl^- and SO_4^{-2} show similar apparent viscosity behavior as a function of the cation concentration, and appear to be reaching the same final plateau.

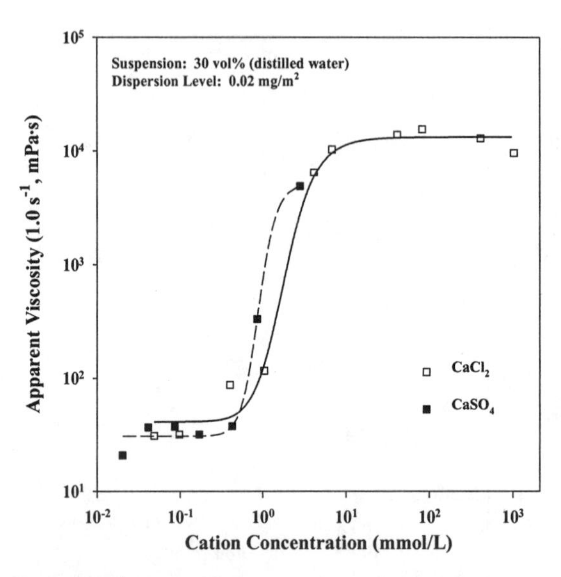

Figure 6. The coagulating efficiency of $CaCl_2$ and $CaSO_4$ appears similar over the range of compositions evaluated.

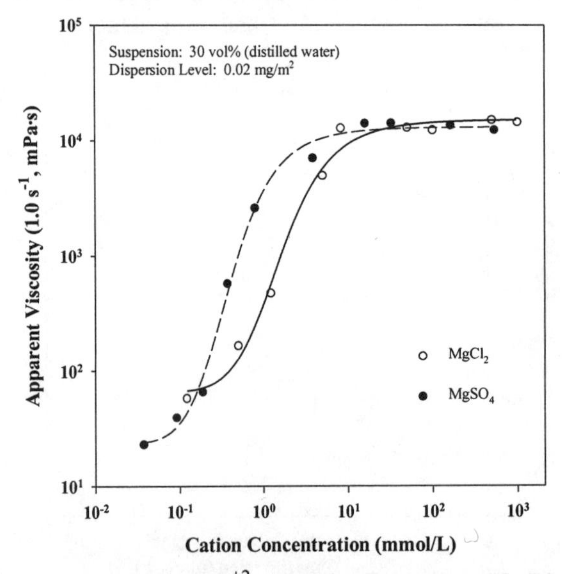

Figure 7. The curves for Mg^{+2} obtained from the chloride and sulfate salts ($MgCl_2$ and $MgSO_4$) showing similar final plateau apparent viscosity values. The $MgSO_4$ appears to be more effective at inducing coagulation indicating a minor difference in specific ion affinity.

The data presented in Figures 2-7 have been normalized to cation concentration, and the similarity of the trends supports the argument that the cation is controlling the rheology. To further support this hypothesis the ionic additions have been normalized to the anion concentration in Figure 8. If the anion was responsible for coagulation, the data from the different salts should overlap. It is clear this does not occur.

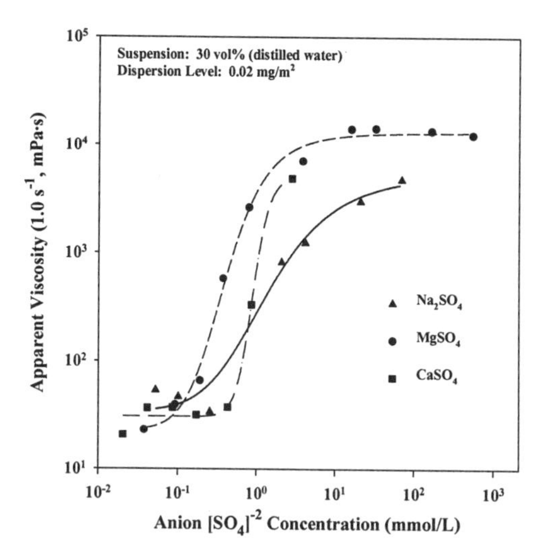

Figure 8. Normalizing the concentration to that of the anion can assess the debate regarding the controlling species, cation or anion.

The Effect of Dispersant Level

With all six salts, three dispersion levels were evaluated: 0.00, 0.02, and 0.05 mg/m^2. Figure 9 shows the effect of dispersant level and the effect of $CaCl_2$ and Na_2SO_4 additions. As the dispersant addition increases, the charge on the particle surface increases (becoming more negative), thus reducing the suspension viscosity.[3] As surface charge increases, a higher ionic concentration is necessary to coagulate the suspension. It is clear in Figure 9 that the CCC is shifting to a higher concentration with increasing dispersant addition, consistent with theory. In both cases presented in Figure 9, the Region III plateau viscosity level is identical for the three dispersant levels. Although the Region III plateau has not been obtained for the Na_2SO_4 case, the trend appears to be approaching a similar viscosity level as obtained with $CaCl_2$. Similar trends were observed for the other four salts.

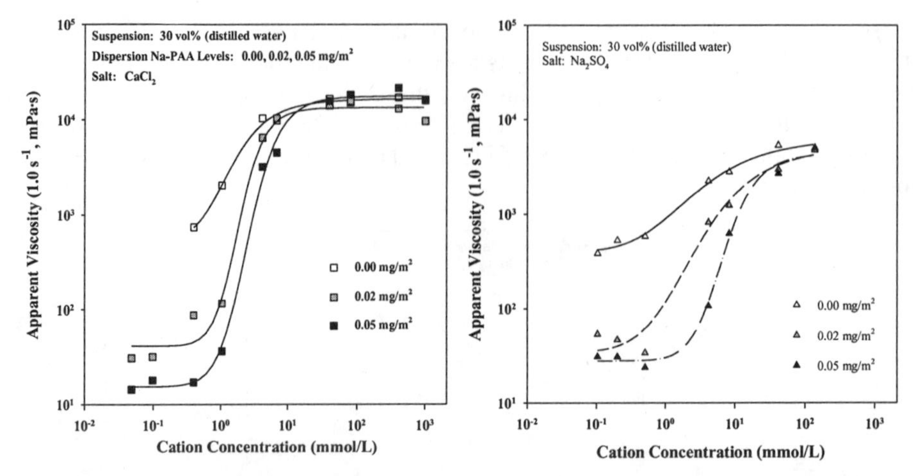

Figure 9. Increased dispersant additions decreases the suspension viscosity and also increases the concentration of salt necessary to coagulate the suspension. Similar results are observed for $CaCl_2$ (left) and Na_2SO_4 (right) additions. Although Region III has not been obtained for the Na_2SO_4 suspensions, the trend in the data is similar to that observed for $CaCl_2$.

Correlation with the Calculated Double Layer Thickness

According to colloidal theory, coagulation of a given suspension should occur at a specific double-layer thickness which is a function of ionic concentration and charge of an indifferent electrolyte (i.e., the ions involved have similar affinities for the particle surface). Figure 10 shows the calculated double layer thickness using Equation 2 as a function of the total ionic concentration of the salt solutions for 1:1 and 2:2 symmetrical electrolytes, representing NaCl and $MgSO_4$ (or $CaSO_4$), respectively. The double-layer thickness, at an identical ionic strength, is smaller for the 2:2 electrolytes compared to the 1:1 electrolytes. Therefore, a higher concentration of NaCl would be required to compress the double-layer to the same degree compared to $MgSO_4$.

As illustrated in Figure 11, the viscosity of the suspensions decrease with increasing double-layer thickness. In addition, and most importantly, the viscosity of all three suspensions increases dramatically at the same double-layer thickness, approximately 3.5 nm. The correlation of viscosity with the double-layer thickness further supports the argument against specific adsorption of the ions in solution.

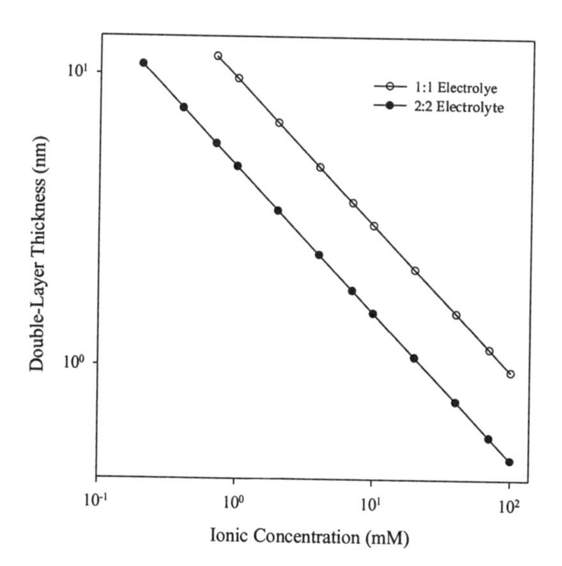

Figure 10. The double layer thickness as a function of the 1:1 and 2:2 electrolyte concentrations.

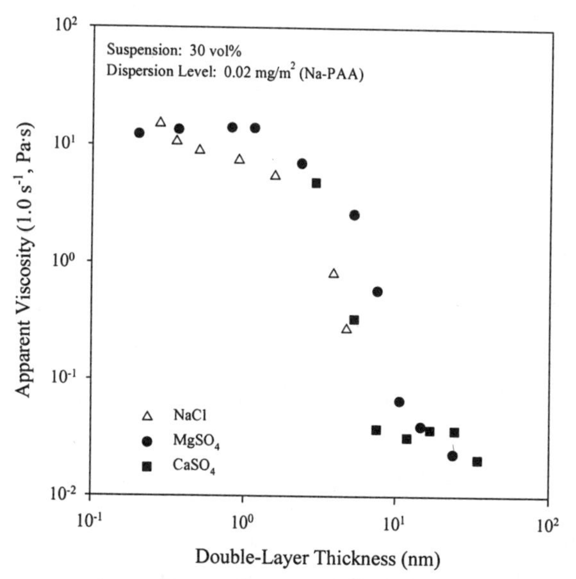

Figure 11. The apparent viscosity as a function of the calculated double layer thickness for the symmetrical electrolytes. The CCC determined from the inflection points for NaCl, $MgSO_4$, and $CaSO_4$ were 3.1, 3.8, and 3.5, respectively.

Correlation with ζ-Potential Measurements

The ζ-potential is defined at the potential at the shear plane separating the strongly adsorbed layer of ions from the diffuse layer of ions that together form the double-layer surrounding a colloidal particle. The magnitude and the sign of the ζ-potential is dictated by the charge at the particle surface (the Nernst Potential), the nature of the suspension medium (i.e., its polarity), and the ionic strength (composed of charge and concentration) of the suspension medium. As stated previously, increasing the ionic concentration compresses the double-layer, which reduces the ζ-potential. Figure 12 shows the ζ-potential of a clay suspension with increasing ionic concentration, measured via acoustophoretic mobility at a dispersant level of 0.2 mg/m^2. The ζ-potential data is superimposed onto the apparent viscosity versus $CaCl_2$ concentration data. At the CCC, the ζ-potential decreases significantly, becoming less negative. This decrease in ζ-potential is consistent with the concept of double layer compression and the rate at which the ζ-potential changes strongly suggests that specific affinity is of minimal concern.

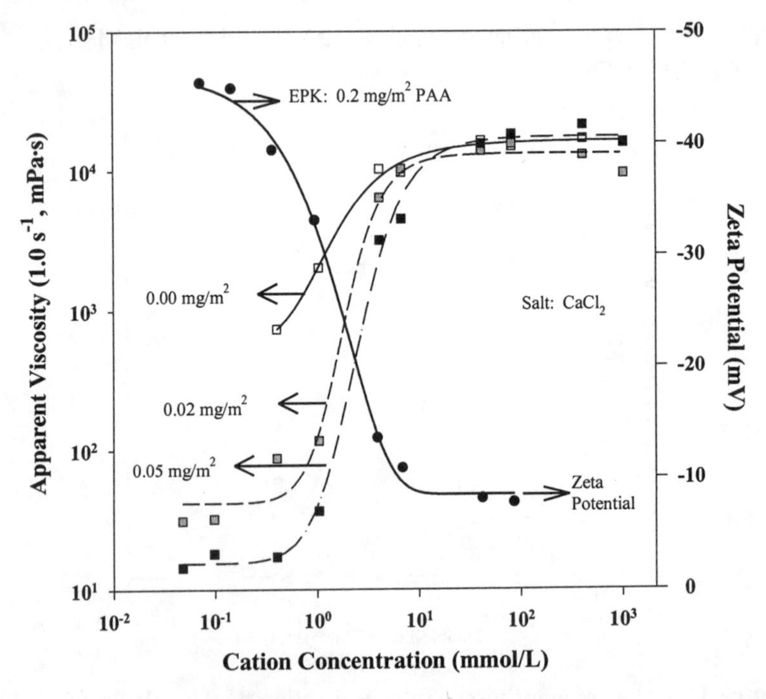

Figure 12. The ζ-potential and the apparent viscosity of a kaolin suspension as a function of cation concentration with varying dispersant levels. This data is consistent with double-layer compression.

SUMMARY AND CONCLUSIONS

Consistent with colloidal theory, and the fact that clay particles are net negatively-charged particles in an aqueous suspension, cations are responsible for the coagulation of whiteware suspensions. Specifically, three observations support this position: First, by correlating the apparent viscosity data with cation or anion concentration demonstrates that rheology behavior differences are clearly cation dependent. Second, when apparent viscosity is plotted as a function of the calculated double layer thickness, the curves overlap. And third, as the ionic concentration increases, the ζ-potential is reduced (i.e., it becomes less negative), consistent with cationic charge compensation, apparently via double-layer compression instead of specific adsorption. The anions – chloride and sulfate – appear to only play a minor role in the viscosity change. Some ionic affinity differences were noted between $MgSO_4$ and $MgCl_2$.

As the salt concentration exceeds the CCC, the viscosity increases by a of 100-10,000, and the magnitude of the viscosity increase ultimately dependent on the dispersant level. At ionic concentrations above the CCC, the apparent viscosity becomes independent of both dispersant concentration and further salt additions.

REFERENCES

1. D. J. Shaw, *Introduction to Colloid and Surface Chemistry*, pp. 210-228, 4th ed., Butterworth-Heineman Ltd., Boston, Massachusetts, 1992.

2. R. J. Hunter, *Introduction to Modern Colloid Science*, pp. 43-53 and 257, Oxford Science Publications, New York, 1993.

3. W. Carty, K Rossington, and U. Senapati, "A Critical Evaluation of Dispersants for Whiteware Suspensions," this volume.

4. D. Lide, ed., *CRC Handbook of Chemistry and Physics*, pp. **B**36-144, 73rd ed., 1992.

EFFECTS OF AGING ON RHEOLOGY OF GLAZE SLURRIES

C.H. Yoon and W.C. LaCourse
NYS Center for Advanced Ceramic Technology—Whiteware Research Center
School of Ceramic Engineering and Materials Science
New York State College of Ceramics at Alfred University
Alfred, NY 14802

INTRODUCTION

Glass frit has been used for many applications, such as glaze coating, sealing glass, and nuclear waste immobilization. In most cases, glass frit can be used in the form of a suspension in water. Rheology is important in numerous ceramic processing operations, including wet mixing/milling, shape forming, and coating/deposition. Proper control over rheological properties is essential for efficiency in processing and also for optimizing the physical properties in the processed material. In the case of glaze coatings, the rheological properties of the glaze (or frit) suspensions used for coating ceramic bodies must fit the application procedure, to ensure correct usage and obtain a defect-free glaze coating. For obvious reasons, the rheological properties involved should remain constant with time. However, gradual changes have been observed in the properties of glaze suspensions as they age. These variations are mainly attributed to the interactions between the glass and water.

The most important reactions, in the sequence in which they typically occur, include: (1) surface charge development, in which the extensively hydrolyzed surface of frits reacts to form positive or negative sites; (2) ion-exchange reactions, in which network-modifier cations such as alkali or alkaline earth ions are replaced by water or hydronium ions; (3) selective dissolution of network formers; (4) complete dissolution of the network.[1] For complex frit compositions, all reactions occur simultaneously. Each reaction influences the kinetics and mechanisms of the other reactions.

The result of frit/water interactions is a changing ion concentration of the suspension, which can strongly influence its rheological behavior. The major consequence is an altered viscosity, leading to changes in glaze thickness, tendency for settling due to the agglomeration, and possible glaze inhomogeneities due to segregation of suspension particles. The state of dispersion in suspension has an important influence on the green microstructure, which develops during powder consolidation and liquid removal.

The objectives of this study are to examine how the interactions between water and frit influence the rheological properties of a glaze suspension and the physical properties and bubble formation of the resulting glaze. Since the dissolution rate of glass by aqueous solution is largely determined by the composition of the glass, the dissolution of several different glass frits was studied. The relationship between chemical durability and frit composition was also investigated. The effects of long-term interaction of water with various frit suspensions were considered in the present study.

EXPERIMENTAL PROCEDURE

One lead-containing frit, Q, and two leadless frits, A and M, were used for this study. For dissolution tests, all frit specimens (or melts) were crushed and sieved below 325 mesh (44 μm). Each frit was added into de-ionized water by keeping 65 wt % of solid content with no correction for the relatively small differences in density. The compositions of starting materials are summarized in Table I.

Table I. Chemical compositions of frits (mol%).

Element	A Frit	Q Frit	M Frit
R'_2O	14	4.9	4.64
$R''O$	11	26.2	16.78
R'''_2O_3	2	13.7	22.14
SiO_2	73	55.4	56.41

R'= Na, K; R"= Ca, Mg, Pb, Sr; R"'= Al, B.

Dissolution tests were carried out in polypropylene bottles. The pH values were measured at each testing interval using a universal glass pH electrode and an ion analyzer.[a] The measured pH values were converted to the concentration of OH^- ions. The pH measurements were repeated several times. For element analysis, each aged frit suspension was centrifuged at 3000 rpm for 30 minutes. After centrifugation, the supernatant was decanted off, using a syringe into a 15-ml polyethylene bottle, and chemical analysis was performed using an atomic absorption spectrometer.[b] The precision of this technique is dependent on the cations being analyzed, roughly within ±5%. The detection limit allowed for analysis was 0.5 ppm for monovalent cations (e.g., Na^+, K^+) and 0.1 ppm for divalent cations (e.g., Mg^{2+}, Ca^{2+}).

Rheological behavior of suspension was studied by a Haake viscometer[c] with MVI sensor system, thermostating the suspensions at 25°C. The suspension was stirred for 5 min at a constant rotation rate. After this, it was immediately transferred to the viscometer, with a time lag of 10 seconds, and its viscosity and shear stress were measured with time under a constant shear rate.

[a] Orion Research Inc.,Boston,MA.
[b] Atomic Absorption Spectrometer Model 360, Norwalk, CT.
[c] Haake, Haake Viscometer, Paramus, NJ

RESULTS AND DISCUSSION

A-frit was prepared from commercial SLS slide glass. The solution pH and ion concentration in A-frit suspension, 65 wt % solid content, is plotted against aging time in Figure 1(a). When A-frit is added to water, the solution pH reaches above 11 within 1 minute. Since the initial solution pH is about 6.5, the initial rise is essentially instantaneous as surface alkali ions are released. H_3O^+ from the water penetrates the glassy network, replacing an alkali or alkali earth ion, which is released into solution, producing an OH^- ion via the reaction.[1,2]

$$(\mathrm{Si}-\mathrm{O}^- \cdots \mathrm{R}^+)_{\mathrm{glass}} + \mathrm{H_2O}_{\mathrm{sol}} \rightarrow \mathrm{Si}-\mathrm{OH}_{\mathrm{glass}} + \mathrm{R}^+_{\mathrm{sol}} + \mathrm{OH}^-_{\mathrm{sol}} \qquad (1)$$

As the consequence of this reaction, the surface becomes depleted in alkali and alkaline ions and the surrounding liquid becomes rich in these species. The pH of the solution also increases due to the formation of OH^- ions. After a certain maximum point, solution pH begins to decrease. Glasses typically dissolve in high-pH solutions. El-Shamy et al.[3] established from equilibrium theory that, as pH exceeds 9.8, dissolution of the silicate network occurs, and $Si(OH)_4$ molecules appear in solution. Since $Si(OH)_4$ ionizes to form $Si\text{-}O^-$, releasing H_3O^+, the appearance of $Si(OH)_4$ in solution can cause the pH to decrease. The combined effect of ions which raise the pH and those which decrease the pH leads to an "equilibrium" pH of the solution after some long time. If the reaction were limited to the surface in contact with the solution, the effects would be limited. However, the above reaction can continue by diffusion of water deeper into the surface. As a result, an increasingly thick layer, rich in Si-OH and deficient in the mobile ions, forms on the surface over time.

Figure 1(b) shows the effect of solids content (surface area per unit volume of water) on the change of pH in A-frit suspension as a function of aging time. As the solid content of suspension is increased, the initial maximum pH is also increased, and consequently the pH reaches, a higher value before decreasing. Generally, the kinetic contribution for chemical durability of glass (or glass frit) is primarily a function of the test geometry.[4] In particular, test parameters such as exposed glass surface area, the solution volume, the aging time, and the test temperature alter the kinetic contribution. As solid content is increased, glass surface area in contact with water is increased and the alkali release rate will be increased leading to a high pH and OH^- concentration.

Ion concentration change in Q- and M-frit suspension is shown in Figure 2(a) and 2(b), respectively. As aging time is increased, alkali ion concentrations (mostly Na^+) in Q-frit suspension increase, whereas the concentration of alkaline earth ions remain stable after 1 hour. In case of M-frit suspension, the leaching rate of alkali ion is relatively faster than that of other divalent or polyvalent ions. After aging for 24 hours, A-frit suspension has about 1100 ppm of Na^+ ion, whereas Q- and M-frit suspension has 500 and 120 ppm of Na^+ ion, respectively. From the results of Figure 2 and Table I, it can be concluded that the concentration of leached alkali ions in frit suspension mainly depends on the original glass composition. Changing ion concentrations in a suspension can strongly influence its rheological behavior.

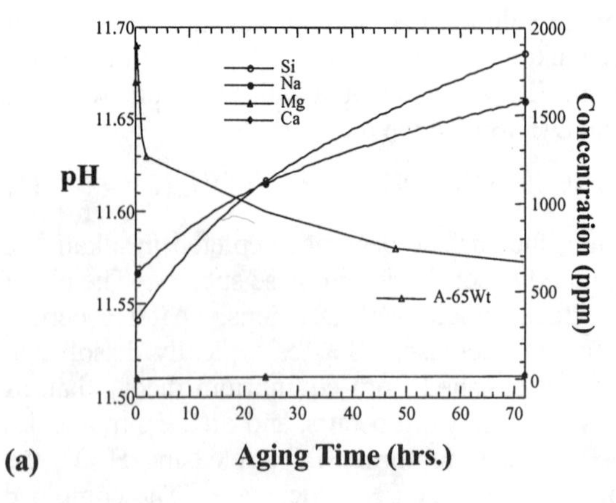

Figure 1. The change of (a) ionic concentration and (b) pH in A-frit suspension (solids content: 65wt % and initial pH ~ 6.5 -- data obtained from a "blank" run).

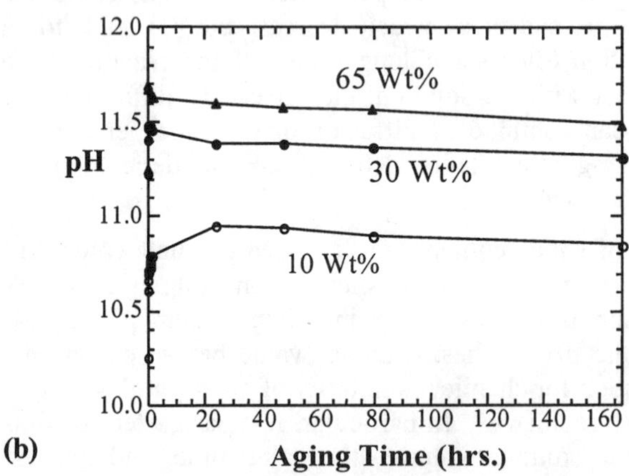

In industry, oversprayed glaze is removed from spray booths and reconditioned for use. The method by which the frit suspension is recycled can play an important role in the suspension rheology and in the firing behavior of the glaze. To observe "reclaim" effects, the suspension was aged for 24 hours and was then filtered, dried, and re-dispersed by keeping 65 wt % solid content. Results of A- and Q-frit are shown in Figure 3(a) and 3(b), respectively. In the case of A-frit, pH slowly increased as aging proceeded. I n contrast, solution pH of Q-frit after 72 hours aging doesn't have the same "equilibrium pH" as that of an original suspension. The possible reason for this result is that the leached ions are removed from the original suspension, and more soluble ions (alkali or alkaline earth) have already been removed from the surface layer, so further reaction needs more time. The remaining frit is thus rich in network formers, leading to a substantial increase in the glaze viscosity and Tg.

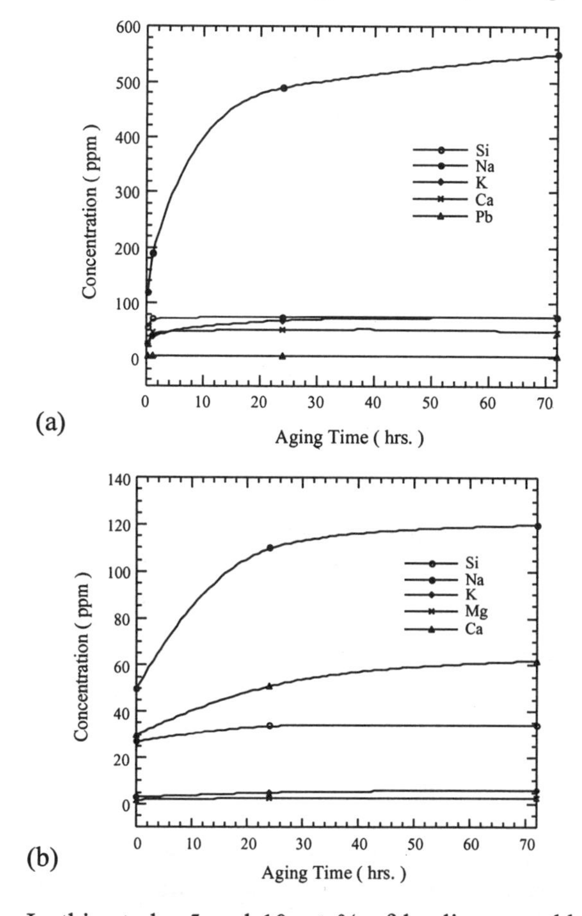

Figure 2. Change in ionic concentration of suspensions of (a) Q-frit and (b) M-frit as a function of aging time.

In this study, 5 and 10 wt % of kaolin was added into an A-frit suspension to control rheological properties. Its effect on suspension pH is shown in Figure 4(a). When the mixture of A-frit with 5wt % kaolin is added to water, the solution pH is increased slowly, and it reaches above 11 after 5 hours, whereas the suspension that contains only frit reaches its maximum solution pH within 1 minute. There was no apparent maximum solution, and it reaches its maximum point after aging for 48 hours. When the content of kaolin is increased, the resulting maximum and equilibrium solution pH is decreased. The ion-exchange process occurring on the surface of kaolin might explain this result.

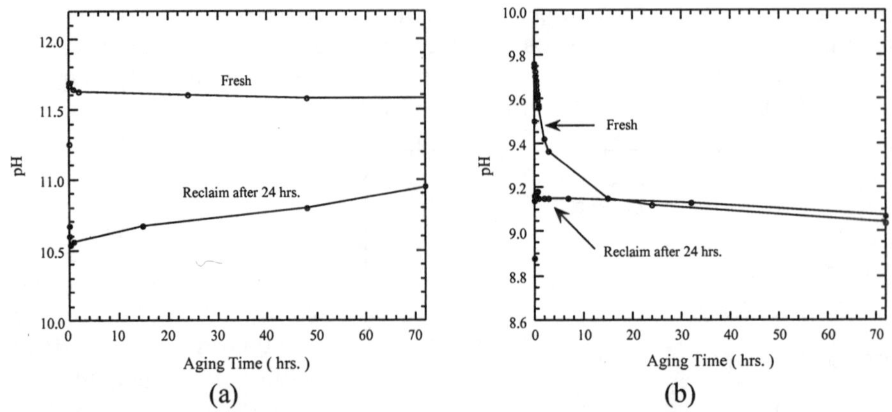

Figure 3. The effect of reclaim on pH change in (a) A-frit and (b) Q-frit suspension as a function of aging time (65 wt% of solid content).

Surface charge of particles can be controlled by the change of pH of solution. When pH is greater than the isoelectric point of the surface, a negative surface charge will be formed. Figure 4(b) shows the zeta potential of silica, alumina, and kaolin with pH. From the figure, it can be said that the net electrophoretic charge of kaolin is always negative independent of pH. Kaolin particles in suspension are negatively charged. Their charge is balanced by an equal and opposite charge carried by the cations Na^+, Ca^{2+}, etc., in the surrounding liquid. The exchange process occur predominately at the edge of crystals. Na^+ ions are adsorbed on oppositely charged surfaces and as a result of this reaction, Na^+ ions neutralize the surface charge, and these act as a buffer.[5] The influence of EPK kaolin addition to A-frit may be represented by the following proposed reactions.[6]

$$\mathrm{ClayOH - H^+ + Na^+ = ClayOH - Na^+ + H^+} \tag{2}$$

$$\mathrm{ClayOH - H^+ + Ca^{2+} = ClayOH - Ca^{2+} + H^+} \tag{3}$$

If suspensions have a high enough concentration of Na^+ ions in solution, the reaction will proceed to the right in equations (2) and (3). Alkali ions are much more mobile than alkali earth ions (divalent ions). Since the A-frit has a high alkali concentration, reaction will proceed to the right in equation (3) and consequently reduce the reaction rate of solution pH.

Changing ion concentrations in a suspension can strongly influence its rheological behavior. The major consequences of these rheological changes will be an altered viscosity leading to changes in glaze thickness, tendency for settling due to agglomeration, and possible glaze inhomogeneities due to segregation of suspension particles.[9] The suspension's flow curve was determined 10 minutes, 24 hours, and 72 hours after initial preparation. Ion concentration of these suspensions after aging strongly depends on the frit composition. These were shown in Figures 1(a), 2(a) and 2(b). Figure 5 shows the flow curves of A-frit suspension. From the shapes of the curves, it may be deduced that the suspensions after 10 minutes and 24 hours exhibited relatively dilatant behavior at lower shear rates, whereas at the higher shear rates they have approximately Newtonian behavior. As aging time is increased, the apparent viscosity is slightly increased. After 72 hours had elapsed, the suspension has a higher shear rate compared to that of 10 minutes and 24 hours aging. Consequently, it has high apparent viscosity. It might be explained by the formation of a microgel or large extent of agglomeration due to bridging action of adsorbed ions. Ion-concentration data in Figure 1(a) indicate a relatively high amount of alkali ions were removed from A-frit. Since leaching and dissolution of frit particles proceeds at their surfaces, the interfacial regions can become gelatinous, which results in sticking of interfacial regions of frit particles.[10] As a result, the apparent viscosity is significantly increased, and the dependence of apparent viscosity on the flow rate is also altered. The 72-hour aged suspension shows shear thinning behavior which can be explained by interparticle forces. If the particles suspended in a slip tend to agglomerate as a result of interparticle attractive forces, the viscosity then varies with the rate of shear, i.e., the agglomerates are broken by the shearing action and later recombine when the shear rate is reduced.[11] This thixotropic behavior, also referred to as shear thinning, is often observed in concentrated suspensions.

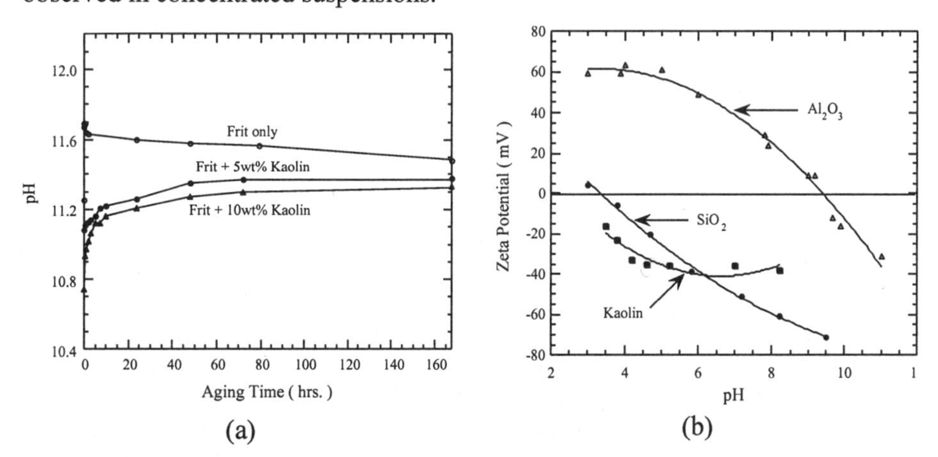

Figure 4. The effect of kaolin additions on the pH change in A-frit suspension as a function of aging time (65 wt% of solid content) and (b) Zeta-potential measurement with pH (silica[7], kaolin[6], α-alumina[8]).

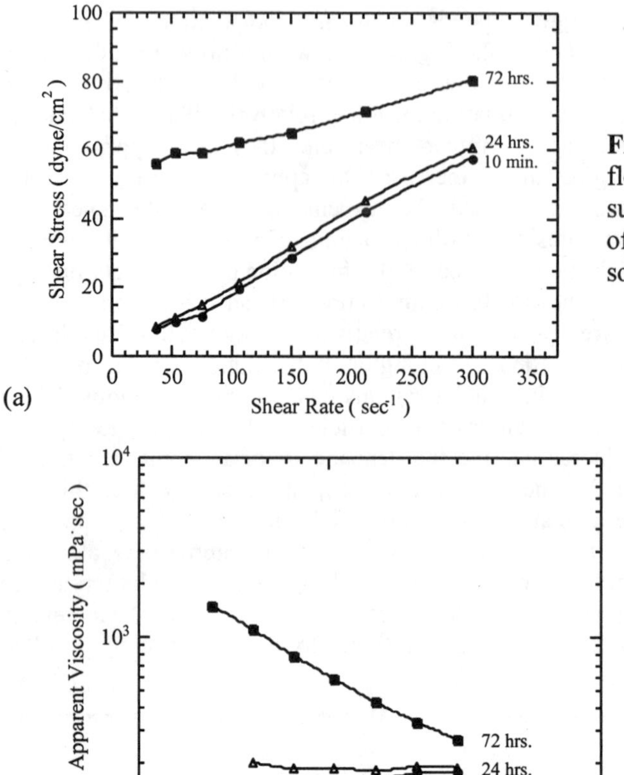

Figure 5. The change of flow property of A-frit suspension as a function of aging time (65 wt % solid content).

Figure 6 shows the flow curves and apparent viscosity of Q-frit suspension as a function of aging time. It may be observed that the suspension rheology changed as the suspension aged. Initially it had a Newtonian behavior, which became slightly shear thinning, with a drop in viscosity. In the cases of 24- and 72- hour aging, both shear thickening and shear thinning behavior were observed. The balance between shear thickening and shear thinning effects resulted in slight shear thickening behavior at lower shear rates, whereas at higher shear rates, the behavior was shear thinning. Apparent viscosity was decreased with aging time. From the results of Figure 2(a), it may deduced that alkali ion concentrations (mostly Na^+) are increased as aging time is increased, whereas the concentration of alkaline earth ions are stable and do not change much after 1 hour. Coagulation has been considered to be the result of van der Waals attraction, which draws two particles together at the moment of collision, unless opposed by a hydration barrier layer or by the electrostatic repulsion forces between similarly charged particles, or both.[12] In the case of Q-frit suspension, the solution pH is about 9, and consequently the frit has negative ionic charge on the surfaces. The surrounding cloud of positive Na^+ ions forms the "double layer," and it retards coagulation of frit suspension. Similar results were reported by Negre et al.[13]

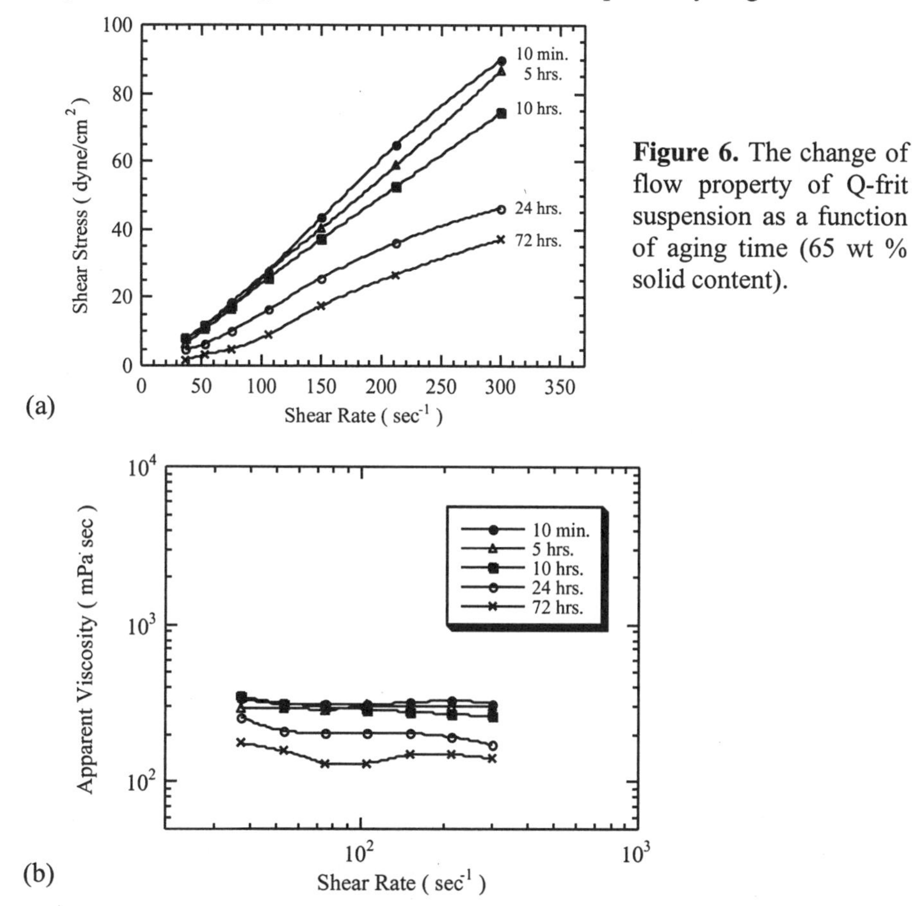

Figure 6. The change of flow property of Q-frit suspension as a function of aging time (65 wt % solid content).

The flow behavior of the M-frit is shown in Figure 7. Initially it had a Newtonian character. The apparent viscosity decreased after 24 hours but became stable after 24 hours. In the initial stage, the concentration of alkali ion increased rapidly, and the concentration of alkaline earth ions, mostly Ca^{2+}, is also increased. The alkali concentration in the M-frit suspension is 3-4 orders of magnitude less than that of the Q-frit suspension. From the results of Figures 2(a) and 2(b), it may be deduced that, in the case of M-frit suspension, alkaline earth ions dominate the rheological behavior. Generally, a characteristic of a divalent cation is that, when it is adsorbed on the surface of glass frit, only a single negative charge is neutralized; that is, one hydrogen ion is released, at least in the first stage. At about pH 9, the divalent ion can act as a positive charge site. Coagulation occurs long before the glass surface is saturated with divalent ions, because the divalent ion can act as a bridge by reacting with two particles at their points of contact. Iler verified this behavior of calcium in studying the flocculation of silica particles of different sizes.[14]

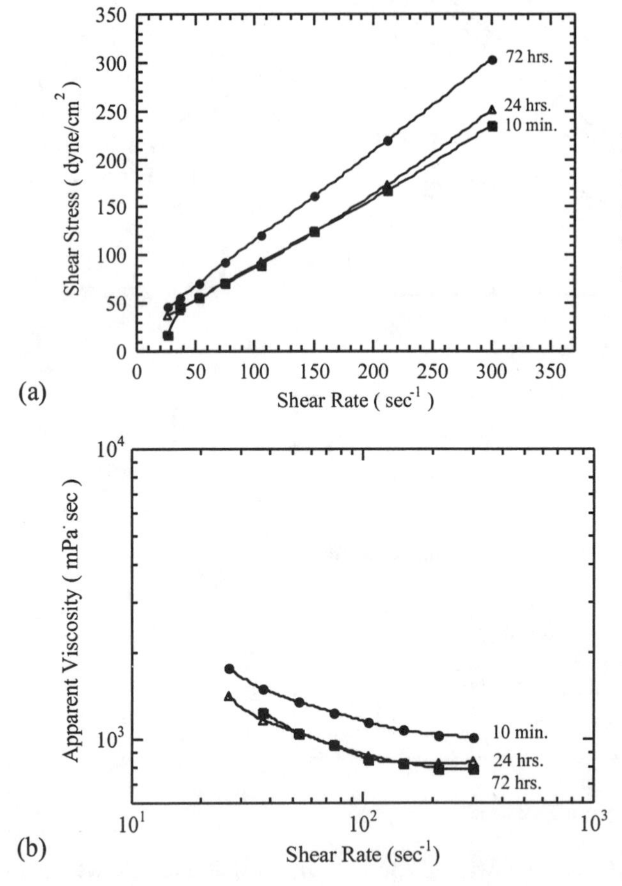

Figure 7. The change in flow property of M-frit suspension as a function of aging time (65 wt % solid content).

CONCLUSIONS

The dissolution study shows that the change of pH and ion concentrations in a suspension depends on the frit composition, additive, and solid content of the frit suspension. Higher solid content of the frit suspension results in higher solution pH. A large amount of alkali ions were removed from the original frit surface, leaving a silica-rich surface in a high-alkali-bearing frit composition (A-frit). The OH^- and water species also increased with aging time. The different frit composition results in different "equilibrium solution pH". The change of suspension rheology is associated with ion concentration of the frit suspensions. The surface layer as a result of dissolution has an influence on the rheology and time-dependent change of frit suspensions.

REFERENCES

1. W. C. LaCourse and W. Mason, "Glaze Problems from a Glass Science Perspective," pp. 339-356 in *Science of Whitewares;* Edited by V. E. Henkes, G. Y. Onoda, and W. M. Carty, The American Ceramic Society, Westerville, Ohio, 1995.
2. R. J. Charles, "Static Fatigue of Glasses: I," *J. Appl. Phys.,* **29** [11] 1549-53 (1958).
3. T. M. El-Shamy, J. Lewins, and R. W. Douglas, "Dependence on pH of Decomposition of Glasses by Aqueous Solutions," *Glass Technol.*, **13** [3], 81-87 (1972).
4. C. M. Jantzen, "Prediction of Glass Durability as a Function of Glass Composition and Test Conditions: Thermodynamics and Kinetics," pp. 24.1-24.17 in *Advances in the Fusion of Glass*. Edited by D.F. Bickford, et al. American Ceramic Society, Westerville, OH, 1988.
5. D.J. Shaw, *Introduction to Colloid and Surface Chemistry*; pp. 225-28, Butterworth, London, 4th Edition, 1991.
6. C. Lee, "The Characterization of Plasticity in Clay-Based Systems," M.S. Thesis, Alfred University College of Ceramics, Alfred, New York, 1995.
7. M.D. Sacks, "Rheological Science in Ceramic Processing," pp. 522-38 in *Science of Ceramic Chemical Processing,* Edited by L.L. Hench and D.R. Ulrich. Wiley, New York, 1986.
8. R.J. Pugh, "Dispersion and Stability of Ceramic Powders in Liquids," pp.146 in *Surface and Colloid Chemistry in Advanced Ceramics Processing*, Edited by R.J. Pugh and L. Bergstrom. Marcel Dekker Inc., New York, 1994.
9. F. Andreola, A.B. Corradi, T. Manfredini, G.C. Pellacani and P. Pozzi, "Concentrated Glaze Suspensions," *Am. Ceram. Soc. Bull.*, **73** [10], 75-78 (1994).
10. J.D. Vienna, P.A. Smith, D.A. Dorn, and P. Hrma, " The Role of Frit in Nuclear Waste Vitrification," *Ceram. Trans.* **45**, 311-325 (1994).
11. V.K. Marghussian and H. Sarpoolaky, "Rheological Behavior of Vycor Silica Glass Slips," *Br. Ceram. Trans.*, **96** [4] 149-154 (1997).
12. F. Moore, *Rheology of Ceramic Systems*; pp. 27-45, Maclaren and Sons Ltd., London, England, 1965.
13. F. Negre, C. Feliu, A. Moreno, E. Sanchez, and E. Bou, "Effect of Frit Cation Release on Rheological Behavior of Glaze Suspensions," *Br. Ceram. Trans.*, **95** [2], 53-57 (1996).
14. R.K. Iler, *The Chemistry of Silica*; pp. 213, Wiley-Interscience, New York, 1979.

PLASTICITY REVISITED

William M. Carty, Katherine R. Rossington, and Douglas S. Schuckers
NYS Center for Advanced Ceramic Technology—Whiteware Research Center
School of Ceramic Engineering and Materials Science
New York State College of Ceramics at Alfred University
Alfred, NY 14802

ABSTRACT

At the first Science of Whitewares Conference, we introduced the High Pressure Annular Shear Cell, or HPASC, as a new plasticity characterization technique based on the direct shear testing concepts derived from soil mechanics. Since that meeting, we have tested hundreds of whiteware-based samples, as well as some from the advanced ceramics arena. Throughout these tests, we have learned the benefits and limitations of the instrument. In addition, and perhaps of critical importance to the whiteware manufacturing community, our results indicate that the divalent cation concentration can profoundly change the plasticity of a whiteware body, and as such, may be the root cause of aging in plastic bodies and may have important implications for drying.

INTRODUCTION

The High Pressure Annular Shear Cell Concept

Historically, plasticity probably represents one of the most difficult rheological behaviors to characterize, in spite of its critical role in the whiteware forming operations. Borrowing from soil mechanics, a high pressure annular shear cell (HPASC) was developed to allow the direct measurement of shear forces within a plastic body under an externally applied pressure, mimicking the environment that a plastic body would experience in a forming process such as jiggering or extrusion. A more detailed discussion of the HPASC is presented elsewhere,[1-3] but it is reasonable to briefly revisit the concepts of the HPASC approach.

The underlying concept of the HPASC is that plastic bodies behave as Bingham fluids, that is, Newtonian-like with a yield stress. As the pressure increases, the yield stress increases linearly with the applied normal pressure, as illustrated in Figure 1. Figure 2 shows the proposed dependence of the slope (pressure dependence) and intercept (cohesion) of the yield stress versus applied pressure for three hypothetical samples. Note that in this example, ***A*** and ***B*** have the same cohesion and ***B*** and ***C*** have the same pressure dependence.

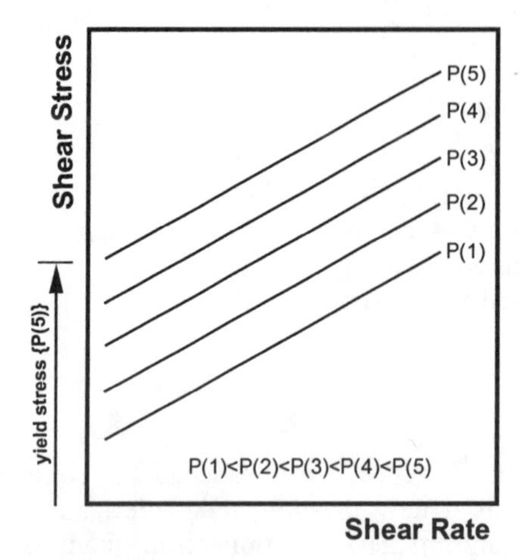

Figure 1. Schematic illustration of the effect of pressure on a Bingham fluid. Note that yield stress scales with increasing pressure.

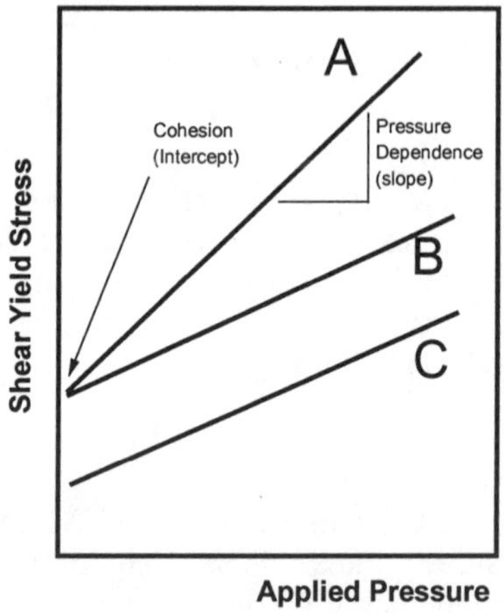

Figure 2. The definition of cohesion and the pressure dependence (the intercept and the slope, respectively) for the examples, A, B, and C.

The HPASC is represented schematically in Figure 3, and consists of a donut-shaped test cell that holds 75-100 grams of sample. Pressure is applied via a dead-weight loading system with a hydraulic amplifier to the top of the test fixture and the test cell design allows the sample to compact during testing. Rotation rate and applied pressure can be altered independently, allowing a broad range of testing conditions. A photograph of the test cell, showing the torque transfer and hydraulic loading setup, is presented in Figure 4. The torque required to prevent the top from rotating is measured and used to calculate the shear stress within the sample. The surfaces of the test fixture are roughened to prevent slippage between the sample and the test chamber.

In the most general sense, three simple categories of behavior can be imagined: dry, plastic, and wet. The shear yield stress versus pressure behavior of these materials is illustrated in Figure 5. The yield stress of the dry sample would be highly dependent on the applied normal pressure in contrast to the wet sample, in which the shear stress is nearly independent of the applied pressure. The plastic sample ideally lies somewhere in between wet and dry. Figure 6 illustrates the behavior of a series of clay samples with differing water contents and an example of a typical whiteware body formulation.

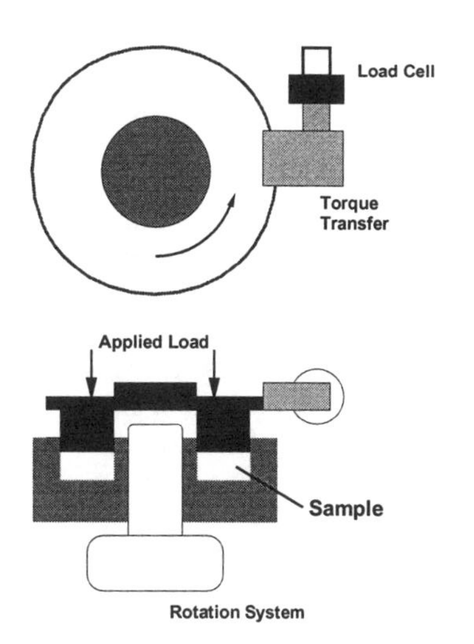

Figure 3. Schematic illustration of the HPASC. Pressure is applied to the top of the test fixture and the sample can compact during the test.

Figure 4. A photograph of the HPASC testing fixture with the testing chamber as illustrated schematically in Figure 4.

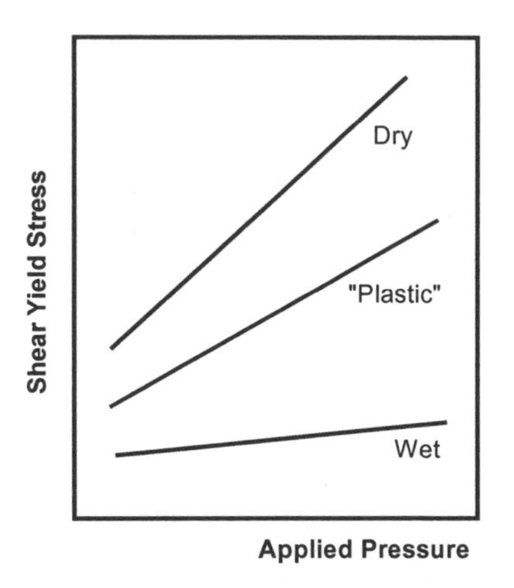

Figure 5. Schematic illustration of the change in slope and intercept of yield stress versus applied pressure for wet, plastic, and dry samples.

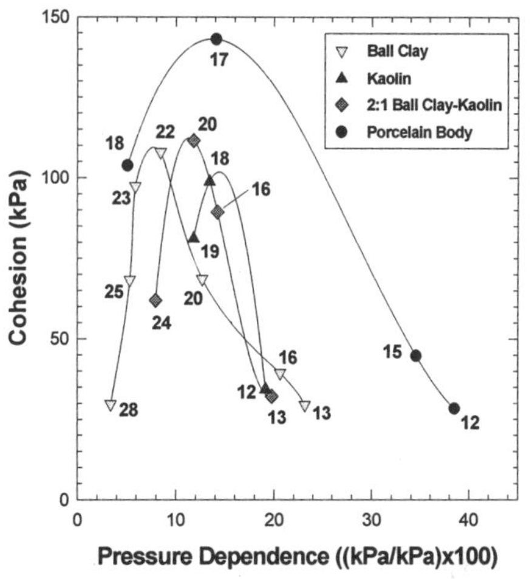

Figure 6. HPASC results of kaolin, ball clay, 2:1 mixture of the clays, and a standard porcelain body[1] showing the dependence of rheology on water content. The water content (d.w.b.) is noted next to each data point.

Factors Affecting Rheology and Plasticity

There are five factors that control suspension rheology: 1) particle-particle interactions; 2) particle concentration; 3) particle size and distribution; 4) particle morphology; and 5) rheology of the suspension medium. It is further proposed that these five factors also control plasticity, but the problem of demonstration has been hampered by the difficulties associated with plasticity measurement. One of the largest issues surrounding plastic body behavior has been the substantial change in body properties due to aging. Prior to the use of refined clays and plentiful potable water, these effects were attributed to bacteria growth. We propose that bacteria growth is a relatively minor, if not altogether insignificant, factor in modern industrial factories, although aging of whiteware systems persists. Based on experimental results on a wide variety of bodies, we propose that aging is due to raw material dissolution.[4] The resulting increase in ionic strength changes the particle-particle interactions in the plastic body, altering the plasticity as measured using the HPASC. (Raw material dissolution also increases the ionic strength in suspensions, but the resulting cation concentration is below that necessary to create a measurable rheology change.)

Consistent with colloidal theory, cations compress the electrical double layer surrounding net negatively-charged clay particles, reducing the repulsive force between particles, allowing particle agglomeration.[5] Also consistent with colloidal theory, divalent cations were approximately eight times more effective at altering the suspension behavior than monovalent cations. To demonstrate that the plasticity of industrial porcelain bodies change with time, several industrial bodies were characterized as a function of moisture content, in both an aged and un-aged form. To test the hypothesis that divalent cations can specifically alter the behavior of plastic bodies, $CaCl_2$ was incrementally added to a standard porcelain body composition and the plasticity measured at a moisture content of 16% (dry weight basis, or d.w.b.) using the HPASC. The results of the $CaCl_2$ additions were consistent with the aging studies and with effect of divalent cations on suspension rheology.[5]

EXPERIMENTAL PROCEDURE

Plastic Body Composition

The plasticity of two whiteware bodies was evaluated, *Body A* and *Body B*. *Body A* is a generic triaxial porcelain composed of clay, quartz, and feldspar. The composition of *Body A* is listed on a dry weight basis (d.w.b.) in Table I along with the corresponding specific surface area measurements for each raw material.[6] *Body B* (listed in Table II) is an industrial triaxial porcelain composition prepared in an industrial setting.

Plastic Body Chemistry

Body A was prepared using distilled water to control batch chemistry. The ionic concentration of *Body A* batches was adjusted using saturated $CaCl_2$

solutions diluted to the desired concentration. The cation concentration of the saturated $CaCl_2$ solution (6.3 M) was verified by Inductively Coupled Plasma (ICP) analysis (ACME Analytical Laboratories Ltd., Vancouver, British Columbia, Canada).

Body B was prepared with regular process water received from the local municipal water treatment facility. The dispersant and coagulant used in production are sodium polyacrylic acid (Dispex®, Ciba Specialty Chemicals, Basel, Switzerland, formerly Allied Colloids) and $CaCl_2$, respectively. The specific chemistry of the suspension medium was not determined.

Table I. The composition *Body A* (d.w.b.) of the standard porcelain body.

Component	Brand and Supplier	d.w.b.	Surface Area (m^2/g)
Kaolin	Edgar Plastic Kaolin (EPK) Zemex Minerals, Inc., Monticello, GA	50.0	26.9
Quartz	325-mesh Flint Ogledbay Norton; Glenford, OH	25.0	0.9
Feldspar	K-200 Zemex Minerals, Inc., Monticello, GA	25.0	1.8

Table II. The composition for *Body B* (d.w.b.) of the industrial porcelain body.

Component	Brand and Supplier	d.w.b.	Surface Area (m^2/g)
Kaolin (blend)	Kingsley Rogers K-T Clay Co., Mayfield, KY	10.0	13.5 27.8
Ball Clay	Weldon/Victoria (blend) United Clays, Inc., Brentwood, TN	34.0	23.3
Quartz	325-mesh Quartz Ogledbay Norton; Glenford, OH	23.0	0.9
Feldspar	K-200 G-200 Zemex Minerals, Inc., Monticello, GA	33.0	1.8 1.1

Plastic Body Preparation: *Body A*

Individual batches were prepared at 16% water content (250 grams of dry batch plus 47.6 grams of water + salt solution). The range of $CaCl_2$ level was 0.0131 mM to 132 mM. All dry materials were placed into 'sigma' mixer (Plastosizer Prep-Mixer, C. W. Brabender Instruments, Inc., South Hackensack, NJ) and mixed with twin "sigma" blades at 60 rpm. The liquid was added gradually and the plastic body was mixed for 15 minutes. The sample chamber was water cooled to minimize sample heating and evaporative water loss during mixing.

Plastic Body Preparation: *Body B*

Two varieties of *Body B* were prepared. Samples were removed from the production process after the filter pressing operation. One sample 'As-Received' was tested within hours after filter pressing and the other sample 'Aged' was stored for two weeks in a sealed container prior to testing.

To determine the change in plasticity with changing moisture content, *Body B* samples were carefully dried to reduce the water content. *Body B* samples were received from the production process with approximate 19% water. Water content values were obtained using the procedure outlined in ASTM method C324-82, but with a smaller sample size. Water contents for HPASC testing were obtained by drying 300 gram samples in an oven, then homogenizing the samples using the sigma mixer. Drying times were varied to generate a range of moisture levels. The moisture content was again measured after mixing prior to HPASC testing.

Plasticity Measurement Method

The plasticity of all samples was measured using the HPASC. The *Body A* samples were tested immediately after mixing so the effects of aging the plastic body would be minimized. The *Body B* samples were tested upon receipt from production facility except the 'Aged Standard' *Body B* sample. Each test required 75 g of a plastic body to be placed into the test fixture. All tests were conducted at a shear rate of 0.5 rpm with incremental weight additions: 22.9, 34.0, 45.4, 56.7, and 68.0 kg (50.0, 75.0, 100, 125, and 150 lbs.). Between each weight addition, there was a five minute equilibration period prior to data collection. The data typically exhibited linear yield stress versus applied pressure with linear correlation coefficients (r^2) of >0.97. (All of the data presented here are the slope (pressure dependence) and intercept (cohesion) values of those data sets. In most cases, the pressure dependence is ignored due to limited variation between the samples, so only the cohesion values are presented.)

RESULTS AND DISCUSSION

Typical Behavior and the Effect of Time on Plasticity

A typical example of the effect of water content on HPASC results of cohesion and pressure dependence is illustrated in Figure 7 for *Body A*. The maximum cohesion occurs at a water content of approximately 15%, although, as will be discussed below, both the magnitude and the water content at which the maximum cohesion occurs changes with body composition and water chemistry. The pressure dependence decreases linearly with increasing water content, up to the water content at which the maximum cohesion occurs, at which point the pressure dependence changes. It is proposed that the maximum cohesion obtained is a function of particle packing, capillary forces due to partially filled pores, and van der Waals attractive forces. The relative contributions of these three forces is unknown.

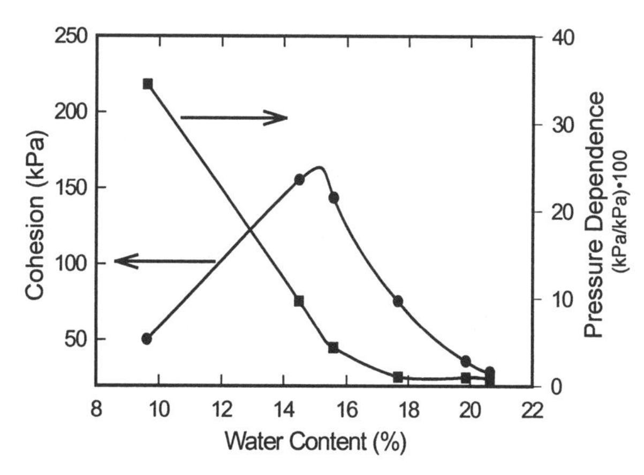

Figure 7. Typical HPASC behavior as a function of water content for *Body A*.

As shown in Figure 8, the maximum cohesion, and the corresponding water content change with aging, both shifting to lower values. While the change in the water content that the maximum occurs is small (and of questionable validity), the maximum cohesion is reduced by nearly 25%. In addition, *Body B* exhibits a much greater cohesion than either the 'as received' and nearly twice that of 'aged' *Body A*.

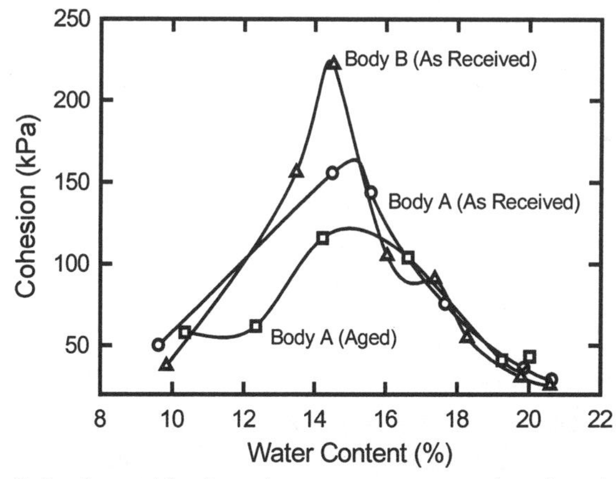

Figure 8. Cohesion with changing water content showing the difference between *Body A (as received)*, *Body A (aged)* and *Body B(as received)*.

It is proposed that drying performance and green strength are inversely related to the peak cohesion values. A higher cohesion value creates a strength differential in the body associated with the drying front. A moisture content difference of approximately 2% could create a strength difference of several hundred kPa, potentially resulting in cracking. (Anecdotal evidence seems to support the correlation of green strength with cohesion strength – it was observed in the industrial setting that *Body B* exhibited lower strength than *Body A*, although specific values were not available.) The change in the peak cohesion for the two bodies can be partly attributed to changes in chemistry, but more probably, and with greater importance, to changes in particle packing. It has long been argued that particle packing is essential for body development,[7] and the substitution of one clay for another will certainly result in a change in particle packing. It is clear, however, that there is a large amount of work necessary to determine the specific influences of particle size on cohesion and green strength.

To demonstrate that these observations are both reasonable and repeatable, Figure 9 shows the effect of time on the plasticity of a ball clay-water sample.[1] The cohesion decreases with increasing time, with a minor increase in the pressure dependence.

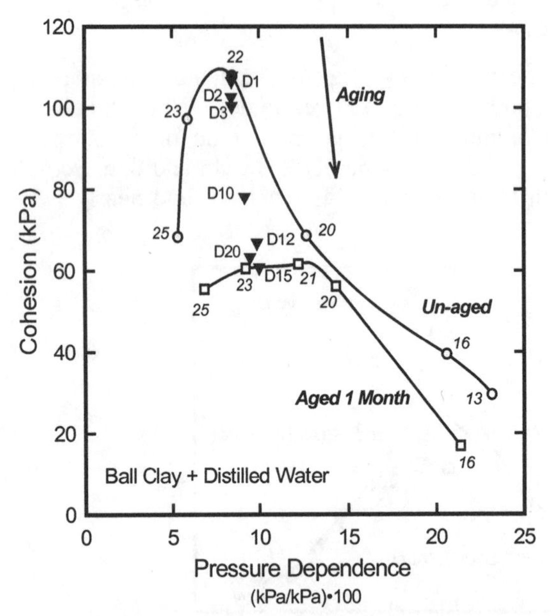

Figure 9. The plasticity of a ball clay-water body showing the change with time.[1] The water content levels are noted in italics (d.w.b.) and the change in plasticity for the 22% water content sample over 20 days is denoted by the ▼. After one month, the maximum cohesion has decreased by approximately 45%.

When the data in Figure 9 was first generated, it was unclear what caused the reduction in cohesion with time. In addition, the time frame for a ball clay sample to achieve a steady-state was nearly two weeks, or about three to four times longer than required for a porcelain body composition.[1,2] Subsequent work indicated that cation dissolution was a function of composition and closely related to the raw materials.[4] It was also clear that water played an important role, as dissolved salts in the water could "jump-start" the aging process.

The impact of initial water quality and the suspension solids loading on divalent cation concentration generated by the dissolution of the raw materials, is presented in Figure 10. The cation dissolution level, after an initial, relatively short (on the order of <24 hours) incubation period, becomes constant, and is equivalent to approximately 1.5% of the bulk cation concentration level in the batch.[4] Because the cation dissolution level is constant (i.e., in these results it amounts to 1.5% of the bulk ionic concentration), the total cation level increases with increasing particle concentration, and is non-linear with solids concentration because the water content decreases with increasing solids loading. (Also noted was a deviation of the measured cation level with the calculated level – this is currently the subject of another investigation.) The dissolution of cations from ball clay in water, without kaolin or feldspar, was demonstrated to be slower, thus indicating a possible explanation for the differences in aging behavior.

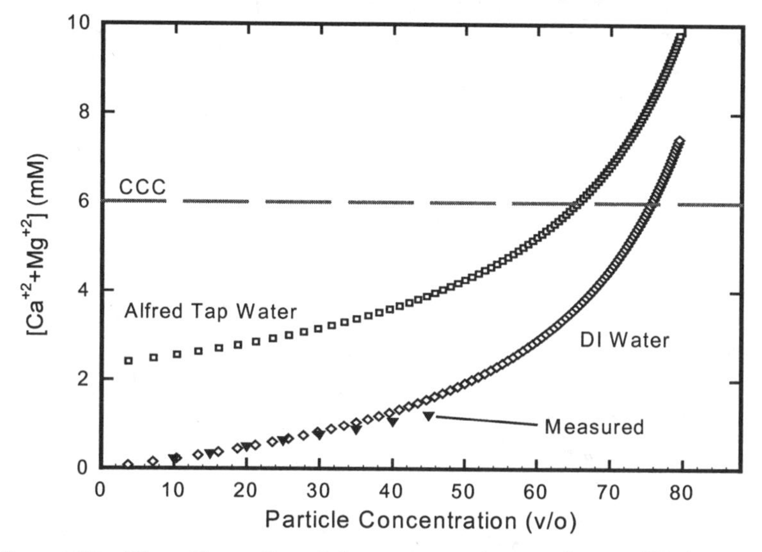

Figure 10. The effect of particle concentration on the equilibrium cation dissolution level (i.e., in the time-independent dissolution range), for a porcelain body slip prepared with de-ionized (DI) water and with Alfred, NY tap water (containing 2.5 mM dissolved divalent cations initially). The critical coagulation concentration (CCC) determination is presented elsewhere,[5] and is denoted by the dashed line.

As shown in Figure 10, at a solids loading of ~65v/o (~20% d.w.b.) and ~75v/o (~13% d.w.b.), for bodies prepared with DI water and Alfred water, respectively, reach the CCC, resulting in a noticeable change in suspension properties. In the concentration range, these samples will exhibit plastic character, but depending on the cation concentration, the plastic character will differ. Based on the results from the plasticity measurements, and from the cation dissolution behavior, it is proposed that raw material dissolution is responsible for aging by changing the dissolved cation levels in plastic bodies. Furthermore, as demonstrated by Figure 10, it is clear that suspension may not exhibit aging effects because the cation concentration due to dissolution is below the CCC at moderate solids loadings commonly used for suspension preparation.

In an industrial process, however, the solids loading is not changed in a continuous manner. Any process which employs de-watering, such as slip casting or filter pressing, causes an abrupt shift in the solids loading at a constant cation level, as illustrated in Figure 11. The cation concentration in the expressed water is identical to the cation concentration remaining in the plastic body, thus providing an opportunity to characterize the chemistry of the water within the plastic body. Also, in slip casting operations, changes in chemistry may not be obvious in the relatively low concentration casting slip, but become apparent in the cast piece which possesses plastic body solids loading. Once the de-watering step is complete, drying causes an increase in dissolved ions due to the removal of water, assuming the salt concentration remains below the solubility limit.

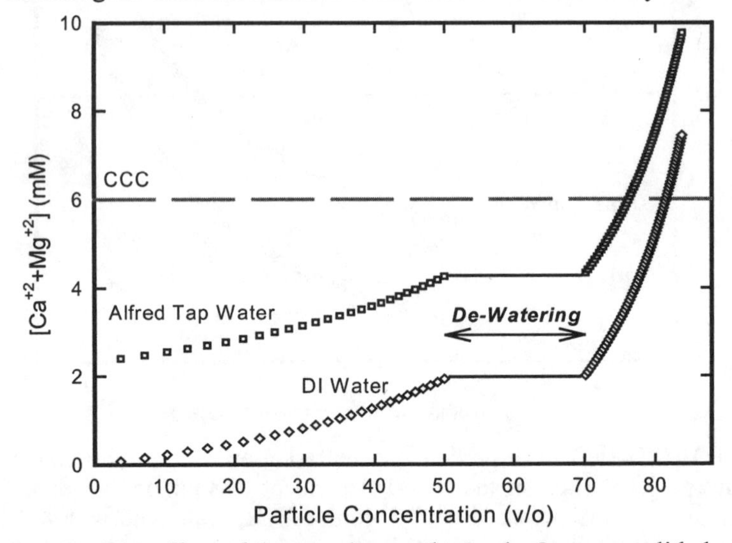

Figure 11. The effect of de-watering on the [cation] versus solids loading for a porcelain body, as illustrated in Figure 10. De-watering simply shifts the curve to the right, but does not change the dissolved cation levels.

As a logical extension of these arguments, it is proposed that the property changes associated with aging are due to the change in cation levels with time due to raw material dissolution. These changes cause the plasticity to change, but may not be observed in lower solids content samples, such as slips. If cations are responsible for the change in plasticity, it should be possible to demonstrate this change by adding cations to a plastic body. To do this, the cohesion of Body A was evaluated as a function of divalent cation (Ca^{+2}) concentration, as shown in Figure 12.

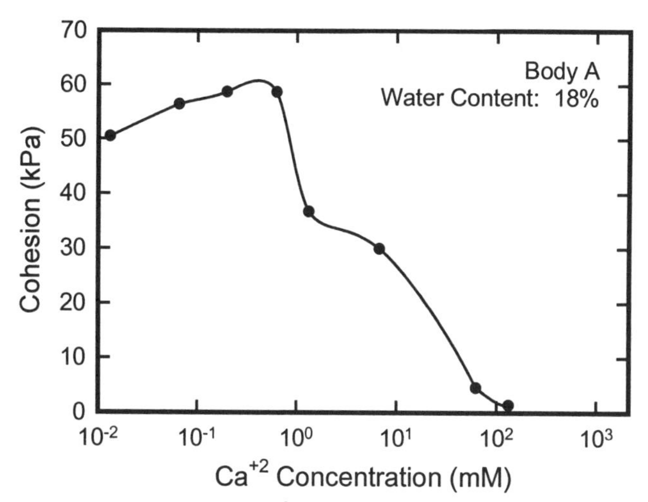

Figure 12. The effect of Ca^{+2} cation concentration on the cohesion of *Body A*, showing an substantial decrease in cohesion above 1.0 mM.

To demonstrate that the decrease in cohesion occurs at a reasonable cation concentration, the results from previous studies on ionic strength effects[5] are superimposed on Figure 12 to generate Figure 13. As is obvious, the CCC values for both systems are practically identical, indicating that the colloidal concepts are applicable even in highly loaded systems, as in plastic bodies.

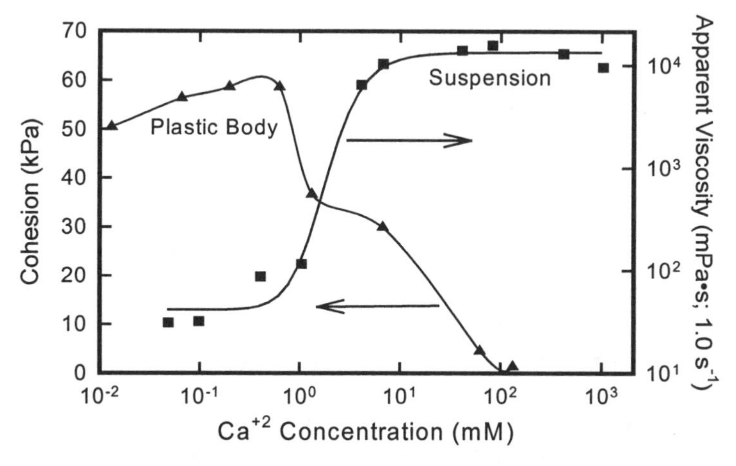

Figure 13. Comparison of plastic body behavior as cohesion via the HPASC and apparent viscosity from porcelain body suspensions showing that the critical coagulation is nearly identical and therefore independent of the water content.

Figure 13 shows that when the cation level increases in a suspension, the viscosity increases, but in a plastic body, the cohesion decreases. If cohesion is considered to be parallel to deformability, higher cation levels would produce softer clay, and in many processes, easier forming.

It appears from this data that drying can be assisted by increased cation levels by reducing the stresses developed in the body during the drying process. Increased cations levels result from aging, but the transition from a dispersed condition below the CCC to a coagulated condition above the CCC may be the critical step in the drying process. If higher ionic concentrations are present in the bodies in the suspension preparation stage, higher cation concentrations remain after de-watering, and therefore, the amount of water removed during drying to reach the CCC level is reduced, allowing the dispersed-coagulated transition earlier in the drying process.

SUMMARY AND CONCLUSIONS

Increasing the divalent cation concentration in a plastic body, whether by aging or by direct salt additions, decreases the cohesion stress measured by the HPASC. The cation level necessary to reduce the cohesion stress in the plastic body is consistent with the measured critical coagulation concentration in suspensions, demonstrating that chemistry can also used to control plasticity. Furthermore, it is proposed that a dispersed to coagulated transition occurs during drying, as a result of dissolved ions in the suspension and when that transition occurs may directly impact the drying performance of porcelain bodies.

REFERENCES

1. C. Lee, *The Characterization of Plasticity in Clay-Based Systems*, M.S. Thesis, Alfred University, Alfred, NY, 1995.
2. W. M. Carty and C. Lee, "The Characterization of Plasticity," *Science of Whitewares* (*Proc. Sci. Whitewares Conf.*), Eds., V. Henkes, G. Onoda, and W. Carty, 89-101 (1996).
3. P. A. Nowak, *Correlation of Extrusion Behavior with High Pressure Shear Rheometry*, M.S. Thesis, Alfred University, Alfred, NY, 1995.
4. H. Lee, *The Effect of Mixing Route on the Properties of Whiteware Suspensions*, M.S. Thesis, Alfred University, Alfred, NY, 1997.
5. K. R. Rossington and W.M. Carty, "The Effects of Ionic Concentration on the Viscosity of Clay-Based Suspensions," this volume.
6. K. R. Rossington, W. M. Carty, and U. Senapati, "A Critical Review of Dispersants for Whiteware Applications," this volume.
7. D. R. Dinger, "Influences of Particle Size Distribution on Whiteware Properties and Processing," *Science of Whitewares* (*Proc. Sci. Whitewares Conf.*), Eds., V. Henkes, G. Onoda, and W. Carty, 105-115 (1996).

DISPERSION OF ALUMINA

Brian R. Sundlof and William M. Carty
New York State College of Ceramics at Alfred University
2 Pine Street
McMahon Building
Alfred, NY 14802

ABSTRACT

The rheological properties of two α-alumina (Al_2O_3) powders dispersed via polymeric and electrostatic (pH) stabilization were addressed during this investigation. The polymers examined included the sodium (Na^+) and ammonium (NH_3^+) salts of polymethacrylic acid (PMAA) and polyacrylic acid (PAA). Acids (HCl and H_2SO_4) and bases (NaOH and NH_4OH) were used to adjust suspension behavior via pH. Rheological evaluations revealed dual minima in some cases of polymeric stabilization for the alumina containing MgO and higher levels of sodium. Washing the powder resulted in elimination of one viscosity minimum. Shrinkage and bulk density measurements were made on cast pellets fired to 1000°C.

INTRODUCTION

Most ceramic processing necessitates the complete dispersion of the raw materials in a medium to obtain the highest degree of homogeneity within the system. Stability of these suspensions is required to obtain dense, uniform, green microstructures that can be sintered with ease. To disperse a suspension of colloidal particles, a long-range repulsive force must be provided. The long-range force must be at least as strong as, and comparable in range to, that of the Van der Waals attractive (VDW) forces. [1-6]

Several different means may be used to overcome the long-range VDW attractive forces. Of the three mechanisms responsible for dispersion (electrostatic, electrosteric, and steric), electrostatic and electrosteric stabilization will be addressed. In electrostatic stabilization, a coulombic repulsive force is provided by adjusting the pH of the suspension.[1,2,7] During steric hindrance polymer molecules are adsorbed or attach to the surface of the particle, physically preventing the particles from getting close enough to allow London forces to be effective. Electrosteric stabilization provides both a physical barrier and an electrostatic contribution from either a charge on the particle surface and/or charges associated with the polymer. The polymeric charge contribution is through the attachment of an electrolyte (e.g., Na^+ or NH_4^+) to the polymer.[8] In this case, the polymers are referred to as polyelectrolytes. A more comprehensive review of the theories can be found in the literature.[1-4,6,9-11]

PROCEDURES

Two powders were used in this investigation, APA-0.5[a] and A-16 S.G.[b] Table I illustrates the differences in chemical composition between the two aluminas. The most evident variations are the sodium and MgO levels.

Table I. Alumina powder characteristics. (*AR* denotes "as received".)

SAMPLES	SiO_2 %	Al_2O_3 %	MgO %	CaO %	Na_2O %	LOI %	Surface area (m^2/g)	Mean Particle Size (μm)
A-16 S.G. (AR)	0.14	98.4	0.07	0.08	0.12	0.9	8.9	0.5
A-16 S.G. Washed*	0.35	98.7	0.07	0.09	0.07	0.7	9.4	0.5
APA-0.5 (AR)	0.11	99.2	0.03	0.07	0.05	0.4	7.8	0.5

Two acids and two bases were used as pH adjusting dispersants. Polymeric dispersants included the sodium and ammonium salts of both polymethacrylic acid (PMAA) and polyacrylic acid (PAA).[c] A summary of the dispersants and their properties can be found in Table II.

Table II. Reported information for dispersants used during this investigation.

Acid	Normality (N)
HCl	12.1
H_2SO_4	36.0
Base	
NaOH	3.2
NH_4OH	14.8

Polymethacrylic acid	Chemical Name	Reported MW_{ave}	% Active Solids
Darvan C	PMAA-NH_4	10-16,000	25
Darvan 7	PMAA-Na	10-16,000	25
Polyacrylic acid			
Darvan 821A	PAA-NH_4	6,000	40
Darvan 811	PAA-Na	3,500-6,000	43

Samples of varying pH were prepared by titrating acid or base into 200ml suspensions of 35vol% solids loading. The pH was measured and 20ml of suspension removed for every 0.1ml of acid or base added. Viscosity

[a] Ceralox Corporation, Tucson, AZ
[b] Alcoa, Pittsburgh, PA
[c] R. T. Vanderbilt Co., Inc., Norwalk CT

measurements were made on the portion removed using a stress-controlled rheometer.[d] Pellets were cast in Lexan molds on gypsum blocks. Following drying overnight, the pellets were fired to 1000°C at a rate of 1.7°C/min and a hold time of 1hr. The furnace was cooled to room temperature at a programmed rate of 5°C/min. Diametric shrinkage and bulk density (modified ASTM standard C20-87) measurements were performed.

RESULTS AND DISCUSSION

Acid/Base Dispersed

Two acids and two bases were evaluated for their ability to disperse the two aluminas and their effects on densification and shrinkage. Figure 1 displays the pH of the A-16 SG suspensions as a function of acid/base addition. The acid/base additions were normalized to milli-equivalents/gram of powder to allow direct comparison of acid or base efficiency. From the data it is obvious that both HCl and H_2SO_4 have equal ability at pH adjustment (pH~1), NH_4OH had little effect on the pH (pH~10), and the NaOH quickly reached a pH of ~12.

The apparent viscosity as a function of shear rate ($\dot{\gamma}$) was measured for each acid/base addition, with apparent viscosity values extrapolated to $\dot{\gamma}$=1.0s^{-1} for comparison (Figure 2). The suspensions dispersed with HCl showed a significant decrease in viscosity (over three orders of magnitude). Although the H_2SO_4 displayed similar pH behavior, the suspension viscosity was unaffected. Neither base could substantially alter the pH from the powder's isoelectric point (pH~9.8). Additions of acid/base increase the electrolyte concentration in solution, compressing the double layer. At some critical coagulation concentration (CCC), the further addition of ions to a suspension compresses the double layer to the point at which short range attractive forces dominate, allowing particles to interact (i.e., coagulation). The estimated CCC of a 35vol% alumina suspension for monovalent and divalent ions is 43-60mM and 0.3-0.7mM, respectively.[2] (Post presentation calculations indicate that the first addition of H_2SO_4 contributes enough divalent counterions to the suspension to achieve the CCC.)

---

[d] DSR 200 Rheometric Scientific, Inc., Piscataway, NJ

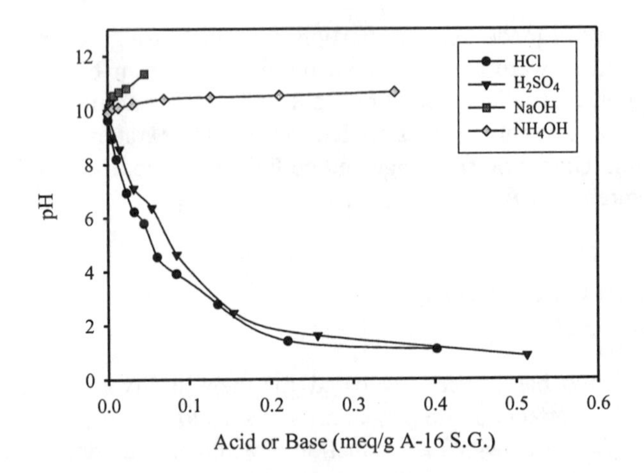

Figure 1. pH of A-16 SG suspensions as a function of acid/base addition.

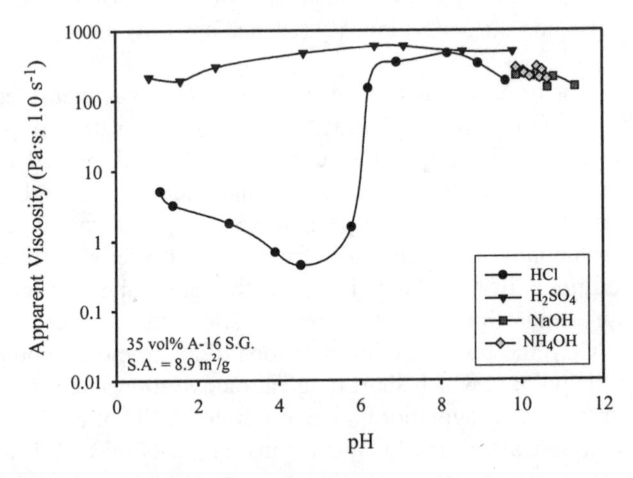

Figure 2. Apparent viscosity measured at $\dot{\gamma}$=1.0 s^{-1}, as a function of pH.

The APA-0.5 powder was also evaluated for rheological behavior using acids and bases to control pH. The suspension pH as a function of acid/base addition is presented in Figure 3. Similar trends were seen for the APA-0.5 as were seen in the A-16 SG suspensions, with the acids leveling off at a pH of 1.0, NaOH at 12, and NH_4OH at 11. Apparent viscosity measurements on these suspensions revealed a slight difference in the HCl plot (Figure 4). The viscosity of the A-16

SG suspension dispersed with HCl drops off steeply at a pH of 6 to ~1.0Pa·s. The APA-0.5 suspension has a more gradual change in apparent viscosity, with a minimum of 10mPa·s occurring at a pH of 3.5. The initial pH of the A-16 SG suspensions was higher than that of the APA-0.5 (i.e., 9.8 versus to 8.5). This data is consistent with that of previous researchers, in which ion exchange between Na^+ impurity ions on the particle surface and H^+ ions in suspension is believed to change the suspension pH and particle mobility.[7] The tendency of MgO to dissolve in water and form a basic solution is also a contributing factor in the pH and surface chemistry of the A-16 SG suspensions.[12]

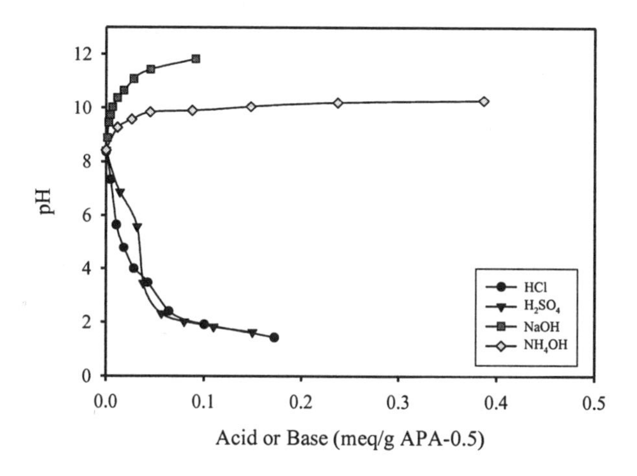

Figure 3. pH as a function of acid/base additions for APA-0.5.

Shrinkage data indicates less than 1.0% linear shrinkage from the green diameter measurement. Average values for the APA-0.5 and A-16 SG samples were 0.82 and 0.53%, respectively. Bulk density calculations were performed based on a theoretical density for α-alumina of 3.98g/cm^3. Plots of bulk density as a function of pH for A-16 SG (Figure 5) and APA-0.5 (Figure 6) show a significant increase in bulk density for HCl dispersed cases only. The H_2SO_4, NaOH, and NH_4OH dispersed suspensions produced pellets that achieved an average of 55% of the theoretical. HCl dispersed cases showed bulk densities of 67% of the theoretical. This data correlates well with that of the viscosity data, in which the HCl suspensions displayed a large decrease in apparent viscosity, compared to the other acid/base dispersants. The bulk density increases sharply in the A-16 SG samples, which is in contrast to a more gradual increase in the APA-0.5 samples. A minimum in bulk density is seen at the isoelectric point for both powders.

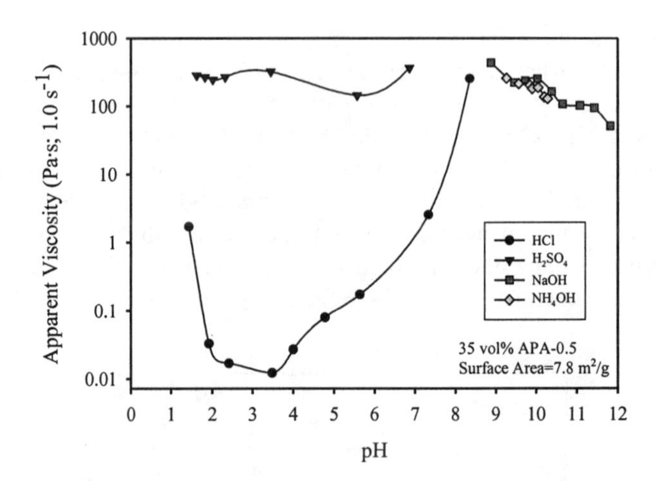

Figure 4. Apparent viscosity measured at $\dot{\gamma}$=1.0 s^{-1}, as a function of pH.

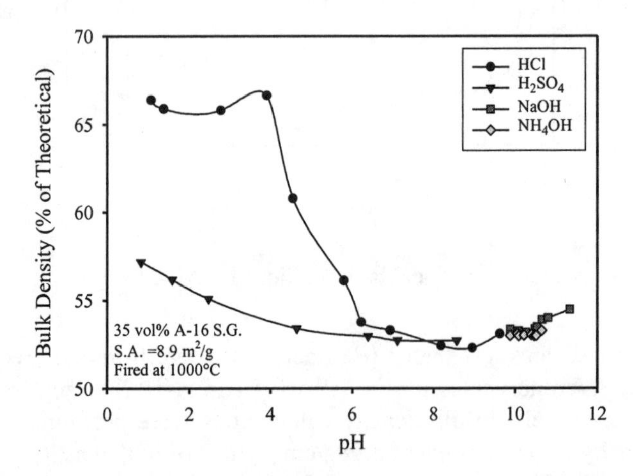

Figure 5. Bulk density of A-16 SG pellets fired to 1000°C at a rate of 1.7°C/min and a hold time of 1hr.

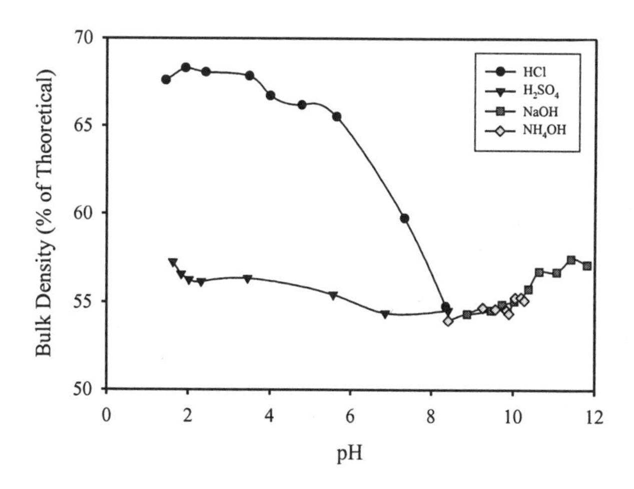

Figure 6. Bulk density of APA-0.5 pellets fired to 1000°C at a rate of 1.7°C/min and a hold time of 1hr.

Polyelectrolyte Dispersed

Polyelectrolyte dispersant levels were represented in mg/m^2, to eliminate effects due to variations in Darvan's active solids and powder surface area. All of the polyelectrolytes were found to significantly reduce the apparent viscosity at $\dot{\gamma}=1.0s^{-1}$. Darvans 7 and 811 displayed minima of 50mPa·s near $0.3mg/m^2$ of dispersant (Figure 7). Darvan C also had a viscosity minimum of 50mPa·s, but a higher concentration of dispersant was required ($0.4mg/m^2$). Darvan 821A's minimum viscosity was 100mPa·s at a dispersant level of $0.6mg/m^2$. A definite shift to lower dispersant levels can be observed with the Na-polyelectrolytes.[8]

The suspensions prepared with A-16 SG displayed somewhat similar results to the APA-0.5, with two major exceptions (Figure 8). The first exception is the degree of dispersion. The apparent viscosity at the lowest minima are a factor of five greater for the A-16 SG suspensions (0.1-0.5Pa·s). The second deviation involves the presence of two minima in plots for Darvan 7, 811, and 821A (Hereafter referred to as the "A-16 SG anomaly").

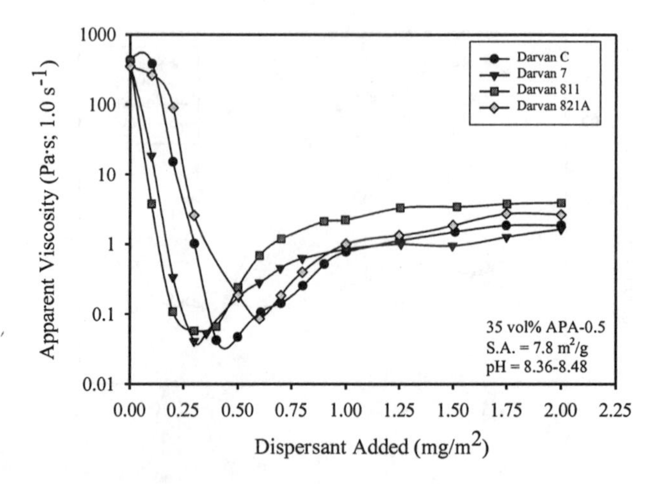

Figure 7. Apparent viscosity at $\dot{\gamma}=1.0s^{-1}$ of APA-0.5 suspensions, as a function of dispersant level.

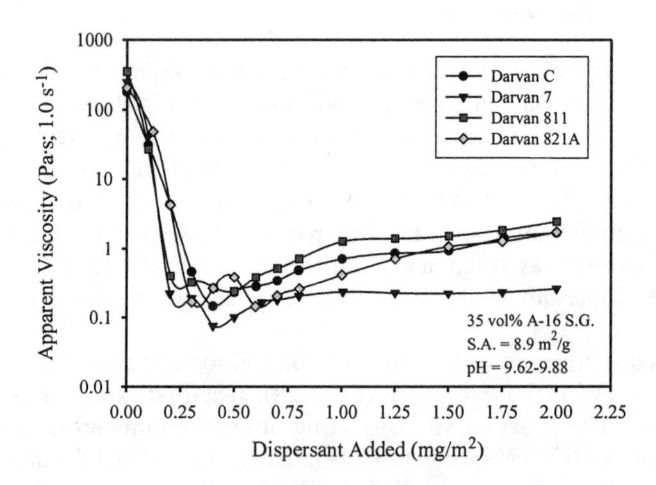

Figure 8. Apparent viscosity at $\dot{\gamma}=1.0s^{-1}$ of A-16 SG suspensions, as a function of dispersant level.

A-16 SG Anomaly

In order to explain the anomaly, several tests were run on A-16 SG powder washed with distilled water. Washing involved mixing 500g of powder for 2hr in

3500ml of distilled water, allowing the powder to settle, removing the supernatant, then repeating a second time before drying the powder. Water samples were taken from the two washings, along with samples of the dried powders, and ICP[e] analyses performed. The results of the dry powder analysis were seen in Table I, while the water analyses can be found in Table III. Of particular interest are the sodium contents. Washing of the A-16 SG resulted in the removal of significant amounts of Na (~6.7mM).

Table III. Chemical analyses of A-16 SG wash supernatant.

SAMPLES	Al ppm	B Ppb	K ppm	Mg Ppm	Mo ppb	Na Ppm	Si Ppm	W ppb
A-16 S.G Wash 1	7.2	64	0.7	0.1	23	112.1	0.22	21
A-16 S.G Wash 2	3.6	36	0.2	1.6	7	42.2	0.14	< 10

The rheological behavior of the washed powder was evaluated using Darvan 821A as a dispersant (Figure 9). Dispersing the washed A-16 SG resulted in only one viscosity minimum and a shoulder. The removal of Na from the powder and the resultant rheological behavior correlates with an ionic contribution to stabilization for the first viscosity minimum and polymeric (electrosteric) stabilization for the second. Researchers have found that a small increase in the surface adsorption of polymer to the particles can be correlated with a decrease in the near-surface Na_2O content, while others have indicated an increase in polymer-electrolyte dissociation in the presence of higher ionic concentrations.[13] Either case could lead to a minimum in viscosity at dispersant levels less than the optimum.

Surface chemistry is a major controlling factor in the dispersion of alumina in water. Potentiometric titrations[f] of suspensions containing $5m^2$ of powder were carried out by initially bringing the suspensions to a pH of 12 by adding NaOH (1N). HCl (3N) was then added, and the pH recorded, until a pH of 2.0 was achieved (Figure 10). Titrations of as-received A-16 SG and APA-0.5 reveal a higher number of active acid (negative) sites on the APA-0.5 powder. When the A-16 SG powder is washed both Na and Mg are seen in the supernatant (Table III). Titration of the washed A-16 SG powder shows an increase in the number of active acid sites. This correlates to a larger number of active sites on the powders surface.[8,12]

---

[e] Acme Analytical Laboratories LTD., Vancouver, B.C., Canada
[f] ABU900, Radiometer, Cleveland, OH

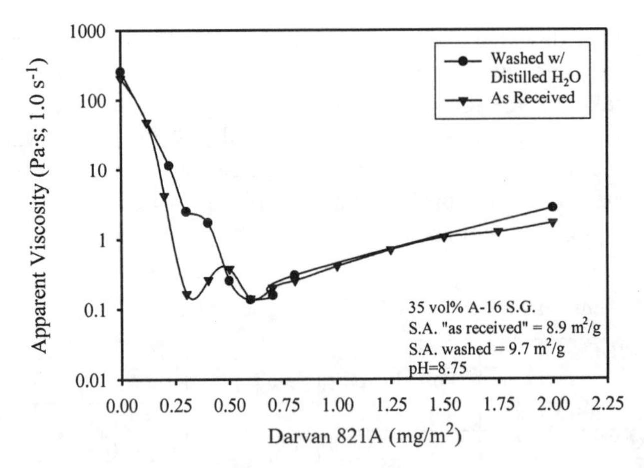

Figure 9. Washing A-16 SG powder with distilled water removed sodium and magnesium. One minimum is observed in the washed powder.

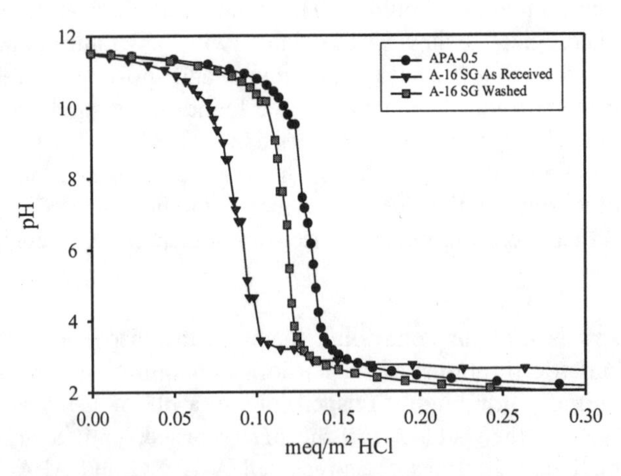

Figure 10. Titration curves of APA-0.5, and the as received and washed A-16 SG powders reveal variations in surface charge/characteristics.

Shrinkage and Bulk Density

Shrinkage and bulk density measurements were also carried out on the polymer-dispersed samples. The results are summarized in Table IV. The bulk density varied little between samples (~56-61% of theoretical). Linear shrinkage was small (<1.0%), although the A-16 SG dispersed with Darvan C showed significantly less shrinkage than Darvan C dispersed APA-0.5.

Table IV. Bulk density and shrinkage measurements of pellets fired to 1000°C.

Dispersant	**Average % Theoretical Density**		**Average % Linear Shrinkage**	
	A-16 SG	***APA-0.5***	***A-16 SG***	***APA-0.5***
Darvan C	57	61	0.23	0.71
Darvan 7	57	58	0.36	0.45
Darvan 811	58	60	0.49	0.55
Darvan 821A	56	58	0.43	0.52
Washed 821A*	59	-----	0.48	-----

*Pellets cast from samples prepared with A-16 SG powder washed twice with distilled water.

CONCLUSIONS

When dispersing alumina using an electrolyte (i.e., acid or base), the most effective dispersant was HCl. Important parameters are the valency of the counterions in suspension. This was prevalent when dispersion was attempted with H_2SO_4. The concentration (number of ions times the valence) in solution influences the size of the diffuse double layer. High concentrations can cause compression of the double layer and subsequently, coagulation. The bases were poor dispersants.

Polymeric stabilization with polyelectrolytes of PMAA and PAA resulted in two observations that are connected with the electrolyte and its concentration. The minimum viscosity in suspensions dispersed with PMAA- and PAA-Na displayed a shift to lower dispersant levels. Samples prepared using A-16 SG bore two viscosity minima. Washing the A-16 SG with distilled water resulted in the elimination of the first minimum. Subsequent work with titration revealed a larger number of active surface sites on the washed powder. This information coupled with the chemical analyses of the raw materials and wash supernatant leads to the belief that two types of stabilization are occurring. The first minima are related to an ionic contribution to the stabilization. The second corresponds to the more stable electrosteric stabilization of the polymer.

ACKNOWLEDGEMENTS

The authors would like to thank Peter Kupinski and David Clark for their efforts in data collection. The funding of this project by the Corning Foundation and the New York State Center for Advanced Ceramic Technology was greatly appreciated.

REFERENCES

1. James, R.O., *Characterization of Colloids in Aqueous Systems*. Advances in Ceramics. Vol. 21. 1987. 349-410.
2. Hunter, R.J., *Introduction to Modern Colloid Science*. 1993, Oxford: Oxford University Press. 194-298.
3. Overbeek, J.T., *Recent Developments in the Understanding of Colloid Stability*. Journal of colloid and Interface Science, 1977. **58**(2): p. 408-422.
4. Overbeek, J.T., *How Colloid Stability Affects the Behavior of Suspensions*. Journal of Materials Education, 1985. **7**(3): p. 401-427.
5. Horn, R.G., *Surface Forces and Their Action in Ceramic Materials*. Journal of the American Ceramic Society, 1990. **73**(5): p. 1117-1135.
6. Napper, D.H., *Polymeric Stabilization of Colloidal Dispersions*. 1983, London: Academic Press. 138-413.
7. Kelso, J.F. and T.A. Ferazzoli, *Effect of Powder Surface Chemistry on the Stability of Concentrated Aqueous Dispersions of Alumina*. Journal of the American Ceramic Society, 1989. **72**(4): p. 625-627.
8. Cesarano III, J. and I.A. Aksay, *Stability of Aqueous α-Al_2O_3 Suspensions with Poly(methacrylic acid) Polyelectrolyte*. Journal of the American Ceramic Society, 1988. **71**(4): p. 250-255.
9. Lyklema, J., *Double Layer Dynamics and Colloid Stability*. Materials Science Forum, 1988. **25-26**: p. 1-16.
10. Napper, D.H., *Steric Stabilization*. Journal of colloid and Interface Science, 1977. **58**(2): p. 390-407.
11. Sushumna, I., R.K. Gupta, and E. Ruchenstein, *Effective Dispersants for Concentrated, Nonaqueous Suspensions*. Journal of the Materials Research Society, 1992. **7**(10): p. 2884-2893.
12. Wei, W.C.J., S.J. Lu, and B. Yu, *Characterization of Submicron Alumina dispersions with Poly(methacrylic Acid) Polyelectrolyte*. Journal of the European Ceramic Society, 1995. **15**(2): p. 155-164.
13. Incorvati, C.M., D.H. Lee, and R.A. Condrate Sr., *Obtaining Dispersible Bayer-Process Aluminas in Water*. The American Ceramic Society Bulletin, 1997. **76**(9): p. 65-68.

Forming

IMPROVED CONSISTENCY OF PLASTER MOULDS

Nigel J. Leak
CERAM Research
Stoke on Trent
UK

ABSTRACT

A major factor in producing consistent moulds is accurate prediction of the time at which a plaster blend should be poured. Work at CERAM Research using a laboratory scale continuous plaster blender and scaled up models of the devise, identified that pouring after a constant change in viscosity produced more consistent moulds than pouring at a constant viscosity or constant blending time. A devise for measuring change in blend viscosity, which can be attached to most batch blenders, is also described.

INTRODUCTION

The time at which plaster water blends are poured after mixing has an important influence on such mould properties as setting time setting expansion. The ability to judge the exact time to pour a plaster blend, the creaming point, is an important factor in achieving consistent mould behaviour.

The aims of this paper are to:

(i) Briefly review the factors which can cause the judgement of the creaming point to be unpredictable.

(ii) Describe an investigation carried out by CERAM Research into methods of accurately determining the creaming point.

(iii)Describe a devise for determining the creaming point which can be attached to most exiting batch blenders.

FACTORS WHICH INFLUENCE THE DETERMINATION OF THE CREAMING POINT

The principle causes of difficulties in predicting the creaming point are variations in the following properties:

1 The water demand of plaster.
2 Plaster to Water Ratio.
3 Blending Speed.
4 Blender Paddle Configuration.
5 Batch Size.

6 Slurry Temperature.
7 Water Quality.

CONTINUOUS VISCOSITY MEASUREMENT

In order to accurately monitor the changes which occur during the blending process, CERAM Research developed a continuous viscosity recorder.

A schematic outline of the operation is shown in Figure 1. The mixing vessel, carrying the plaster water blend, stands on a turntable. During blending, the rotating paddle exerts a torque through the blend, onto the vessel, which causes the turntable to rotate. The rotating turntable makes contact with a spring loaded transducer and recorder, which monitors the viscosity. At the set mixing speed of 136 rpm, the blade configuration is designed to maintain movement of the whole blend while minimising the amount of turbulence.

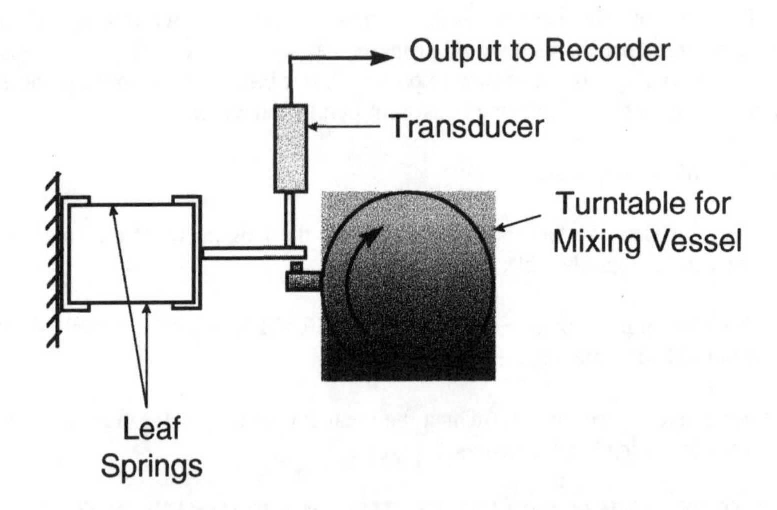

Figure 1. Schematic of a Continuous Viscosity Recorder

When calibrated the instrument is able to monitor the continuous change of blend viscosity with blending time. The blending times on industrial scale blenders are shorter because of the greater amount of turbulence. However it is possible to obtain correlation between the blending times from this laboratory scale blender with the larger scale blenders, which makes it possible to predict the blending times for industrial size blenders.

Figure 2 shows a series of blend curves for a typical pottery plaster in a range of conditions (age) and plaster to water ratio's. A key observation is that at the end of blending, which is taken as the fixed viscosity of 450 cp, the viscosity at which it is generally considered a mould maker would judge that a blend was ready for pouring,

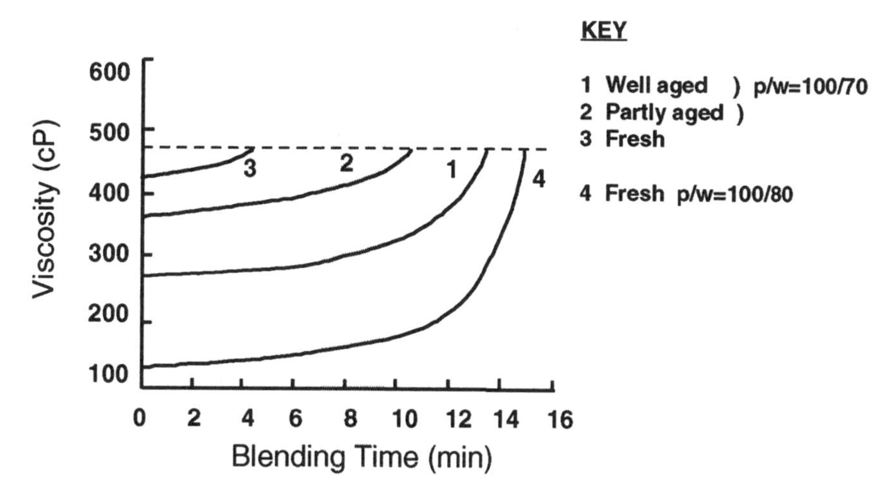

Figure 2. Continuous blending curves for a series of pottery plasters

the rate of change of viscosity is not uniform. These observations illustrates the difficulties experienced by mould makers when attempting to judge the creaming point of blends of fresh plaster. It is not the pouring of highly viscous blends which is the primary problem but being able to detect the rate of change of viscosity.

DEVELOPMENT OF CONTROLLED VISCOSITY BLENDERS

Trials have been undertaken at CERAM Research to incorporate continuous blending curve measurements into blending machines which can automatically predict the creaming point.

Initially, two devices were made and tested. The first was a scaled up version of the technology described in previous section. The second device, had the drive motor suspended from a freely rotating bearing, which then pushes against the torque measuring unit mounted above the bearing. This device measures the changing drag between the plaster blend and the mixing propellor.

In both devises the torque measuring unit consisted of a leaf spring, which detects torque, transducer and recorder. Both devices gave satisfactory measurements of torque providing the conditions in the mixing vessel were non-turbulent. These conditions were achieved by using low rotational speeds and impellers with flat, vertical, radial blades rather than "marine" type axial flow impellers.

Three different methods of viscosity control were investigated:
(i) Pouring at a constant viscosity.
(ii) Pouring at a constant rate of change of viscosity.
(iii) Pouring after a constant increase in viscosity.

The three methods are illustrated on viscosity blending times curves in Figure 3.

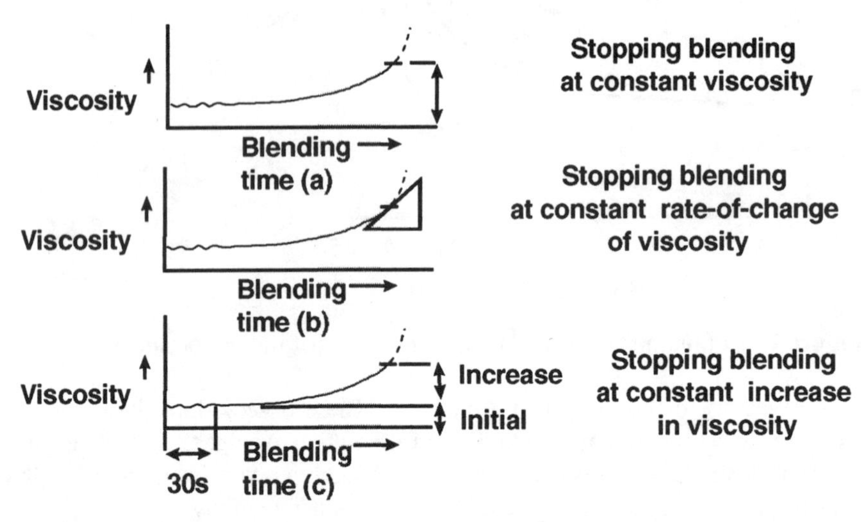

Figure 3. Various control methods

During the initial trials it was quickly established that the equipment developed at that time had insufficient sensitivity to allow for reliable control by the rate of change of viscosity (Figure 3b). Extended trials were carried out comparing the remaining two control methods with pouring after a fixed time. To achieve the maximum consistency throughout the trial, the blend size, consistency, filling procedure and soak time prior to mixing (standardised to 30 seconds) were kept as constant as possible. The parameters (Figure 4), demonstrate that constant viscosity increase gives the most precise prediction of mould properties.

Mean Values, Deviation from Mean and Range of Values for Setting and Set Properties of Plasters Poured According to Three Different Criteria Special Potters Plaster			
Properties	**Constant blending time**	**Constant viscosity**	**Constant increase of viscosity**
Blending time (min) Range (min)	7± 0	7.47 ± 1.48 7.57	7.50 ± 1.30 6.10
Setting time (min) Range (min)	9.63 ± 3.02 11.77	9.15 ± 1.35 5.82	8.45 ± 1.00 3.82
Permeability (CERAM units) Range (CERAM units)	2.80 ± 1.10 3.90	2.60 ± 0.60 2.80	2.50 ± 0.60 2.40
Strength (lbf/in^2) Range (lbf/in^2)	676 ± 51.7 206	687 ± 45.1 175	696 ± 51.5 191
Hardness penetration (1/10 000 in) Range (1/10 000 in)	17.8 ± 2.80 9.40	17.6 ± 2.80 13.30	Mean Values, Deviation from Mean and Range of Values for Setting and Set Properties of Plasters Poured According to Three Different Criteria Special Potters Plaster

Figure 4. Comparison of control methods

The mould permeability, measured using a meter developed by CERAM Research, accurately characteristics blending time where high permeability values indicate early pouring and low permeability values late pouring.

FACTORY TRIALS

Prototype Blender

When consideration was given to factory trials it was found that batch blenders used for tableware and sanitaryware were unsuitable to accommodate either of the viscosity measuring devises. A prototype plaster blender was developed based on the model where the measuring devise was above the motor. This method was chosen as the most practical means of keeping the measuring devise clean. Other modifications included:

1 Vibration storage hopper.
2 Weighing devise.

3 "Swing-in" platform and washing station.
4 Microprocessor control.

A diagram of the prototype blender is shown in Figure 5.

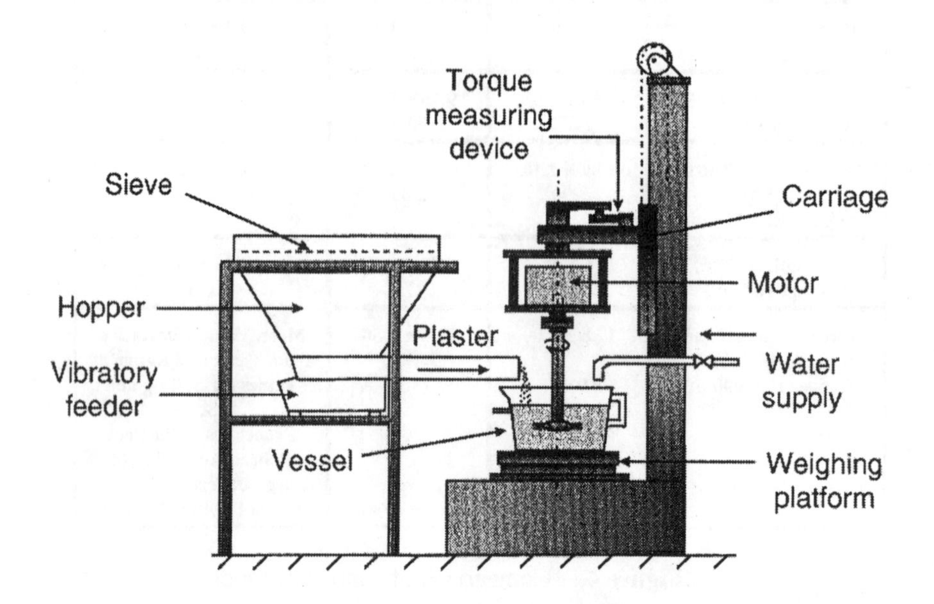

Figure 5. Prototype production blender

Factory Trials

Moulds produced using the prototype blender, and several copies of the machine, were trialed in production against moulds made at the same period using the current mould making facilities, at several tableware factories. The mould permeability values which were measured are summarised in Figure 6. In all cases a narrower range of values was achieved from the controlled viscosity blenders. In many cases improved ware quality, lower losses and longer mould life were reported.

NEW DEVICE

SL Electrotech have recently developed a device to monitor changes in blend viscosity which can be attached to most batch blenders (Figure 7). The instrument simply gives a signal to tell the mouldmaker when a blend is ready to pour.

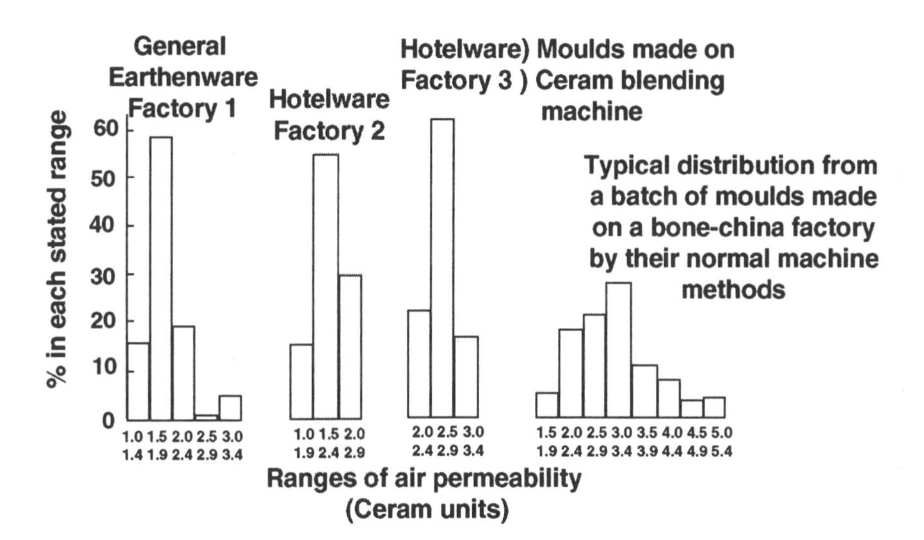

Figure 6. Distribution of mould permeabilities obtained during factory trials

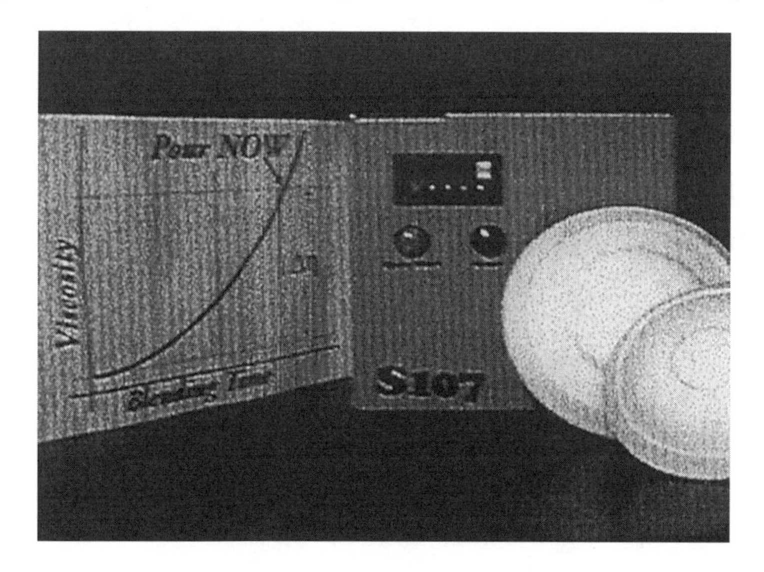

Figure7. Blender control instrument

DEGRADATION OF GYPSUM MOLD MATERIALS

Brett M. Schulz* and William M. Carty
NYS Center for Advanced Ceramic Technology—Whiteware Research Center
New York State College of Ceramics
Alfred University, 2 Pine Street, Alfred, NY 14802

ABSTRACT

Gypsum molds are used for the dewatering and forming of clay suspensions in whiteware plants. During the use of the mold inconsistencies develop over time, due to mold degradation, that eventually result in the mold being discarded. To determine the cause and rate of gypsum mold degradation, gypsum samples were immersed in salt solutions of various concentrations over a range of times. The samples were suspended from an analytical balance allowing the change in weight to be monitored as a function of time. It was determined that an increase in either the concentration or the valence charge of the ions had the greatest effect on the gypsum dissolution, thus indicating a potential route for reducing degradation.

INTRODUCTION

Gypsum molds are used in the whiteware industry for the dewatering of clay slips and the forming of plastic masses. During the lifetime of the gypsum mold the casting behavior of the clay slip, or the forming behavior of plastic body, becomes inconsistent, at which point the mold is removed from the process. This change in mold properties is often variable causing unpredictable losses in production time and raw materials. Since millions of pounds of plaster are used in the whiteware industry each year, an increase in the lifetime of the mold by 10 percent would result in a tremendous savings for whiteware companies. It was hypothesized that dissolved ions accelerate the dissolution of gypsum, resulting in the change in the casting behavior. So the goal was to determine specifically what augmented mold degradation and, as a consequence, what would allow the lifetime of the mold to be extended.

Figure 1 shows the typical type of degradation that is seen in the use of gypsum molds. Show are SEM micrographs taken from a fracture surface of a gypsum mold prepared for use in a jiggering process. Figure 1A shows the microstructure of the mold in the new, as cast, condition and Figure 1B is the microstructure after jiggering 80 pieces. It is obvious that the microstructure has changed dramatically during the use of the mold. The gypsum needles are clearly visible in the as cast microstructure, but there is almost no evidence of gypsum needles in the used jiggering mold.

---

* Now with MMCC, Inc., 101 Clematis Ave., Unit 1, Waltham, MA 02154

(A) (B)

Figure 1. SEM micrographs of (**A**) an unused gypsum jiggering mold and (**B**) a jiggering mold after 80 casts. The microstructure of (**B**) has obviously coarsened during use. (The size bar = 10 μm.)

BACKGROUND

The solubility of gypsum is low, about 0.014 moles per liter (0.241g/100 ml) in de-ionized water at 20°C. The solubility is enhanced by dissolved ions that are present in the water, providing that the ions in solution are not common to the gypsum (i.e., no Ca^{+2} or SO_4^{-2} are present in the solution). The enhanced solubility is due to the ions in solution increasing the polarity of the water and making the water act as a better solvent. The theoretical solubility of gypsum in the presence of a salt solution can be calculated by using a modified Debye-Hückel equation for strong electrolytes:

$$pK_{sp(c)} = pK_{sp^0} - 0.51\{a(z_c)^2 + b(z_a)^2\}\left(\frac{(\mu)^{1/2}}{1+(\mu)^{1/2}}\right) \quad (1)$$

The term K_{sp^0} is the solubility product constant, or the equilibrium constant for the solubility of a slightly soluble or nearly insoluble ionic compound. Likewise, the term K_{sp} is the solubility constant of the slightly soluble species in the presence of a more soluble salt in solution. The prefix 'p' on the solubility constants is used to designate the negative logarithm of the constant. The constants a and b are determined by the following equation:

$$Ca_a(SO_4)_b \Leftrightarrow aCa^{+2} + b(SO_4)^{-2} \quad (2)$$

In the case of calcium sulfate the values for 'a' and 'b' are one. The term 'μ' is the ionic strength of the salt solution and is calculated by:

$$\mu = \frac{1}{2}\left\langle [\text{cation}](z_c)^2 + [\text{anion}](z_a)^2 \right\rangle \quad (3)$$

The variables z_c and z_a are used to denote the valence state of the cation and anion of the species being dissolved, respectively. The square brackets denote the concentration of the ion in solution (in millimoles/liter). Using Equation 1 the theoretical solubility of gypsum can be calculated. The higher the valence charge on the ions, the higher the solubility of the gypsum for the same concentration salt solution and that as the concentration of the salt solution is increased the solubility of gypsum is increased. Figure 2 shows the effect of increasing ionic strength on the solubility of gypsum using Equation 1.

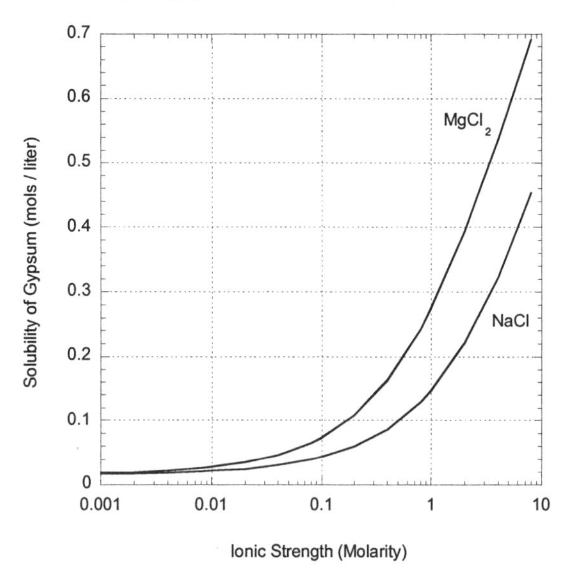

Figure 2. Plot of the theoretical solubility of gypsum as a function of the ionic strength for uncommon salt solutions. The data is plotted for both a monovalent and a divalent cation.

If a common ion salt is used, the common ion will hinder the dissolution of gypsum. If the dissolved common ion concentration levels are in excess of the solubility limit of gypsum, there should be no noticeable dissolution of the gypsum sample. If the concentration of the common ion is below the saturation level for the dissolution of gypsum, the gypsum will dissolve until the solubility limit is reached.

In addition to ions specifically added to a suspension, there are three ion sources in a whiteware batch suspension: 1) dissolved salts in the tap water used for the preparation of the slip; 2) raw materials dissolution (attributed to the aging process); and 3) ion exchange with the clay minerals. Therefore, along with the water entering the mold during the dewatering stage process are any dissolved ions. As stated earlier, the uncommon ions increase the polarity of the water and increase the dissolution rate of gypsum.

EXPERIMENTAL PROCEDURE

The effect of various salts on the degradation of gypsum was determined by performing a dissolution study. The study consisted of suspending samples of gypsum in salt solutions of varying concentration. Right circular cylinders of gypsum were cast at a consistency of 70g of water per 100g plaster using K69 pottery plaster (from Georgia Pacific). After the plaster had set the gypsum samples were dried in a forced air dryer at 100°F (43°C) for 24 hours.

Prior to analysis the gypsum samples were saturated with de-ionized water to create a common starting point for each of the experiments. The saturated gypsum samples were suspended from an analytical balance in a salt solution and the change in weight was monitored with a computer. The change in weight was recorded once per minute using a data collection program (via LabVIEW) to produce the dissolution curves for each of the solutions.

EXPERIMENTAL RESULTS

Effect of Various Salts and Concentration on Gypsum Dissolution

Several different salts, at varying concentration levels (listed in Table I), were used to evaluate the effect of cations and anions on the dissolution rate of gypsum. The effect of concentration for each of the salts studied was determined by preparing a dilute solution, at a concentration roughly equivalent to filter press water, and a higher concentration solution. In some of the collected dissolution curves discontinuities are evident due to the formation of gas bubbles on the surface of the immersed gypsum sample. The formation of gas on the surface resulted in the decrease in the weight of the gypsum sample followed by the release of the bubble. (The data sets were not corrected for "bubble events.")

Table I. List of the salt solutions used in the dissolution study. The salts are grouped by cations, anions and water samples

Salt	**Concentration**		
NaCl	0.014 M	0.854 M	
KCl	0.011 M	0.542 M	
$MgCl_2 \cdot 6H_2O$	0.011 M	0.487 M	
$CaCl_2 \cdot 2H_2O$	0.013 M	0.403 M	
$NaNO_3$	0.012 M	0.520 M	
$Na_2SO_4 \cdot 10H_2O$	0.011 M	0.320 M	
Na-PAA	1.27 wt%	4.50 wt%	10.00 wt%
Na_2SiO_3	0.80 wt%	1.70 wt%	4.25 wt%
De-ionized Water	See Table II		
Alfred Tap Water			
Filter Press Water			

Effects of Water

De-ionized water was used as a baseline for comparison to subsequent tests. In the presence of de-ionized water the sample lost 4.5% of its initial weight in 24 hours. A precipitate also formed in the bottom of the beaker which created a sink for the dissolved material, thereby allowing the reaction to continue indefinitely. The next sample was immersed in tap water, the composition of the tap water can be seen in Table II. The dissolution rate in the presence of the tap water was enhanced due to dissolved ions that were present in the tap water sample, resulting in a 7.7% weight loss over 24 hours. When a gypsum sample was immersed in a filter press water sample from industry, the initial weight loss rate was lower than that of the de-ionized water rate. As the experiment was continued, the sample started to gain weight as a function of time resulting in a 1.9% loss of the initial weight after 24 hours and 0.5% weight loss after 90 hours. (This behavior indicates the presence of common ions in solution or the precipitation of another crystalline species.) These three data curves are presented in Figure 3.

Table II. Chemical analysis of the water samples involved in the degradation of gypsum molds study.

Sample	Na^+ ppm (*mM*)	K^+ ppm (*mM*)	Mg^{+2} ppm (*mM*)	Ca^{+2} ppm (*mM*)
De-ionized Water	<0.1 (*<0.004*)	0.1 (*0.003*)	<0.1 (*<0.004*)	<0.1 (*<0.003*)
Filter Press Water*	220.9 (*9.5*)	53.2 (*1.4*)	1.41 (*0.06*)	0.76 (*0.02*)
Process Water	9.1 (*0.39*)	1.9 (*0.05*)	9.0 (*0.37*)	36.6 (*0.92*)
Filter Press Water**	199.6 (*8.72*)	28.4 (*0.73*)	21.5 (*0.88*)	123.5 (*3.09*)
Alfred Tap Water	57.3 (*2.50*)	2.1 (*0.05*)	19.9 (*0.82*)	77.9 (*1.95*)

* Sample of filter press water prepared from a 12.5 volume percent suspension in de-ionized water, extrapolated to an equivalent specific gravity.
** Sample of filter press water from an industrial dinnerware process.

Cation Effects

The effect of monovalent cations was determined using NaCl and KCl. The solubility of gypsum was slightly enhanced by the dilute salt solutions, resulting in a 4.5% and a 5.7% weight loss in the NaCl and KCl solutions, respectively, within 24 hours of immersion. When the concentration of the salt solutions was increased, the solubility of the gypsum samples immersed in the salt solutions also increased, as illustrated in Figure 4. The sample that was suspended in the higher concentration NaCl solution had a 15.2% weight loss while the higher concentration KCl solution sample had a 16.1% weight loss in 24 hours. It is important to note that the samples immersed in the KCl solutions had a higher rate of dissolution even though the concentration of the solutions used in the study were more dilute than the NaCl solutions indicating that the selection of cation is important.

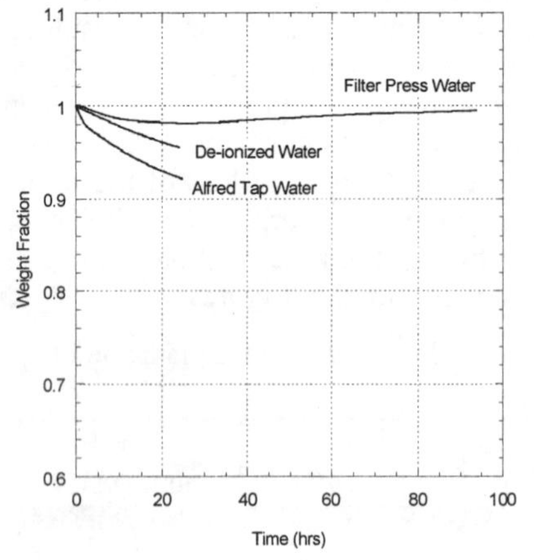

Figure 3. Effect of the various water samples on gypsum degradation.

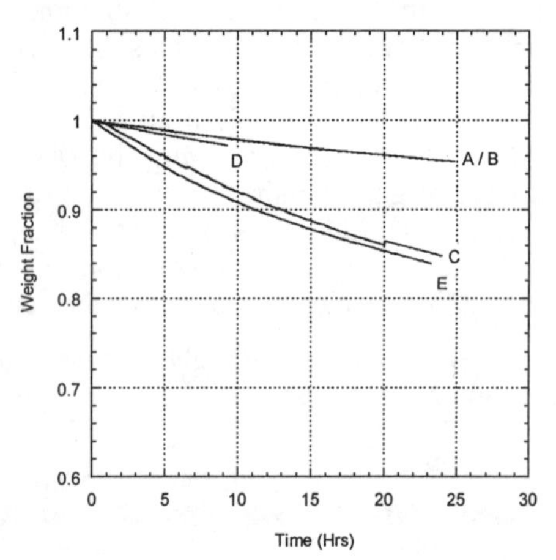

Figure 4. Effect of monovalent cations on the degradation of gypsum. Shown are the effect of **(A)** de-ionized water, **(B)** 0.014 M NaCl, **(C)** 0.854 M NaCl, **(D)** 0.011 M KCl and **(E)** 0.542 M KCl. **(A)** and **(B)** overlap.

Two divalent cation salts were tested in this study, one uncommon salt, $MgCl_2$, and a common ion salt, $CaCl_2$. In the presence of the dilute salt solutions, the gypsum samples lost 8.7% and 5.7% of their initial weight within 24 hours for the $MgCl_2$ and $CaCl_2$ salts, respectively. The dilute $CaCl_2$ solution was prepared at a concentration just below the solubility limit of $CaSO_4$, which allowed the dissolution of gypsum in addition to the material that was precipitated within the beaker. When the concentration of the uncommon ion salt was increased, the dissolution rate greatly increased, resulting in a 27.5% weight loss in 24 hours. When the concentration of the common ion salt was increased, the weight of sample gradually increased as material was precipitated within the gypsum structure followed by a slow weight loss as the gypsum structure degraded due to the dissolution precipitation reaction coarsening the gypsum needles. After immersion for 24 hours in the higher concentration common ion salt, the sample had gained 0.3% of its initial weight, the results are plotted in Figure 5.

Anion Effects

The effects of Cl^-, NO_3^-, and SO_4^{-2} anions were tested using a series of Na-salts. Two dispersants, sodium silicate and a sodium polyacrylic acid (Na-PAA) solution (Dispex A40, Allied Colloid Inc., Suffolk, VA), were also tested to to evaluate the effect of polyvalent anions commonly used as dispersants.

As previously stated the presence of NaCl resulted in a 4.5% weight loss in the dilute solution and 15.2% weight loss at a higher salt solution concentration (Figure 4). The use of $NaNO_3$ resulted in a 3.7% weight loss after 24 hours of immersion in the dilute solution, as illustrated in Figure 6. When the concentration of the $NaNO_3$ salt solution was increased, the gypsum sample lost 15.7% of its initial weight. The presence of the common ion salt Na_2SO_4 caused an initial weight gain in both the dilute and the higher concentration solutions followed by a gradual weight loss in the dilute solution. After 24 hours of immersion in the Na_2SO_4 solutions the sample in the dilute solution had lost only 1.6% of its weight and the sample from the higher concentration solution had gained 2.0% of it's initial weight.

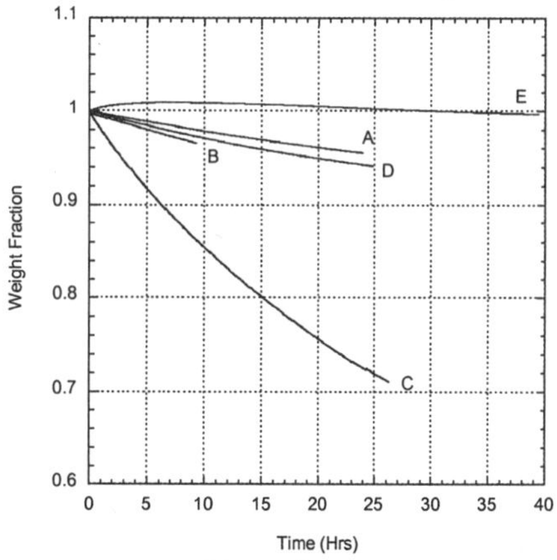

Figure 5. Effect of divalent cations on the degradation of gypsum. Shown are the curves for **(A)** de-ionized water, **(B)** 0.011 M $MgCl_2$, **(C)** 0.487 M $MgCl_2$, **(D)** 0.013 M $CaCl_2$ and **(E)** 0.403 M $CaCl_2$.

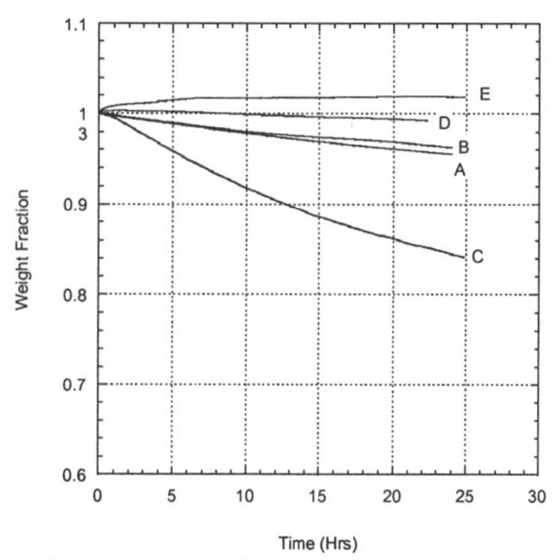

Figure 6. Effect of anions on the degradation of gypsum molds. Shown are the curves for **(A)** de-ionized water, **(B)** 0.012 M $NaNO_3$, **(C)** 0.520 M $NaNO_3$, **(D)** 0.011 M Na_2SO_4 and **(E)** 0.320 M Na_2SO_4.

When the samples were immersed in Na_2SiO_3 solutions, over the concentration range of 0.8 to 4.25 weight percent, each sample showed a weight gain as a function of time. The samples immersed in the 0.8 and 1.7 weight percent solutions gained 1.0% of their weight while the 4.25 weight percent solution gained 5.5% of it's initial weight, as illustrated in Figure 7.

In the presence of Na-poly acrylic acid (Na-PAA), the gypsum samples demonstrated a rapid weight loss presumably due to the high affinity of the calcium ion for the carboxylic acid groups in the polymer chain, resulting in a calcium salt-bridged polymer that precipitates from solution, i.e., a Ca-PAA compound. As the concentration of the polymer in solution increased, the dissolution rate of the gypsum was increased. After 24 hours in the Na-PAA solutions, the 1.27 weight percent solution demonstrated a 14.2% weight loss, the 4.5 weight percent Na-PAA demonstrated a 28.9% weight loss, and the 10 weight percent solution had a 32.5% weight loss, as presented in Figure 7.

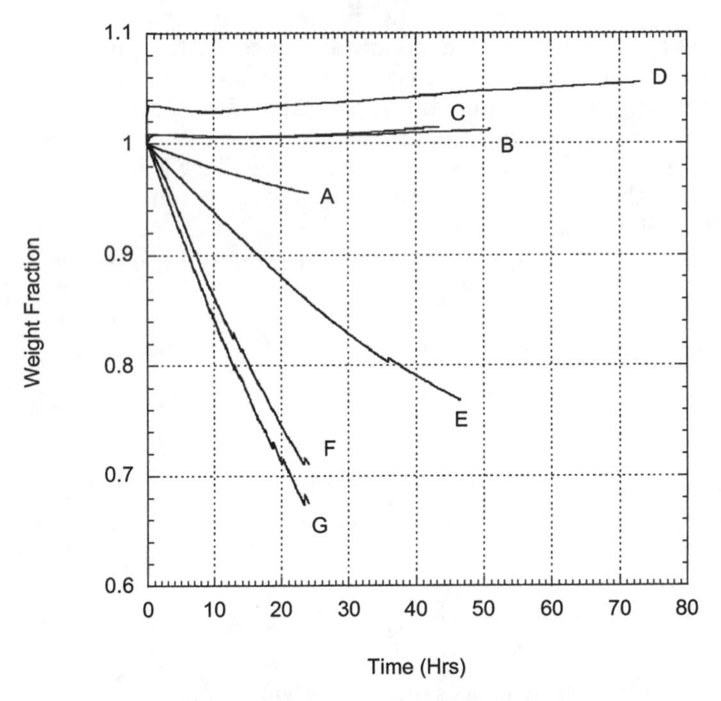

Figure 7. Effect of dispersants on the degradation of gypsum molds. Shown are the curves for **(A)** de-ionized water, **(B)** 0.80 wt% Na_2SiO_3, **(C)** 1.70 wt% Na_2SiO_3, **(D)** 4.25 wt% Na_2SiO_3, **(E)** 1.27 wt% Na-PAA, **(F)** 4.50wt% Na-PAA and **(G)** 10.0 wt% Na-PAA.

It is important to recognize that these addition levels shown in Figure 7 are far above those normally used in an industrial process, and that these results probably do not, in any way, indicate how Na-PAA would behave in an industrial process. The benefits of Na-PAA are well established, and the need to avoid excess dispersant additions, from the suspension control perspective, is obvious.

IDENTIFICATION OF PRECIPITATED MINERAL PHASES

To determine the phases present in the gypsum samples after immersion in the salt solutions, the samples were dried at room temperature for two weeks and crushed and ground to a powder. The samples were then analyzed using powder X-ray diffraction. Only the samples from the higher concentration solutions were tested, since those samples had the highest probability of containing a secondary phase. The secondary phases identified in the gypsum microstructure are listed in Table III.

Table III. List of the precipitated phases as identified using powder X-ray diffraction for each of the salts tested in the dissolution study

Salt	Concentration	Precipitate
NaCl	0.854 M	Na_2SO_4
KCl	0.542 M	KCl
$MgCl_2$	0.487 M	$MgSO_4$
$CaCl_2$	0.403 M	$CaCl_2$
$NaNO_3$	0.520 M	Na_2SO_4
Na_2SO_4	0.487 M	Na_2SO_4
Na-PAA	10.0 wt%	None Found*
Na_2SiO_3	4.25 wt%	$CaSiO_3$
De-ionized Water	Table II	None Found
Tap Water	Table II	None Found
Filter Press Water	Table II	Na_2SO_4

* Ca-PAA is not a crystalline compound that can be identified via powder XRD.

DISCUSSION OF RESULTS

The dissolution studies produced results consistent with the concepts outlined in the introduction. The most obvious observation is that the presence of a common ion salt hinders the dissolution of gypsum until the concentration of the ion is reduced to the solubility limit of the mold. Also, an increase in the ionic strength of an uncommon ion results in the enhanced dissolution of gypsum consistent with the theory of increased polarity of the solution medium. When the valence charge of the uncommon ion is increased, the effect increases. In this study the most serious degradation effects were observed in the presence of Mg^{+2} ions and excess dispersant additions. When these materials are present in solution they greatly enhance the degradation of the gypsum mold. Therefore, to increase the lifetime of a gypsum molds the salts added to affect rheology should be carefully chosen and excess dispersant should be avoided. The best choices for salt additions would be Na_2SO_4, $MgSO_4$, or $CaSO_4$ (ignoring solubility limits).

COMPARISON OF MICROSTRUCTURE DEGRADATION

As a final comparison to the degradation that is seen in industry, a sample of gypsum that had been immersed in a 0.2 M $MgCl_2$ solution for 144 hours was analyzed in the SEM. The microstructure of the gypsum sample was compared to the jiggering mold sample that was obtained from industry. Figure 8 (top) is the used gypsum sample from industry and Figure 8 (bottom) is the sample from the salt solution. There are some similarities between the samples, both microstructures have been obviously altered compared to the "as cast" gypsum sample. It is proposed that the coarsening of the microstructure is due to the dissolution of the finer needles with the growth of the larger needles through a solution-reprecipitation mechanism. Further research is necessary to clearly identify the responsible mechanisms.

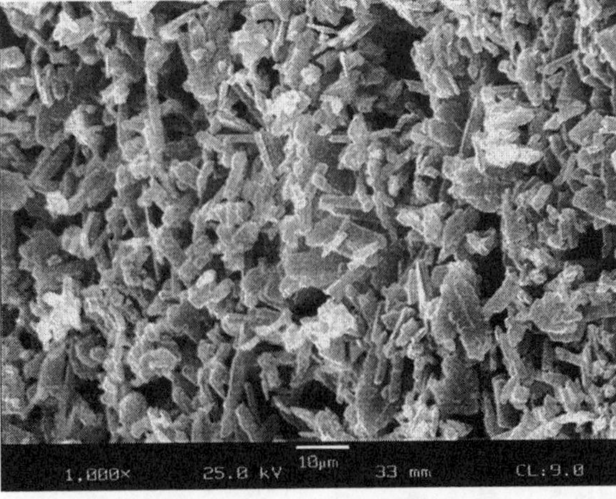

Figure 8. Comparison of the degradation that is seen in industry to the degradation that was observed in this study. The top image is the SEM image from Figure 1B (a sample from an industrial jiggering mold after 80 cycles). The bottom SEM photomicrograph is of a mold immersed in 0.2 molar $MgCl_2$ solution for 144 hours. (The size bar = 10 µm.)

CONCLUSIONS

The dissolution of gypsum in an aqueous environment depends on three factors: 1) the type of ions (i.e., common or uncommon) present in the aqueous solution; 2) the ionic concentration; and 3) the valence of the dissolved ions. When common ions are present in the solution above the solubility limit of gypsum, the dissolution rate of gypsum is slowed until the concentration of the common ion falls below the gypsum solubility limit. For an uncommon salt, the dissolution rate increases with ionic concentration and the valence of the ions in solution. An increase in the dissolved ion valence increases the polarity of the water making it a better solvent for gypsum. Based on these findings, if the gypsum mold lifetime is related to the dissolution of gypsum, increased gypsum mold lifetimes can be obtained by keeping uncommon ion concentrations as low as possible. Therefore, excess dispersant additions should be avoided, and $MgSO_4$ or $CaCl_2$ should be used as coagulants, instead of $MgCl_2$ or NaCl, to adjust suspension rheology.

REFERENCES

Excerpted from *Characterization of Gypsum Mold Materials*, M.S. thesis, Brett M. Schulz, Alfred University, 1996.

For information on excess dispersion effects on suspension rheology and the impact of ionic concentration on rheology of whiteware suspensions:

W. M. Carty, K. R. Rossington, and U. Senapati, “A Critical Review of Dispersants for Whiteware Applications” (this volume), and

K. R. Rossington and W.M. Carty, "The Effects of Ionic Concentration on the Viscosity of Clay-Based Suspensions" (this volume).

THE CHARACTERIZATION OF PRESSURE CASTING

Shiraz H. Khan and William M. Carty
NYS Center for Advanced Ceramic Technology—Whiteware Research Center
School of Ceramic Engineering and Materials Science
New York State College of Ceramics at Alfred University
Alfred, NY 14802

ABSTRACT

A new instrument, based on the Baroid filter test, has been constructed for the characterization of pressure casting systems. This instrument, the pressure filtration cell, measures the casting rate and cake moisture retention level of samples prepared under conditions similar to those observed in pressure casting. During the test, the instrument monitors the suspension height reduction as a function of time. Upon test completion, the final suspension height is the cake thickness, allowing water retention level to be calculated automatically by the computer program by knowing the initial solids content of the suspension and by using the starting and ending points of the suspension height. With this information the goal was to correlate suspension rheology with pressure filtration behavior and provide an alternative means of characterizing whiteware bodies. For industrial applications, the goal is to achieve the fastest possible casting rate with low moisture retention in the cast layer. Future experiments will assess the effects of applied pressure, viscosity, solids content, dispersant type, and particle size distribution.

INTRODUCTION

This project is based on the conservation of mass equations proposed by Aksay and Schilling,[1] refining the model of Adcock and McDowall.[2] Assuming a one-dimensional casting situation as illustrated in Figure 1, Aksay and Schilling determined that the thickness of the cast layer can be calculated from the change in suspension height during the casting process or by monitoring the saturation level within the mold material. Represented mathematically:

$$X_S^0 - X_S = nX_C = -\varepsilon_m X_m \tag{1}$$

where: X_S^0 ≡ initial suspension height; X_S ≡ suspension height at time "t"
n ≡ system parameter; X_C ≡ cast layer thickness
ε_m ≡ mold void volume fraction; X_m ≡ mold saturation depth.

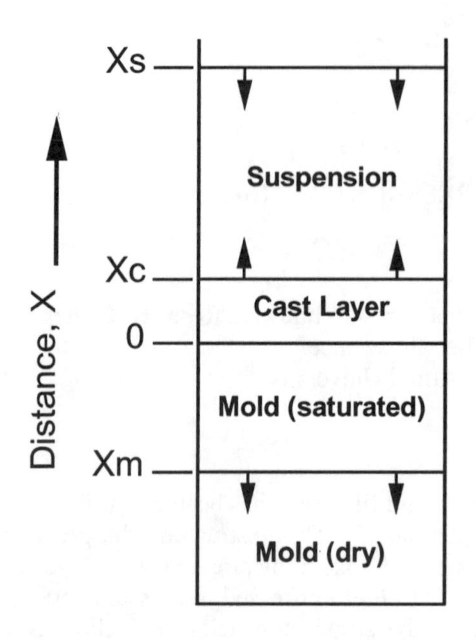

Figure 1. Schematic cross section of the pressure filtration system. As a one-dimensional model this assumes flow only through the cast layer and the mold interface.

The one-dimensional casting model assumes that the suspension is composed of only solid particles and the suspending fluid (i.e., no air bubbles), and thus obeys a simple conservation of mass relationship. The suspension solids loading (x_p), on a volumetric basis, provides the amount of fluid involved. The void fraction in the mold and the volume of mold saturated are directly related to the volumetric decrease in the suspension height at any given time. Two additional assumptions regarding the molds are that the mold is dry, or at least the mold pore volume is known, and that there is negligible deposition of solid particles in the mold material. This second assumption can be reasonably argued based on typical industrial mold lifetimes of 80-100 cycles. (Specific processes and industrial practices however can substantially change the typical number of cycles, but this is a reasonable estimate.) Therefore, the casting system can be adequately (and elegantly) described by following the fluid in the system.

One additional relationship necessary to allow the calculation of the cast layer thickness, is the "system parameter," n:

$$n = \left(\frac{1 - x_p - \varepsilon_c}{x_p} \right) \tag{2}$$

which takes into account the solids loading (x_p) and the void fraction in the cast layer (ε_c). Therefore, upon completion of the casting process, through a conservation of mass relationship, the initial amount of water in the suspension must be equal to the volume of water in the cast layer plus the volume of water in the mold. Also, the volume of particles in the suspension must be equivalent to the volume of particles in the cast layer when casting is complete.

During the casting process, the cast layer acts as an additional filter, but one with a different pore structure compared to that of the mold. Increasing the cast layer thickness results in an increase in the length of the filtration channels through the cast layer, slowing down the casting process with increasing time. In addition, consistent with experimental observation, Aksay and Schilling concluded that the increase in the thickness of the cast layer is directly related to the square root of casting time.[1]

From a practical perspective, the suspension height is a relatively easy interface to monitor, while the cast layer thickness and the mold saturation distance are nearly impossible. Therefore, based on Equations 1 and 2, and the conclusion that casting rate follows a square-root of time dependence, Equation 3 was derived allowing the cast layer thickness (X_C) to be calculated at any time, t, during the casting process:

$$X_C = \left(\frac{m x_p}{1 - x_p - \varepsilon_C} \right) \sqrt{t} \quad \text{or} \quad X_C = \frac{m}{n} \sqrt{t} \tag{3}$$

where "m" is the slope of the change in suspension height versus $\sqrt{t}$.

In industrial casting processes, the problem is two-fold. Competing with a fast casting rates is needed to be able to handle the cast part. In semi-automated pressure casting operations, the total cycle time is critical. The molds are used repeatedly, and unlike gypsum molds in a bench casting process, are saturated with water. The suspension is introduced to the pressure casting machine under pressure, thereby overcoming the capillary pressure in the molds and forcing any water in the molds out with the water from the casting suspension. Typical cycle times in the dinnerware industry are less than three minutes, and include: closing the mold, filling with slip, casting, opening the mold, removing the cast piece, purging the mold, then closing the mold again (thus completing the cycle). When the mold is opened, the cast part must be ready to be handled, if not, the process is delayed. Handleability is directly related to the water retention level in the cast layer, which is also the void fraction in the cast layer (ε_c), and can be represented mathematically as:

$$\varepsilon_C = 1 - x_p - \left(\frac{X_S^0 - X_S}{X_C} \right) x_p . \tag{4}$$

Prior to this work, the instrumentation used to measure pressure casting performance tended to be both incomplete and inefficient, particularly with regards to the needs of pressure casting applications. The objective of this research was to develop an instrument for testing pressure casting slips which could be used to accurately and efficiently test suspensions on a laboratory scale before introduction to a manufacturing environment. The development of this instrument obviously required an iterative process involving testing, comparison to other data, and refinement of both the instrument mechanics and the testing protocols. Additional experiments are still needed to assess the effects of applied pressure, viscosity, solids content, dispersant type and content, and particle size distribution of whitewares on their casting rate. Another area of need is in the testing of various mold materials to encompass those used in an industrial environment.

Construction of the Pressure Casting Instrument

Six units have been produced and installed at Whiteware Research Center member companies. Figure 2 is a photograph of the instrument. The instrument consists of a sample chamber, mold, piston (driven by compressed air), a pressure sensor, and an LVDT to measure suspension height by monitoring the piston position.

The cell was also designed to eliminate a common problem associated with previous pressure casting instruments—the incorporation of a bubble at the piston-suspension surface. Using a flow-through system for sample loading, bubble generation has been eliminated. Unexpected problems (such as the sudden release of the test suspension at high velocity through sealing surfaces) were eventually eliminated, as well. One remaining valid criticism of the new instrument is the difficulty of cleaning the test fixture following a test. The cleaning issue restricts the test cycle turn-around time.

To provide for user-friendly data collection, a Windows-based, visual basic computer code was written that asks the operator for specific information. This information is necessary for subsequent calculations, but also to provide for complete sample identification. Each test is automatically assigned a filename to prevent data corruption by file over-writing. Microsoft Access and Excel are used for data analysis and to present the data as a standard data sheet (through the use of a specifically designed template). All run parameters are tabulated on the data sheet, with the raw data automatically filed for future analysis. Previous testing data can also be re-evaluated at a later date and re-saved under a different filename. The computer program collects data from the LVDT and the applied pressure versus elapsed time. The data is displayed in real time in both a linear time format and a square-root of time format. When the suspension height becomes independent of time, usually represented as an abrupt change in slope, the casting process is complete.

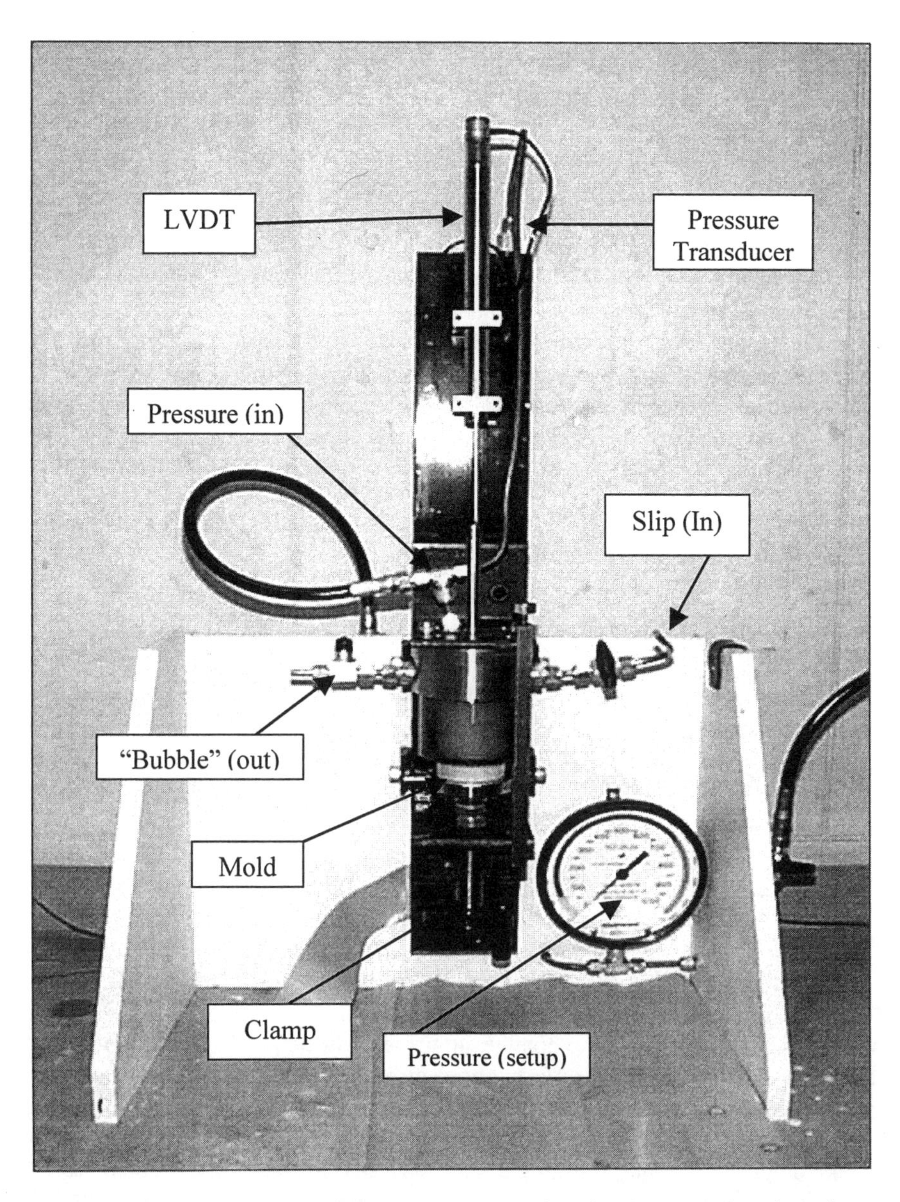

Figure 2. Photograph of the pressure casting instrument showing the general arrangement, the LVDT for measuring the suspension height, the flow-through sample loading system, and the test chamber. A piston, driven by compressed air, applies pressure to the suspension to drive the casting process.

Evaluation of the Pressure Casting Instrument

Once the instrument was fully operational, experiments were conducted to assess the effects of applied pressure on the casting rate of whiteware bodies. To determine the effect of dispersion degree on casting behavior, a typical whiteware body suspension (see Table I) was evaluated at ten dispersion levels, in a random sequence, at 300 psi and a solids loading of 53 v/o. In addition, one suspension was tested three times to check for repeatability and reliability. Another batch was run at three pressure levels (200, 300, and 400 psi) to test for effects of applied pressure on pressure casting rate.

The mold was cut from a polymer-based pressure casting mold for dinnerware production. The specific composition is proprietary, and was not known for these experiments. Several mold samples were prepared at the same time, and the test results indicated they behaved uniformly. (This indicates, however, another use for the instrument—mold evaluation and characterization.)

Table I. The composition (dry weight basis, d.w.b.) for the porcelain composition used in the pressure casting evaluations.

Component	Suppler	%
Kaolin	*EPK* Zemex Minerals, Inc., Monticello, GA	25
Ball Clay	*Marquis* United Clays, Inc., Brentwood, TN	25
Quartz	325-mesh Ogledbay Norton Industrial Sands, Inc., Glenford, OH	25
Feldspar	*G-200* Zemex Minerals, Inc., Monticello, GA	25

RESULTS AND DISCUSSION

Figures 3a and b show the change in suspension height as a function of linear and the square-root of time, respectively. The data is collected and displayed in both forms. The change in suspension height has been normalized and is represented as a positive change, although in actuality, the suspension height is decreasing. When the change in suspension height becomes independent of time, the consolidation process is complete and the suspension height is now the cake height. The data in Figure 3 was generated in a test cell of thickness 12.5 mm (½ inch), so the overall decrease in the suspension height was roughly 40%. Although at a Na-PAA (Darvan 811, R.T. Vanderbilt, Norwalk, CT) dispersant level of 0.7 mg/m^2 (an excess dispersant situation,[3] as illustrated in Figure 4, which obviously does not correspond to any reasonable industrial situations), the curves show typical casting rate behavior for slip casting.

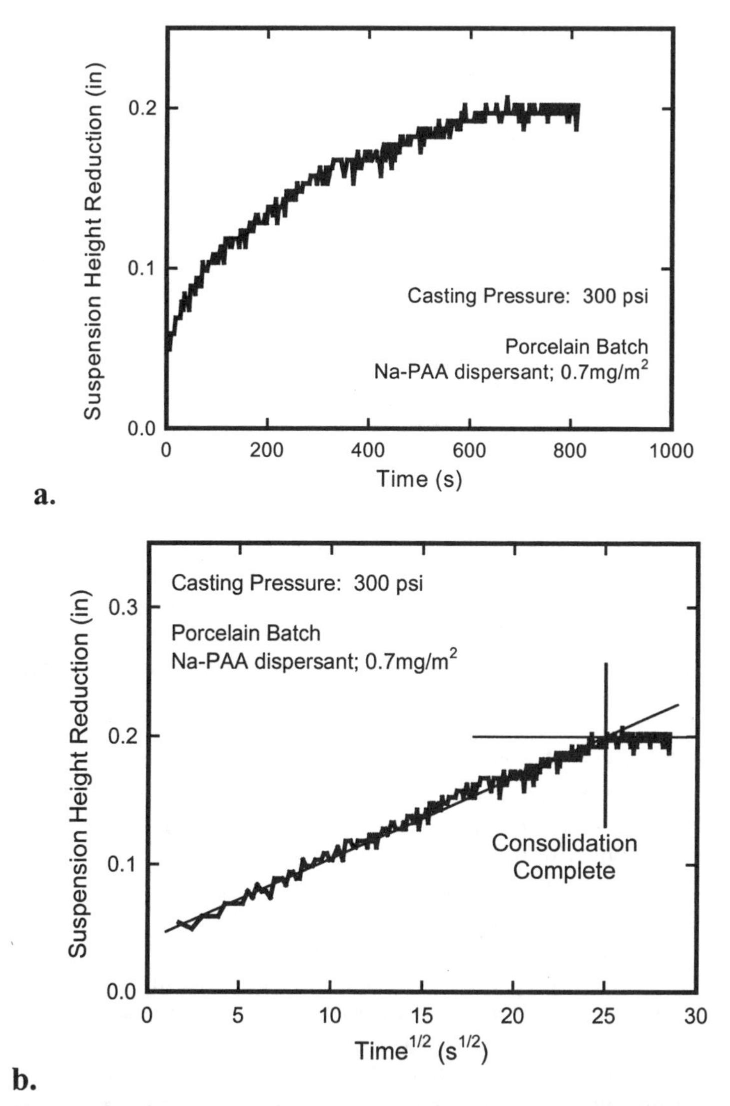

Figure 3. The decrease in the suspension height with (a) casting time and (b) the square-root of time. A dispersant level of 0.7 mg/m^2 was used, and a casting pressure of 300 psi. The point at which the suspension height becomes independent of time indicates that the consolidation process is complete. It is obvious from these two plots that the change is more easily identified when the change in suspension height is plotted against the square-root of time.

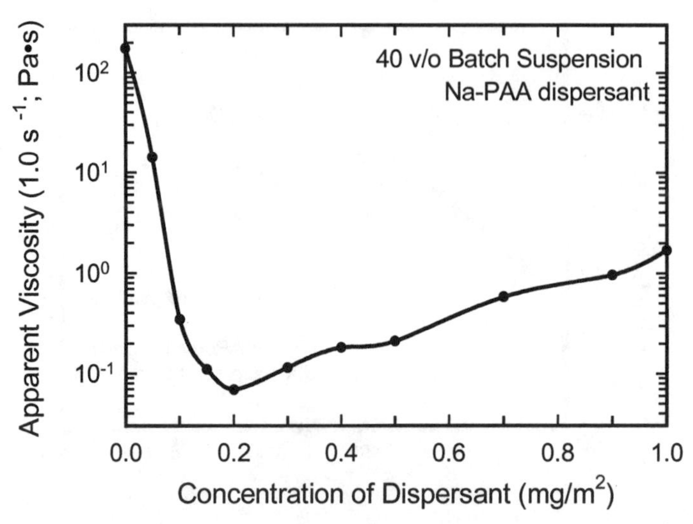

Figure 4. Viscosity as a function of dispersant concentration.[3] These rheology results are for a similar batch composition, but were measured at 40 v/o instead of the 53 v/o used in the casting studies. The exact apparent viscosity values will differ with the change in solids concentration (53 v/o having significantly greater viscosity), however the trends will remain similar.

Correlation of suspension rheology with casting behavior

One of the original goals of this research was to identify a quantitative link between suspension rheology and casting behavior. As demonstrated in previous work,[3] and consistent with industrial observations, the simplest and most reliable means of adjusting the rheology of porcelain suspensions is through the use of dispersants. Therefore, the casting rate was evaluated as a function of dispersant concentration, initially for the reduction in suspension height, as shown in Figure 5 for three Na-PAA dispersant levels: 0.05, 0.40, and 0.90 mg/m^2. Based on the data in Figure 4, 0.05 mg/m^2 is below the dispersant level necessary to create the most fluid suspension (shown to be 0.20 mg/m^2), and is close to the level of dispersant used industrially. Both 0.40 and 0.90 mg/m^2 additions constitute excess dispersant levels, and although not applicable industrially, represent the coagulated state, due to increased cation (Na^+) levels from the dissociation of the Na-PAA.[4]

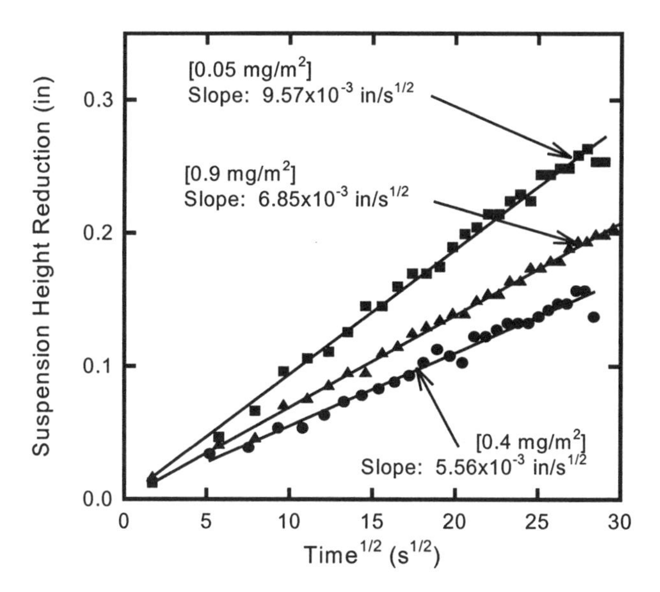

Figure 5. A comparison of change in suspension height as a function of the square root of casting time for three dispersant levels.

The data in Figure 5 is the suspension height reduction, not the rate of build-up of the consolidated layer (the true casting rate). To convert to casting rate, the system parameter, ***n***, is necessary (Equation 2), and that requires knowing, by measurement or by calculation, the void fraction in the consolidated layer, ε_C. Equation 4 shows that ε_C can be calculated knowing the initial and final dimensions of the suspension height and the consolidated layer, and the volume fraction of particles in the suspension. It is also necessary to know the rate of change of the suspension height (denoted ***m*** in Equation 3).

To determine the cast layer pore volume, the cast cake is weighed after the test is completed, dried, then re-weighed. The difference is the amount of water in the cake, and is equivalent to the pore volume in the as-cast cake assuming no air bubbles are present. Measured cast layer pore volumes were used to calculate the cast layer thicknesses as a function of $\sqrt{t}$ presented in Figure 6. The slope of the thickness versus $\sqrt{t}$ curve demonstrates the rate of cake growth, and thus the casting rate. Knowing the casting rate would allow accurate prediction of the wall thickness for drain casting applications.

The change in cast layer thickness is plotted as a function of dispersant addition in Figure 7. This curve clearly illustrates that casting rate correlates strongly with dispersant addition level, showing that flocculated and coagulated suspensions have similar casting rates, while dispersed suspensions have low casting rates.

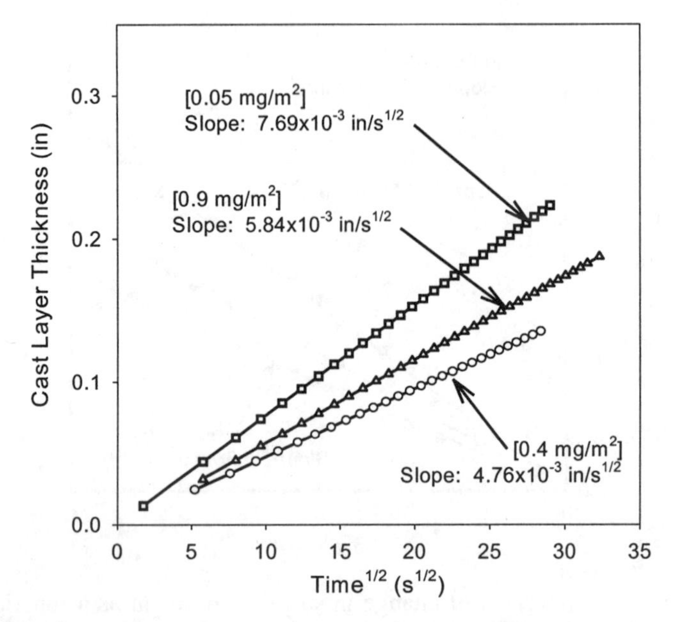

Figure 6. Cast layer thickness, calculated using Equation 3, plotted as a function of the square root of time for 0.05, 0.40, and 0.90 mg/m^2 Na-PAA dispersant addition levels.

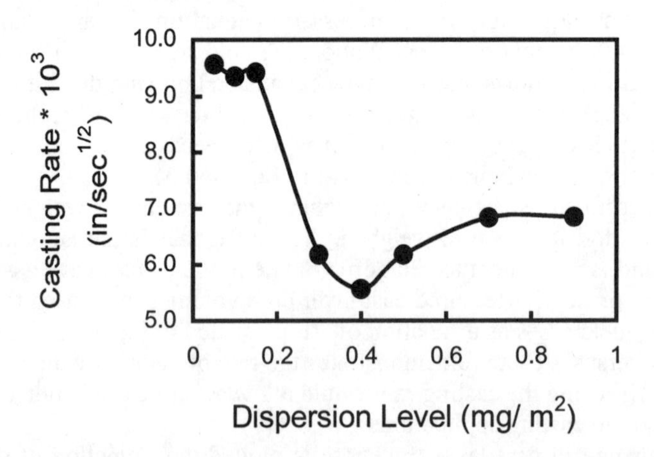

Figure 7. Casting rate is shown as a function of dispersant concentration demonstrating a close similarity to the trend with apparent viscosity presented in Figure 4.

The instrument software was designed to automatically calculate the water retention level in the cast layer, but this approach generated some questionable data, as illustrated in Figure 8. Subsequent analysis revealed that the calculation of the final cake pore volumes was extremely sensitive to the instrument calibration and setup, and work is ongoing to refine the void fraction determination, as other problems may also be present. In addition, it was noticed throughout these experiments that the water retention levels did not correlate well with the change in suspension properties as affected by the change in dispersant levels. It is expected that flocculated suspensions would exhibit fast casting rates, with corresponding high water retention levels, while dispersed suspensions would exhibit slower casting rates with low water retention levels. This discrepancy is also being investigated.

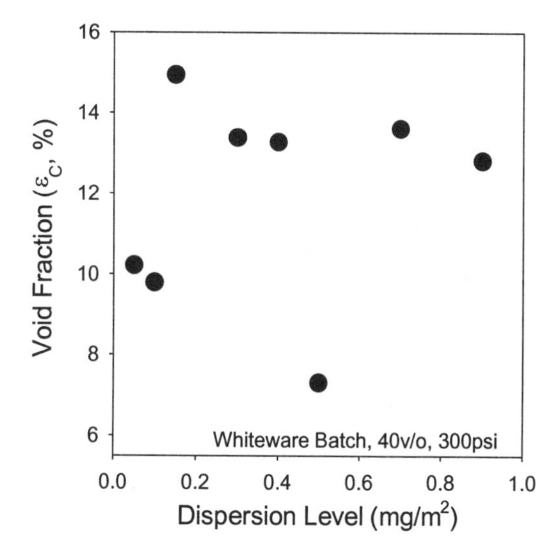

Figure 9. Void fraction (water retention level) is plotted against the dispersion level. The data does not follow the expected trends, specifically that void fraction should be inversely related to casting rate. Additional work is in progress to refine the model and the calculations.

Effect of pressure on casting rate

Intuitively, it was expected that casting rate would increase with applied pressure. To evaluate this, samples were cast at different pressures, with the results presented in Figure 10. The effects of applied pressure were studied for one porcelain body suspension at a Na-PAA dispersion level of 0.40 mg/m^2. Three pressures were evaluated, 200, 300, and 400 psi. (All previous data was collected at 300 psi.) A direct relationship was found between applied pressure and casting rate, a linear rate of increase was observed.

(According to discussion following the presentation at the conference, it was stated that the data should not necessarily exhibit a linear dependence on pressure, and that this result needs to be further evaluated.[5] It is probable that over the range of pressures tested, a linear approximation may be reasonable.)

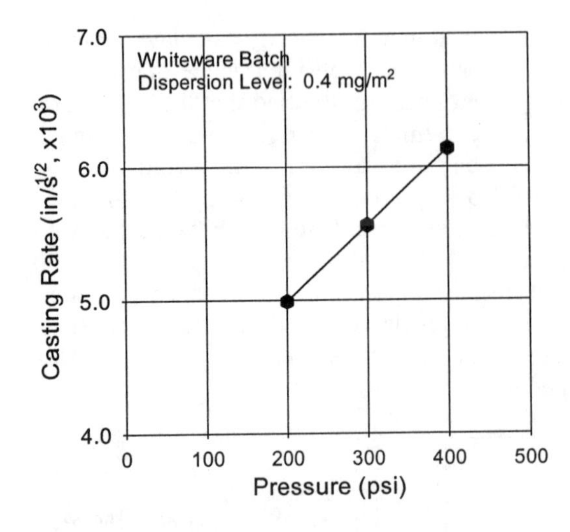

Figure 10. Casting rate exhibits a linear dependence on applied pressure. Discussion at the conference indicated that this would not necessarily be the expected result and warrants further investigation.

CONCLUSIONS

A laboratory test instrument was developed to characterize samples for pressure casting applications. The casting rate and the water retention level are automatically calculated from the data sets using the instrument software. Although the initial data is promising, additional work is necessary to account for discrepancies in the cast layer water retention level compared to the expected trends, and the pressure dependence needs to be further investigated. Other experiments to be conducted include the evaluation of mold materials, addition evaluation of the correlation with suspension rheology, and the effect of other parameters such as solids loading and temperature.

REFERENCES

1. I. A. Aksay and C. H. Schilling, "Mechanics of Colloidal Filtration," Advances In Ceramics, *Ceramic Fabrication Processes*, American Ceramic Society, **9,** 85-93 (1984).
2. D. S. Adcock and I. C. McDowall, "The Mechanism of Filter Pressing and Slip Casting," *J. Amer. Cer. Soc.*, **40** [10], 355-62 (1957).
3. W. M. Carty, K. R. Rossington, and U. Senapati, "A Critical Review of Dispersants for Whiteware Applications;" this volume.
4. K. R. Rossington and W.M. Carty, "The Effects of Ionic Concentration on the Viscosity of Clay-Based Suspensions," this volume.
5. B. Kellett, discussions following the presentation, *Science of Whitewares II*, June 1, 1998.

OEDOMETER EXPERIMENTS TO DETERMINE SLURRY PROPERTIES

B. J. Kellett and D. N. Ravishankar
Department of Materials Science and Engineering
University of Cincinnati
Cincinnati, OH 45221-0012

ABSTRACT

Slip casting, pressure casting, drying, and other dewatering processing operations are ultimately defined by permeability and compressibility. An instrumented oedometer is described which measures compressibility, yield curve, and permeability of cakes and slurries. These cakes are formed within the oedometer from slurries, under conditions similar to slip casting and pressure casting. Experiments with flocculated α-Al_2O_3 slurries are used to demonstrate these concepts. A simulation of the cake growth process is shown, using permeability and yield curve characteristics determined from the oedometer experiment. It is shown that large density gradients can develop during dewatering.

INTRODUCTION

Most ceramic materials are fabricated by pressureless sintering. For many applications dimensional tolerances must be maintained to less than 0.5% to preclude costly and time consuming diamond grinding Non-uniform powder packing causes either shape change or creates large crack like flaws during sintering.[1,2] Either outcome must be avoided in a reliable, low cost processing operation. Shape change causes a loss in dimensional tolerance while large flaws greatly reduce component strength. A complete understanding of the underlying physics of powder consolidation and forming operation can be obtained by a granular mechanics approach. Using this approach, a test method is described for measuring relevant slurry characteristics which determine the powder packing uniformity of slip cast cakes. Variations of this approach has been discussed elsewhere.[3-7]

Slurries are commonly characterized by average particle size, viscosity, pH and density. While these characteristics have important effects on processing behavior, they are at best secondary parameters. While reliable relationships between these characteristics and process behavior (i.e. cake uniformity and porosity distribution) may exist, they are at best empirical, for they do not control the underlying physics of cake formation.

SEDIMENTATION THEORY OF CAKE GROWTH

In the absence of shear, the two characteristics which govern the consolidation behavior of granular materials are powder consolidation behavior and permeability. This is apparent from the governing equation of porous media based on Darcy's law:[8]

$$\vec{q} = -\frac{K}{\eta}\nabla u_e \tag{1}$$

The volume flow rate of liquid divided by the cross sectional area (the liquid flux q) proportional to an "excess" liquid pressure gradient (∇u_e). The liquid flow is down the gradient in excess liquid pressure. The liquid pressure is the excess of the absolute liquid pressure, which includes the hydrostatic component produced by the self weight of the permeant and suspended particles. For this article zero pressure is ambient. The proportionality constant of equation 1 defines the permeability (K) divided by the viscosity of the permeant (η). Only a pressure gradient in "excess" of that created by the weight of the liquid phase will generate a liquid flux q. The total liquid pressure gradient includes both the Darcian pressure drop, which is the excess liquid pressure gradient ∇u_e, and the static liquid pressure gradient ∇u_s equal to the weight of the permeant (liquid):

$$\nabla u = \nabla u_e + \nabla u_s = -\frac{\eta}{K}\vec{q} + \rho_l \vec{g}_o \tag{2}$$

where ρ_l is the density of the liquid. The liquid pressure increases in the direction of gravity and decreases in the direction of the liquid flux and gravity respectively. This self weight term is usually negligible and is self compensated when a differential pressure gauge is used. For low flow rates and consolidation measurements as discussed in this article a differential pressure mode is important. For example, for an aqueous slurry $\rho_l g_o$ is about 10 kPa/m, while ∇u_e is on the order of 100 kPa/m for the flow rates and permeability of typical slurries ($K=10^{-14}$ m^2).

For a slurry undergoing unimpeded sedimentation, sometimes called 'plug' sedimentation, the liquid pressure gradient includes the weight of suspended powder particles:

$$\nabla u = \rho_s \vec{g}_o = (\rho_l + \phi(\rho_p - \rho_l))\vec{g}_o \tag{3}$$

where ρ_s is the density of the slurry, ϕ is the solid volume fraction and ρ_p is the density of the solid particles. For a settling powder the excess liquid pressure is thus

$$\nabla u_e = \phi_s(\rho_p - \rho_l)\vec{g}_o \tag{4}$$

This applies to a powder which is not supported or connected to the container. For a 20 vol.% alumina aqueous slurry, du_e/dz = 6 kPa/m. As a reference the pressure gradient produced by water at 20°C is $du/dz = \rho_l g_o$ = 9.8 kPa/m. From the reference frame of the container, this liquid pressure gradient produces an upward liquid flux, which can be related to a particle sedimentation velocity.[9]

$$\vec{q} = -\phi(1-\phi)^p \frac{2}{9}(\rho_p - \rho_l) r_p^2 \frac{\vec{g}_o}{\eta} \quad (5)$$

Stokian settling is assumed, with the hindrance factor p, as discussed by Buscall and White.[10] The calculated liquid flux for an aqueous alumina slurry ϕ_s=0.2, r_p=0.25μm, and no hindrance p=0, is q=0.33 μm/s. For hindered slurries with p=5, q=0.11 μm/s.

The above results apply to dispersed powders which are suspended within the liquid. As the powders settle they become supported by an interconnected network to the bottom of the container or filter. Both the weight of the powder particles in the cake and the Darcian liquid pressure gradients are supported by these interparticle forces. It is the action of these forces, and how these forces increase that leads to the post deposition consolidation of the cake. These forces consolidate the cake as dictated by the cake's consolidation properties. These stresses within the cake are conveniently discussed in terms of an average 'effective' stress σ', defined as the 'equivalent' stress over the entire cross sectional area of the cake. The total stress is divided into two parts: one that is supported by the particle skeleton network (σ') and the other, the liquid pressure (u) as follows

$$\sigma = \sigma' + u \quad (6)$$

Following the convention in the solid mechanics literature, the total stress, the liquid pressure and effective stress are positive and compressive. Darcian flow leads to an effective stress gradient in the direction of liquid flow. The gradient in the total stress in the direction of fluid flow varies with the specific weight of the slurry:

$$\nabla\sigma = \nabla\sigma' + \nabla u = (\rho_l + \phi(\rho_p - \rho_l))\vec{g}_o \quad (7)$$

which leads to

$$\nabla\sigma' = \frac{\eta}{K}\vec{q} + \phi(\rho_s - \rho_l)\vec{g}_o \quad (8)$$

where K is the permeability, η is the liquid viscosity.
This effective stress gradient is the root cause of non uniform consolidation. The cake is consolidating from the stresses applied to the cake from the liquid pressure gradient. This can be understood from the perspective of the interaction forces of isolated particles which are brought closer by the increasing liquid pressure drop. As the distance between them decreases, the inter particle forces of repulsion increase and create stresses which counteract the liquid pressure. After a certain stage, the particles cannot be brought any closer and the increase in pressure drop is fully balanced by these inter particle forces and a steady state is reached.

From the above equations it is clear that these two characteristics--permeability and particle packing density--are central to understanding ceramic forming behavior. Equations based on Darcy's law require these two macroscopic characteristics to be known over a range of conditions. These two characteristics are now discussed.

PERMEABILITY AND YIELD CURVE

Theoretical expressions for permeability fall within two models: capillary models and drag models. The Kozeny-Carman model is derived from capillary flow models and is defined by the specific surface area per unit volume of solid (S_V) and the solid volume faction of the cake (ϕ_C):

$$K = \frac{1}{\zeta S_v^2} \frac{(1-\phi)^3}{\phi^2} \tag{9}$$

ξ is the tortuosity constant (typical values between 3 and 6) which is a measure of the increased distance liquid must travel to pass through the cake. For monosized spheres S_V is equal to 3/r. For a distribution in particle size, $S_v = \Sigma 3/r_i$. Using an average particle size to calculate S_v can lead to a serious overestimation of the permeability. For example the A-16 S.G. (ALCOA) powder used in this study has an S_v=33 μm^{-1} and an average diameter of 0.48μm as determined by a Sedigraph (Micromeritics Inc.).

Permeability equations are functions of particle packing density. Darcy's law defines permeability (K), which is a central issue in the study of flow through porous media. The central issue for ceramic processing is powder packing uniformity. The yield curve defines the relationship between particle packing density and effective stress and is central to understanding ceramic process behavior. Thus the yield curve is a central connecting element to determining processing behavior and the ultimate uniformity of a ceramic compact formed from slurries.

The consolidation behavior of powders is highly nonlinear and frequently found to increase to the log of the effective stress:

$$\phi = \phi_O + \phi_\sigma \log\left(\frac{\sigma_Z'}{\sigma_O}\right) \tag{10}$$

where ϕ is the solid volume fraction, and σ'_Z is the uniaxial stress. Kellett and Lin have used the following yield curve with both high and lower density limits:[5]

$$\phi = \left(\phi_m\left(\frac{\sigma'_Z}{\sigma_O}\right)^n + \phi_1\right)\left(1+\left(\frac{\sigma'_Z}{\sigma_O}\right)^n\right)^{-1} \tag{11}$$

The stress σ_o marks the point of the yield curve in which the particle packing density increases most rapidly. While the stress σ_O also corresponds to the stress at the average solid fraction $(\phi_m+\phi_l)/2$, the exponent n determines the sharpness of the transition from ϕ_l to ϕ_m. The stress σ_o would represent an apparent strength of flocs, suggesting, from a processing point of view, that one would like the full compact to be well above this effective stress to produce a high density and uniform powder compact.

The derivative of the yield curve is defined as the compressibility in soil mechanics literature. The m_v compressibility has the following definition:

$$m_v = \frac{1}{\phi}\frac{d\phi}{d\sigma} \tag{12}$$

For yield curves 10 and 11:

$$m_v = \frac{\phi_\sigma}{e}\frac{1}{\phi\sigma} \tag{13}$$

$$m_v = n\left(\frac{(\phi_m - \phi_l)(\phi - \phi_l)}{(\phi_m - \phi_l)}\right)\frac{1}{\phi\sigma} \tag{14}$$

Note that m_v is a slightly different measure of compressibility than a_v coefficients defined in an earlier study from the deformation strains.[5]

CAKE CONSOLIDATION

For flocculated slurries and cakes the classical picture of cake growth is incomplete. This is because once cakes form, pressure gradients develop which lead to further consolidation. This time dependent consolidation behavior is central to the study of soil foundations and a large literature in soil mechanics has developed to analyze the consolidation behavior of powders saturated with fluid acted on by external loads and pressures. This literature is also relevant to ceramic processing.

In the limit of small deformations, solid mechanics theory leads to the simplified Terzaghi equation:[11]

$$\frac{\partial u_e}{\partial t} = C_v \frac{\partial^2 u_e}{\partial z^2} \tag{15}$$

Where the consolidation constant C_v is a common design parameter in soil mechanics:

$$C_v = \frac{K}{\eta}\frac{1}{m_v} \tag{16}$$

As the Terzaghi theory assumes small deformation strains, it is common to approximate the strains as proportional to the effective stress. As $du_e=-d\sigma_z$ equation 15 can be rewritten in terms of deformation strain.[12]

$$\frac{\partial \varepsilon}{\partial t} = C_v \frac{\partial^2 \varepsilon}{\partial z^2} \tag{17}$$

Equations 15 and 17 are equivalent in form to the diffusion and heat transfer equation, and solutions for various boundary conditions are widely available.[13,14] One solution of the Terzaghi theory is of particular interest: The sudden application of a constant applied load. The constant applied load solution applies to the oedometer experiment, a common soil mechanics experiment.[11] In this technique a constant load is suddenly applied to a uniform cake to produce an axial stress of P. For an initially uniform cake of thickness d, the liquid pressure through the cake is given by the following solution:

$$u_e = P \sum_{m=0}^{\infty} \frac{2}{M} \sin\left(M\frac{z}{d}\right) \exp\left(-M^2 T_v\right)$$

$$M = \frac{\pi}{2}(2m+1) \tag{18}$$

$$T_v = C_v \frac{t}{d^2}$$

Neglecting the self weight of the cake, the effective stress is $\sigma_z' = P - u_e$. The particle packing density is found accordingly from the yield curve. The total liquid pressure drop across the cake is found by integrating equation 16 from z=0 to the cake thickness d. It is common to measure total deformation strains as proportional to the effective stress. An equation for total deformation strain:

$$\varepsilon_T = 1 - \sum_{m=0}^{\infty} \frac{2}{M} \exp\left(-M^2 T_v\right) \tag{19}$$

is well approximated by:

$$\varepsilon_T = \sqrt{\frac{4}{\pi} T_v} \quad \text{when } \varepsilon_T \text{ is} <0.6 \tag{20}$$

$$\varepsilon_T = 1 - 0.8108 \exp\left(-2.48 T_v\right) \text{ when } \varepsilon_T \text{ is} >0.6 \tag{21}$$

EXPERIMENTAL PROCEDURE

The equipment used for this study consists of a consolidometer (Figure 1) capable of operating at stress ranges ranging from 0.6 kPa to 200 kPa. It is made up of a plastic tube in which the ceramic slurry is contained between two porous filter plates. The bottom plate is fixed and the top plate is closely machined and floats on top of the slurry.

Fine filter papers of pore size 0.22 μm are fixed on the plates to contain even the most dispersed slurries. These effectively separate the slurry from water that is filled on both sides of the slurry and ensure that, when the syringe pump is on, water alone is transferred while the ceramic solid content is confined within. The syringe pump creates the liquid flux for consolidation by operating at flow rates varying from 0.3 to 30 μm/s. The suction pressure of gypsum mold is of the order of 100 kPa which is within the pressure range of this set-up.

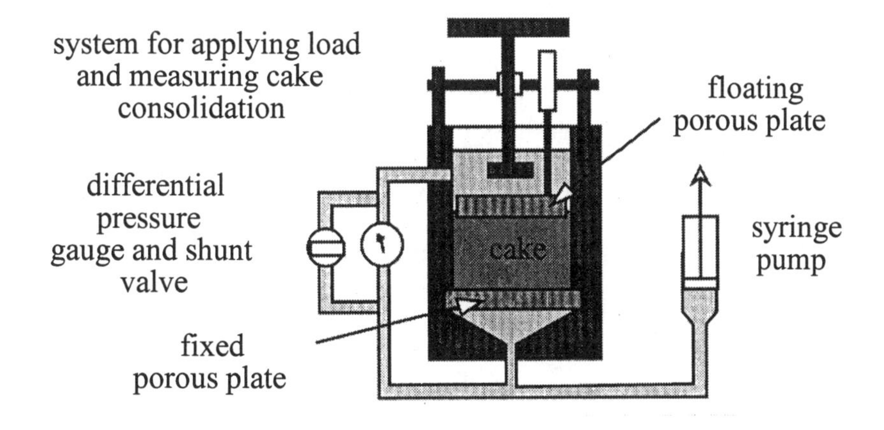

Figure 1. Instrumented oedometer for measuring densification and permeability of cakes under constant flow rate conditions.

Oedometer experiments were undertaken with commercially available alumina powder (A-16 S.G.). Aqueous slurries were prepared in plastic bottles and agitated with high purity alumina media by a SWECO vibratory mixer for about 1 hour to produce compressible slips of 20 vol. % solids. These experiments were conducted with straight DI water and flocculated slurries of viscosity ~1500 mPa·sec. The liquid pressure difference between the top and bottom of the cake was measured using a variable reluctance differential pressure transducer (DP15, Validyne Engineering continuously measured using a linear variable differential transformer, B50, Solartron Metrology, Buffalo, NY). The pressure transducer was connected in parallel with a shunt valve.

EXPERIMENTAL RESULTS

In the first experiment, slurry is poured into the consolidometer to a height of 1.93 cm. The floating top plate was applied to the surface of the slurry. Shown in Figure 2 is a typical pressure/displacement -time response for the self consolidation of a fresh slurry (20 vol.% solids, η=1.5 Pa-s, 1.9cm initial height). The weight of the floating top plate applies an equivalent stress of 0.52kPa to the top of the slurry. The weight of the powder in the slurry produces an additional pressure of 0.056 kPa. The negative pressure shown in Figure 2 implies that the pressure beneath the cake is greater than above. As this is a stable flocculated slurry, the weight of the powder particles and the top plate is already partially self supported by the cake at the beginning of the experiment (about 1 minute after pouring in the slurry). Steady state is reached in about 4 hours at a displacement of 3 mm.

Consolidation slows and stops as the liquid pressure difference across the cake approaches zero. One way of looking at this condition is that the effective stress profile across the cake has reached the yield curve values. The displacement rate, measured by the LVDT, is equal to the liquid flux rising through the cake and passing through the top filter plate. The variation in the liquid flux through the cake will depend on the consolidation characteristics of the cake.

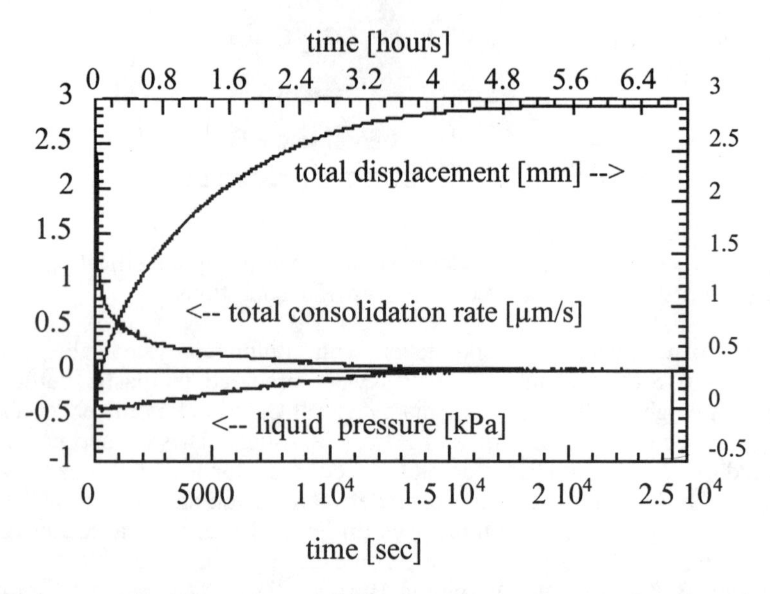

Figure 2. Liquid pressure and displacement response of a flocculated slurry with an imposed stress of 0.6kPa.

Shown in Figure 3 is an analysis of this experiment. As the filter plate pushes into the slurry the particles collect on the filter and form a cake. Also particles may settle out of the slurry onto the bottom filter plate. If the cake is uniform then the liquid flux through the cake would also be uniform throughout and equal to the consolidation rate measured at the top plate (Vp), minus the settling rate of the powder particles in the slurry. The excess liquid pressure gradient is defined by Darcy's law and equation 2. The liquid pressure and excess liquid pressure has been arbitrarily set to zero at the base of the top filter plate. The excess liquid pressure gradient in the slurry is related to the settling rate of the powder particles as given by Equation 4. The static liquid pressure increases by the weight of the slurry as given by equation 5. For a 20 vol.% alumina aqueous slurry $du_e/dz = 6$ kPa/m. The liquid flux for an aqueous alumina slurry ϕ_s=0.2, r_p=0.25 μm, and no hindrance p=0, is q=0.33 μm/s. For hindered slurries with p=5, q=0.11 μm/s. These fluxes would correspond to the total consolidation rate, as shown in Figure 2. The bottom cake formed by powder settling would have a liquid flux of zero, as liquid is not being displaced by moving powder particles. The excess liquid pressure gradient would be zero, and the static liquid pressure gradient would accordingly be the gradient produced by water, which at 20° is $du/dz = \rho_l g_o = 9.8$ kPa/m.

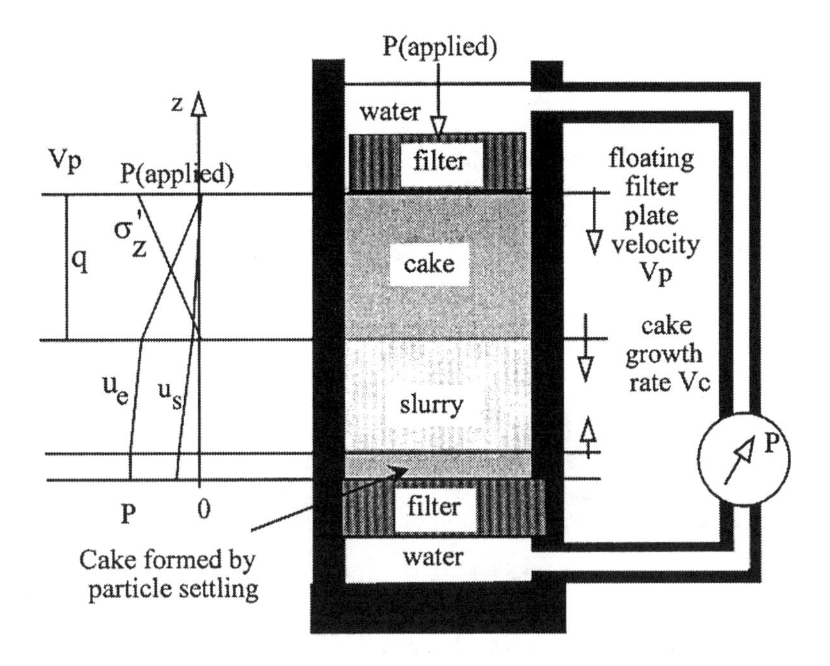

Figure 3. A hard sphere model of cake growth, consistent with the traditional theory of slip casting. Excess liquid pressure (u_e), static liquid pressure (u_s), effective stress (σ_z'). An applied force producing an effective stress of P is applied on the top floating plate.

The liquid pressure across the cell measured by the differential pressure gauge would initially equal the sum of the applied pressure P and the weight of the settling powder particles in the slurry. Initially, the pressure produced by the weight of the top filter plate is supported by the resistance of liquid flow through the top cake. The hard sphere model shown in Figure 3 would show a gradual decrease in differential liquid pressure given by the following formula:

$$P = P_{applied.} + \phi_s \Delta\rho g_o z_s \qquad (22)$$

where z_s is the overall thickness of the slurry. For the experiment shown in Figure 2, this initial pressure would be 0.576 kPa. The pressure would decrease with the slurry thickness z_s until the bottom and top cake contact. At which point the pressure would rapidly decrease from 0.52kPa to zero. At this condition the applied pressure P would be totally supported by the cake which is now resting on the bottom of the cell. As shown in Figure 2, the liquid pressure doesn't follow this behavior.

By this hard sphere model the total consolidation rate, which is the velocity of the top filter plate (V_p), is the sum of the settling velocity of the slurry ($v_{settling}$) and the rate in which liquid flows through the top cake:

$$V_p = \frac{K}{\eta}\frac{P_{applied.} - \Delta\rho\phi_s g_o z_s}{z_c} + V_{settling} \tag{23}$$

The settling rate of the slurry is equal to the liquid flux in the slurry. This would be the rate which the top surface of a slurry (slurry-supernatant) is settling. This rate can be calculated by Stoke's law with hinderance (equation 5), which for the conditions of the experiment shown in Figure 2 is between 0.11 μm/s and 0.33 μm/s depending on the degree of hindrance (value of p in equation 5). The total consolidation rate would decrease as the thickness of the top cake (z_c) increases, approaching the velocity of the settling rate of the slurry ($V_{settling}$). The top plate would abruptly stop when the top and bottom cake contact. This behavior is clearly not seen in our experiments. Further, these flocculated slurries are stable and do not settle. For colloidal dispersions which do not settle and form a bottom cake then $V_{settling}=0$ in Equation 23. Further, if we neglect the $\Delta\rho\phi_s g_o z_s$ term in Equation 23 (which is reasonable for $P_{applied.}>0.5$kPa, then it can be shown that the hard sphere model would predict the following displacement behavior:

$$z_p = C\sqrt{t}$$
$$C = \sqrt{2\frac{K}{\eta}\frac{\phi_c - \phi_s}{\phi_s}P} \tag{24}$$

While the square-root time behavior fits the displacement shown in Figure 2, realistic values for permeability ($K=10^{-15}$ m^2) and cake volume fraction ($\phi_\chi=0.4$) lead to a constant $C=3\text{x}10^{-5}$ m/√s, which is about three orders of magnitude of a best fit value of C=0.0102 m/√s.

We can go further with this hard sphere model and consider the effect of the friction created by the cake sliding over the die wall. A simple slab analysis calculation, assuming a friction coefficient of 0.5, suggests that sliding friction effects may reduce the applied effective stress by about 10%.[15] We will not consider this effect further. Nevertheless, it is clear that the flocculated slurry studied here are not well described by a hard sphere model.

The second approach is to apply the Terzaghi approximation, with the cake forming a continuous skeleton structure and the liquid flux increases from zero at the base to the velocity of the top filter plate, as shown in Figure 4. The effective stress and the "excess" liquid pressure is equal the applied stress P and at the top filter the liquid pressure is zero and the effective stress is equal to the applied stress P. Unlike the hard sphere model, the effective stress is transmitted through the cake to the bottom plate. Thus the total liquid pressure drop across the cake will decrease continuously over time, as found in the experiments (Figure 2).

Shown in Figure 5 is a solution to the experiment shown in Figure 2. For our α-alumina 20 vol.% slurry we find a constant of consolidation C_v=3x10^{-8} m^{2}s^{-1} with a least squares R^2 value of 0.998. The liquid pressure is also well represented by the Terzaghi solution as shown in Figure 5. The solution is based on Equation 21 and the assumption of a linear relationship between liquid excess pressure and consolidation strains. While it would be more consistent, it is slightly more time consuming to solve for the liquid pressure, calculate the local effective stress and then calculate the solid fraction from the effective stress and the yield curve. This approach also requires accurate values for the yield curve parameters, which are frequently not available. When yield curves parameters are known, then solutions are just as well found to fit the experimental data. For the experiment shown in Figure 2 we find a best C_v=9.5x10^{-9} m^{2}s^{-1}. A series of experiments under different applied stress are needed to determine the consolidation coefficient as a function of particle packing density. These more extensive results will be presented elsewhere.[15]

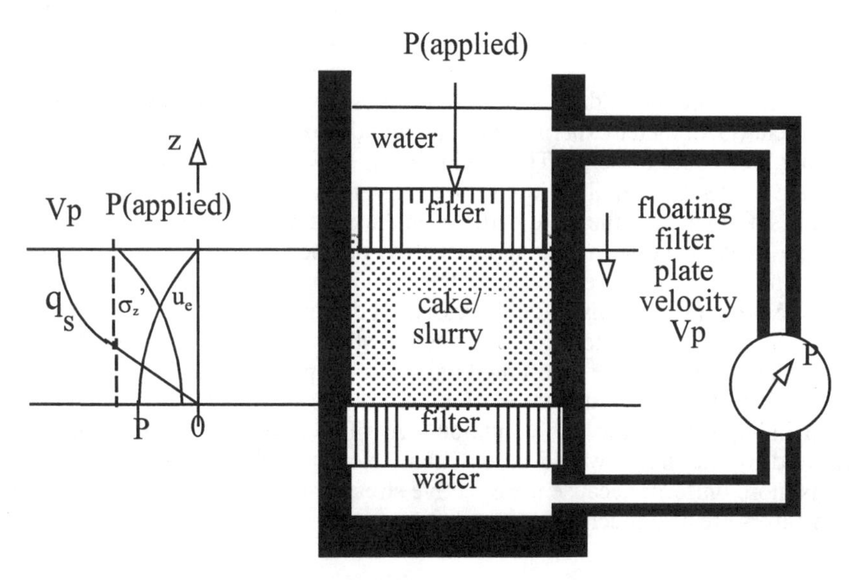

Figure 4. A continuous skeleton cake model, consistent with the traditional theory of slip casting. Excess liquid pressure (u_e), static liquid pressure (u_s), effective stress (σ_z'). An applied force producing an effective stress of P is applied on the top floating plate. For the effective stress profile shown,the measured differential pressure (P) would be less than the stress applied to the top filter ($P_{applied}$).

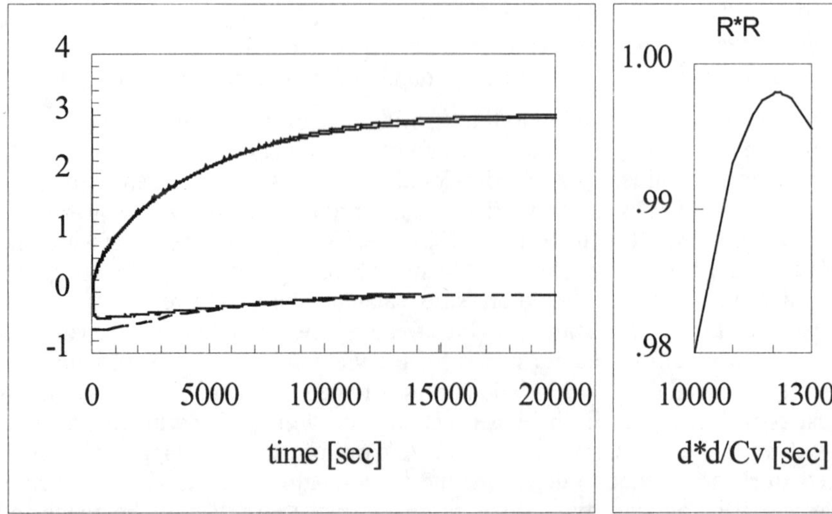

Figure 5. The Terzaghi solution for the experiment shown in Figure 2. The R^2 value for d^2/C_v are also shown. The best fit is for d^2/C_v =12150s which for the cake thickness leads to $C_v = 3 \times 10^{-8} m^2s^{-1}$.

Shown in Figure 6 is the calculated profiles for this constant applied load experiment (Oedometer experiment). The liquid flux is maximum at the top filter plate and decreases to zero. The liquid flux is upward through the cake and the top filter. This is a relative liquid flux. From the reference plane of the die, the liquid is stationary and the solid particles in the cake are moving downward. The effective stress is equal to the applied stress on the top filter, and decreases to zero. The liquid pressure is zero at the top filter and increases to the applied stress 0.56kPa. (The relative liquid pressure is defined as zero at the top plate for convenience.) The gradient in the effective stress creates a non-uniform cake. During the experiment the measured total liquid drop is equal to the applied stress until the effective stress increases from zero at the bottom plate. The profiles shown in Figure 7 show this condition. At 5000 seconds, the liquid pressure measured by the gauge would be 0.42kPa. Note that the cake is thinner. The cake is more uniform because the effective stress gradient has been reduced. The liquid fluxes are also much reduced.

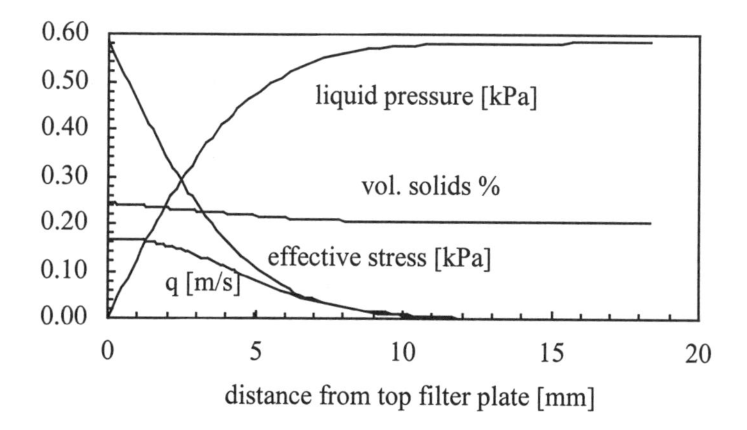

Figure 6. Terzaghi solutions for the excess liquid pressure (u_e), effective stress (σ_z'), liquid flux q and volume fraction solids (ϕ) for experiment 2. C_v=3 x 10^{-8} m^2s^{-1} at t=500s.

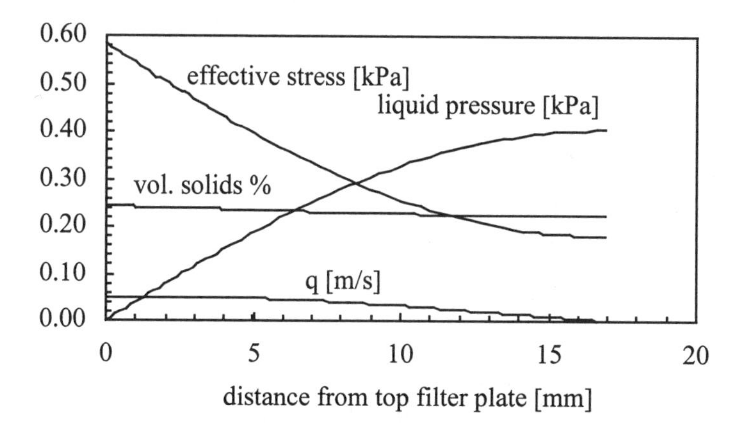

Figure 7. Terzaghi solutions for the excess liquid pressure (u_e), effective stress (σ_z'), liquid flux and volume fraction solids (ϕ) for experiment 2. C_v=3 x 10^{-8} m^2s^{-1} at t=5000s.

FINAL COMMENTS

It has been shown that a Kynch analysis is inappropriate for the flocculated alumina slurries oedometer experiments described here. A Terzaghi analysis based on a continuous cake model appears to well describe the experimental results. Slip and cake characteristics like permeability and particle packing density have been determined from these Oedometer experiments. The pressure readings along with the volume density and flux can be used to compute a range of parameters like permeability, particle packing density and yield curve behavior.

While the calculated results shown in Figures 6 and 7 strictly correspond to the Oedometer experiment, these results are applicable to constant pressure casting with minor changes in boundary condition. The results shown in Figure 6 and 7 that even under these very low applied stresses (0.5kPa) can lead to variations in particle packing density develop during cake formation due to the gradient in effective stress.

There are some practical difficulties encountered in the course of these experiments. The top filter plate has to seal and slide as the cake consolidates. It is difficult to ensure both of these functions, especially for highly dispersed slurries. This makes it more difficult to study the consolidation behavior of dispersed slurry. Long test duration may be uneconomical in an industrial environment. A series of experiments under different applied stress are needed to determine the consolidation coefficient as a function of particle packing density. These results are combined with other measurements to determine the two fundamental characteristics of slurries: permeability (K) and yield curve. These more extensive results will be presented elsewhere.[15]

REFERENCES

1. F. F. Lange, "Powder Processing Science and Technology for Increased Reliability," *J. Am. Cer. Soc.,* **72** [1] 3-15 (1989).
2. C. H. Schilling and I. A. Aksay, "Gamma-ray Attenuation Analysis of Packing Structure Evolution during Powder Consolidation," in *Ceramic Powder Science II*, Eds. G. L. Messing, E. R. Fuller, Jr., and H. Hausner (Am. Cer. Soc. Westerville, Ohio, 1988) pp 800-808l.
3. A. N. Abu-Hejleh and D. Znidarcic, "Estimation of the Consolidation Constitutive Relations," in *Eight International Conference on Computer Methods and Advances in Geomechanics* (IACMAG94), Eds. H. J. Siriwardane and M. M. Zaman, Morgantown, WV, pp. 499-504 (1994).
4. B. J. Kellett, "Application of Granular Mechanics to Ceramic Processing," Ceram. Eng. and Sci. Proc., **16** [3] 85-93 (1995).
5. B. J. Kellett and C. Y. Lin, "The Mechanics of Constant Rate Filter Pressing of Highly Flocculated Slurries," *J. Amer. Ceram. Soc.,* **80** [2] 381-93 (1997).
6. B. Kellett and D. Znidarcic, "Application of Granular Mechanics to Slip Casting, Filter Pressing, and Drying," *Science of Whitewares*, Eds. V. E. Henkes, G. Y. Onoda, and W. M. Carty, pp 133-146 (1996).

7. C.-Y. Lin and B. J. Kellett, "General Observations of Constant Flow Rate Filter Pressing," submitted *J.Am.Ceram.Soc. (1998).*

8. Darcy, H., "Les Fontainer Publiques de la Ville de Dijon," *Dalmont*, Paris (1856).

9. D N Ravishankar and B. J. Kellett, "Application of Kynch Sedimentation Theory to Slip Casting," to appear in *Ceram. Eng. and Sci. Proc.*, (1998).

10. R. Buscall and L. R. White, "The Consolidation of Concentrated Suspensions. Part 1 The Theory of Sedimentation," *J. Chem. Soc. Faraday Trans. I,* **83**, 873-891 (1987).

11. R. F. Craig, *Soil Mechanics*, 6th Edition, E&FN Spon (1997).

12. T. W. Lambe and R. V. Whitman, *Soil Mechanics, SI Version*, John Wiley & Sons, New York (1979).

13. H. S. Carslaw and J. C. Jaeger, *Conduction of Heat in Solids*, London Oxford University Press, 1947.

14. A. V. Luikov, *Analytical Heat Diffusion Theory*, Academic Press, 1968.

15. B. J. Kellett, "Application of Granular Mechanics to Ceramic Processing," *to be submitted J.Am.Ceram.Soc.*

INFLUENCE OF SPRAY-DRIED GRANULE MOISTURE CONTENT ON DRY MECHANICAL STRENGTH OF PORCELAIN TILE BODIES

José L. Amorós, Carlos Felíu, Enrique Sánchez, and Fernando Ginés
Instituto de Tecnología Cerámica. Asociación de Investigación de las Industrias Cerámicas. Universitat Jaume I. Campus Universitario Riu Sec. 12006. Castellón. Spain.

ABSTRACT

Test specimens were prepared at different pressing pressures and moisture contents from spray-dried granules used in porcelain tile manufacture, in order to determine the breaking mechanism in unfired porcelain tile bodies. A study of the effect of the granule's moisture content showed that although porosity considerably affected dry mechanical strength, in compacts with the same porosity, the ones formed from moister granules exhibited higher dry mechanical strength than those formed from drier granules. Finally, a semi-empirical equation is proposed, relating dry mechanical strength to porosity and granule deformability.

INTRODUCTION

A tile body's green mechanical strength is a key property in ceramic tile manufacture. Its importance lies in the fact that the materials must withstand the stresses that arise during pre-firing operations without deteriorating. Although studies abound in the literature on the mechanical properties of fired ceramic materials, few studies have appeared on failure in green tile bodies.[1,2]

The present study addresses the fracture mechanism exhibited by green tile bodies, as well as the effect of pressing variables (pressing pressure and moisture content) and porosity on the compact's dry mechanical strength.

EXPERIMENTAL

Materials

The tests were run on an industrial spray-dried powder agglomerate prepared from kaolinitic-illitic natural clays, potassium-sodium feldspar and feldspathic sand. 90 wt% of the agglomerate size distribution lay in the 200-600 μm range. True density of the material, and specific surface area (BET) were respectively 2.65 g/cm^3 and 12 m^2/g.

Experimental procedure

Powder agglomerate yield pressure was established by monitoring the increase in bulk density of the powder bed with rising pressing pressure (compaction curve). The yield pressure locus was defined as the point in the curve at which the first slope change occurred.[3]

Disk-shaped test specimens (diameter=400 mm, thickness=7 mm) were formed from the pressing powder by uniaxial pressing in a laboratory hydraulic press, at varying pressures (5, 10, 15, 20, 30, 50, 75 and 100 MPa) and different moisture contents (3.0, 5.5, 8.0 and 10.0 wt%).

Test specimen dry bulk density, ρ, was determined by measuring specimen dimensions and weight. This value was then used to calculate specific pore volume V_T (pore volume/solid volume). Dry compact strength was measured by diametrical compression. The measurements were run on a universal testing machine, loading at 1.0 $mm \cdot min^{-1}$. Microstructure was characterized by mercury porosimetry. Certain specimen fracture surfaces were examined by scanning electron microscopy (SEM).

RESULTS AND DISCUSSION

Fracture mechanism in green ceramic bodies

In order to determine how green tile bodies behaved during fracture, prism-shaped test specimens sized 1.5x8x0.7 cm with different porosities were subjected to three-point cross-bending tests, with constant loading at 1.0 mm/min, while continuously monitoring the relation applied load/resulting deformation. Figure 1 shows a typical curve, in which the following three stretches can be distinguished:

a) An initial non-linear area, probably caused by settling of the bending system and plastic deformation at the contact points.
b) A straight stretch, in which deformation was always a linear function of applied load.
c) A slightly curved stretch, in which deformation was not a linear function of load and which ended with specimen failure.

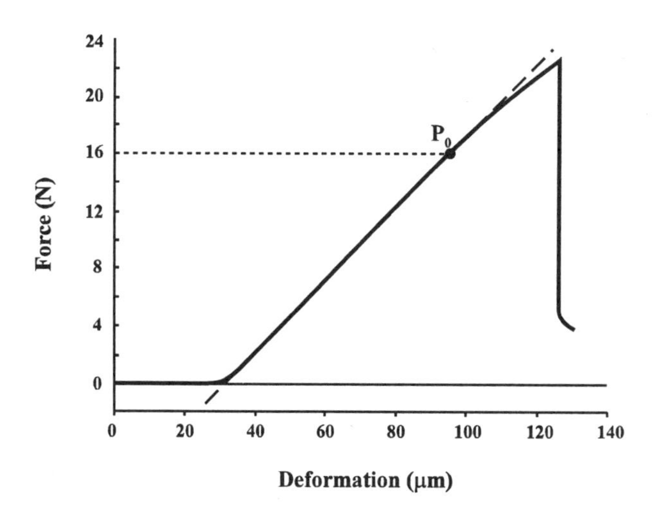

Figure 1. Load/deformation curve in a cross-bending test with specimen failure.

The point at which the values deviate from linearity, P_0, defines the material's elastic limit. On applying a load below this value, the arising permanent deformation was virtually negligible. Beyond this limit, plastic deformation rose appreciably, since on reducing the load the load/deformation curve did not return along the same path.

The elastic limit therefore indicated the point at which crack development and propagation started as a result of plastic deformation at pre-existing defects.[4] In flexural testing, failure occurred at or near the specimen surface subject to tensile stress. Specimen surface texture could therefore noticeably affect test outcomes. On exceeding the elastic limit, an existing surface crack or defect was likely to grow. Such a crack then branched out between agglomerates and particles, producing multiple cracks. Finally, when the crack system reached critical size C, failure occurred.

The material considered in the study did not exhibit purely brittle, but rather semi-brittle behaviour, as propagation of numerous cracks was required to produce failure. These conclusions are consistent with the Weibull modulus values obtained, and high critical deformation observed.[4] The Weibull modulus ranged from 20 to 55 (depending on the forming and pressing powder variables used), while critical deformation was found to lie at 0.4%.

Effect of pressing pressure and powder moisture content on compaction behaviour and resulting microstructure

Amorós et al.[5] derived a compaction equation based on the evolution of a material's porous structure with pressing pressure, which adequately describes the densification of a spray-dried powder bed with pressing pressure:

$$X = a_1\left[1 - exp(-kP)\right] + a_2 \cdot exp\left(\frac{-P_b}{P}\right) \qquad (1)$$

where:

$$a_1 = \left(V_{PG_O} / V_{T_O}\right) \qquad a_2 = \left(\left(V_{PP_O} - V_{PPmin}\right) / V_{T_O}\right)$$

X is a powder bed's degree of packing density on compacting at pressing pressure P.

V_{T_O}, V_{PG_O}, V_{PP_O} are respectively the specific total, intergranular and intragranular pore volumes of the non-compacted powder bed (pore volume/solid volume).

V_{PPmin} is the specific intragranular pore volume in a specimen formed at very high pressure (hypothetically at infinite pressure).

k is a parameter relating to powder compressibility.

P_b is the pressure bound beyond which intragranular porosity starts decreasing.

After verifying the validity of Eq.(1) with the pressing powder studied, the values were calculated for parameters k, P_b and attainable peak relative density of compact $\rho_{r\infty}$ (calculated from the value of a_2), at different moisture contents (H_P). Table I details the resulting values.

Table I.- Influence of H_P, on parameters k, P_b and $\rho_{r\infty}$

H_P (wt%)	k (MPa^{-1})	P_b (MPa)	$\rho_{r\infty}$
3.0	0.45	23.0	0.800
5.5	0.57	17.0	0.805
8.0	0.77	14.2	0.803
10.0	0.94	10.6	0.785

Figure 2 plots the experimental points of specimen relative densities ρ_r together with those computed from Table II data and from Eq. (1) (solid and dashed curves) versus pressing pressure at the foregoing pressing moisture contents. The calculated curves fitted the experimental data well.

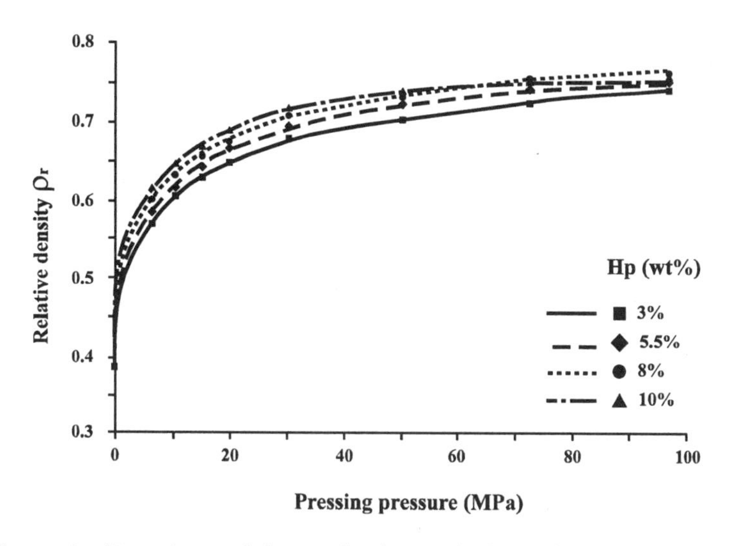

Figure 2. Experimental data and values calculated from Eq. (1) (curves) for specimen relative density.

The effect of pressing variables on packing density was similar to the effect widely described in the literature for non-clayey as well as for clay-containing powders.[3,6] Thus, on raising pressing pressure, relative density rose exponentially at every moisture content. At a given pressing pressure, raising powder moisture content facilitated particle flow, producing improved packing provided that water (a non-compressible fluid) was not found in excess inside the solid's porous structure. However, when this was the case, higher powder moisture content lowered packing density. This occurred at pressing pressures exceeding 50 MPa and moisture contents above 8 wt%.

As Table I shows, the variation of $\rho_{r\infty}$ with granulate moisture content H_p peaked at H_p=5.5 wt%. At lower H_p values, the specimen's excessive after-pressing springback considerably constrained the value of $\rho_{r\infty}$. However, at high agglomerate moisture contents, especially at H_p>8.0 wt%, as indicated above, water virtually saturated specimen pores during pressing at high pressure preventing reaching high densities.

As agglomerate moisture content H_p was raised, the material became more deformable and softer, yielding higher powder compressibility. This was expressed quantitatively by a rise in k and drop in the pressure bound P_b. This rise in powder compressibility with moisture content H_p produced higher packing densities at pressing pressures equal to or lower than industrial pressing pressures (<50 MPa).

Influence of pressing pressure and powder moisture content on dry mechanical strength

Figure 3 plots the variation of dry mechanical strength (σ_s) versus pressing pressure at various powder moisture contents. The variation in dry mechanical strength with pressing variables resembled the variation observed in dry relative density or packing density (Figure 2). Thus, raising pressing pressure at every powder moisture content raised mechanical strength potentially. Similarly, at a given pressure, the specimens formed with a higher moisture content yielded higher dry mechanical strength. As with packing density, at sufficiently high pressing pressures and moisture contents (>50 MPa and >8 wt% respectively), the elastic response of water after compaction produced a drop in relative density (Figure 2) and hence in mechanical strength. This parallelism between relative density and mechanical strength was to be expected as a result of the rise in interparticle and/or interagglomerate contact area on lowering porosity.[6]

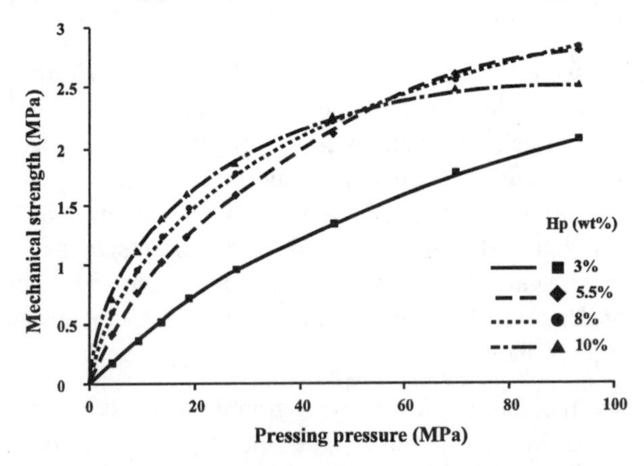

Figure 3. Variation of dry mechanical strength in terms of pressing pressure for powders with different moisture contents.

It can be observed that the effectiveness of water in raising mechanical strength is greater in the 3-5.5 wt% moisture content range than at higher moisture contents.

With a view to identifying the effect that the pressing variables separately have on mechanical strength, σ_s, the values of σ_s have been plotted versus relative density ρ_r in Figure 4.

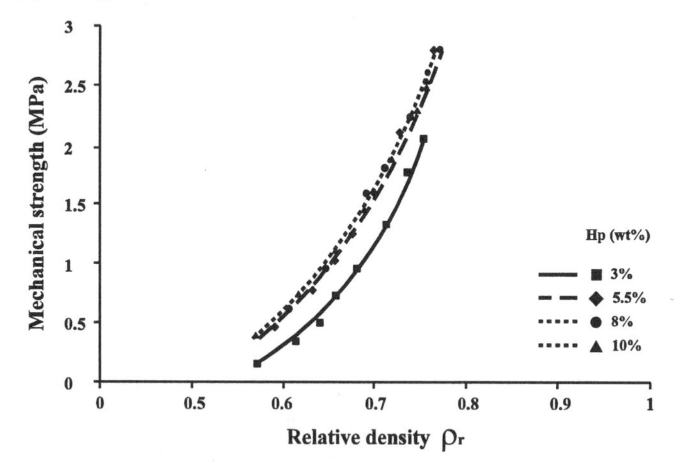

Figure 4. Variation of mechanical strength (σ_s) with relative density (ρ_r) at different pressing moisture contents.

The figure shows that at the same porosity, mechanical strength rose slightly with increased powder moisture content, especially in the moisture content range 3-5.5 wt%, indicating that the relation between dry mechanical strength and porosity was not independent of the forming variables. At moisture contents exceeding 5.5 wt%, mechanical strength hardly increased (5.5-8 wt% range), or exhibited no increase (8-10 wt% range).

The reason why specimens pressed at higher moisture contents exhibited greater dry mechanical strength may lie in water's plasticizing effect. Thus, as agglomerate moisture content rose, the agglomerates became more deformable, facilitating sliding and enhancing particle packing.

Indeed, as Table II reveals, on raising moisture content, agglomerate yield pressure P_Y (an estimation of average agglomerate strength) decreased, thus raising deformability, increasing interparticle contact area and intergranular adhesion. The table also shows that the plasticizing effect of water (lower yield pressure) was much greater in the first moisture content range (3-5.5 wt%) than in subsequent water additions (>5.5 wt%), which matched the findings set out in

Figure 3.

Table II.- Variation in P_Y in terms of moisture content.

Hp (wt %)	Yield pressure (MPa)
3.0	0.30
5.5	0.19
8.0	0.17
10.0	0.16

The foregoing assumption was verified by SEM examination of the fracture surface of two series of specimens with the same relative density formed at different pressing conditions. In the first (ρ_r =0.682), the specimens were formed at 30 MPa and 3.0 wt% moisture content, and 20 MPa and 8.0 wt% moisture content. In the second series (ρ_r =0.740) the following pressing variables were used: 75 MPa and 3.0 wt% moisture content, and 50 MPa and 8 wt% moisture content.

Microstructural differences were qualitatively significant especially at low packing densities, since as a result of the pressing conditions used in this last series of specimens, total granule deformation and/or failure could hardly take place. At higher packing densities, the fracture surfaces in the specimens formed at a moisture content difference of 5% exhibited less evident microstructural differences. However, in both cases, the fracture surfaces of specimens formed at higher moisture contents were always characterized by lower or no intragranular fracture, as a result of greater interagglomerate adhesion.

Equation for determining dry mechanical strength of specimens made from powder agglomerates

Different models can be found in the literature, in which a compact's dry mechanical strength is related to an exponential function of its porosity.[6] However, as shown in the foregoing, these models do not account for the effect of powder moisture content on mechanical strength.

On plotting the logarithm of dry mechanical strength versus specific total porosity (V_T) for specimens formed under different pressing conditions (pressing pressure and moisture content), it can be observed, that the experimental data points fit straight lines that converge at coordinate point (σ_{SS}, V_{TT}).[6] Hence, dry mechanical strength (σ_s) can be described by:

$$\sigma_S = \sigma_{SS}\, exp\left[-m\left(V_T - V_{TT}\right)\right] \quad \textbf{(2)}$$

in which m is a parameter that depends on powder moisture content.

Given the close relationship between agglomerate moisture content and yield pressure (Table II), and between yield pressure and the compact's mechanical strength (Section 3.3), it was thought that parameter m might be a function of agglomerate yield pressure. No equation was found in a literature survey, which, other factors being equal, could relate agglomerate yield pressure, P_Y, to mechanical strength of the compact; however, it is quite likely that this relationship will be of a potential type.

Bearing in mind the above, it has been attempted to fit the experimental data by non-linear correlation to the following equation:

$$\sigma_S = \sigma_{SS}\ exp\left[-AP_Y^B\left(V_T - V_{TT}\right)\right] \qquad \textbf{(3)}$$

The experimental σ_S and V_T values of the specimens formed under all the tested pressing pressures and moisture contents (Section 2.2), and the values of yield pressure, P_Y, (Table II) were fitted in Eq. (3), yielding:

$V_{TT} = 0.243$; $\sigma_{SS} = 3.48$ MPa ; A = 11.75 $MPa^{-0.57}$; B = 0.57

Figure 6 plots the experimental dry mechanical strength data (σ_S) versus total specific porosity (V_T), as well as the values calculated from Eq. (3) (solid curves). It can be observed that a good fit was obtained, substantiating the validity of the equation.

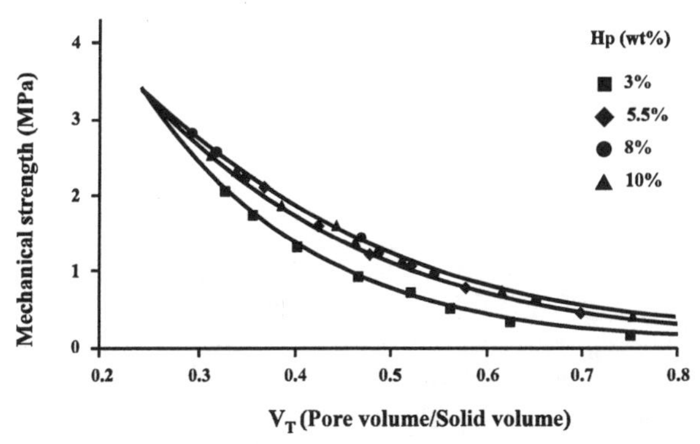

Figure 5. Comparison of experimental dry mechanical strength data (σ_S) and the values calculated from Eq. (3) (solid curves).

CONCLUSIONS

- The green tile compacts studied exhibited semi-brittle behaviour, with fracture starting at pre-existing cracks or defects, which grew microplastically during testing.
- At pressing pressures equal to or lower than the peak pressure typically used in industry (≈ 50 MPa), the rise in powder compressibility with moisture content yielded higher relative density of the compact.
- Dry mechanical strength varied with pressing variables in a way closely resembling the variation observed for relative density or dry packing density. Dry mechanical strength rose considerably on reducing porosity, owing to higher interparticle and/or agglomerate contact area.
- At the typical industrial moisture content range (4-6 wt%), in materials exhibiting the same porosity, the resulting mechanical strength on using moister granules was higher than when drier granules and greater pressing pressures were used. This was because of the effect of moisture content on agglomerate elasto-plastic deformation during pressing, and therefore on interparticle and/or agglomerate contact area and intergranular adhesion.
- Finally, a semi-empirical equation has been proposed and validated, relating mechanical strength to porosity and agglomerate yield pressure.

REFERENCES

[1]K. Kendall, N. Alford; and J.D. Birchall "The Strength of Green Bodies", *Br. Ceram. Proc.*, **37**, 255-265, (1986).

[2]D. Bortzmeyer, G. Langguth, and G. Orange. Fracture Mechanics of Green Products. *J. Eur. Ceram. Soc.*, **11**, 9-16, 1993.

[3]J.S. Reed. *Principles of Ceramic Processing*. Pressing; pp. 418-449, 2nd ed. New York: Wiley, 1995.,.

[4]D.G.S. Davies. The Statistical Approach to Engineering Design in Ceramics. *Br. Ceram. Soc. Proc.*, **21**, 429-451, 1973.

[5]J.L. Amorós; F. Ginés; C. Felíu; and G. Silva. Pressing of Spray-Dried Powders. Derivation of a Compaction Equation. In: *V World Congress on Ceramic Tile Quality*. Castellón: Cámara de Comercio, 1998. p. 49-51.

[6]J.L. Amorós. *Single-fire Floor Tile Bodies. Effect of Pressing Variables on Green Ttile Properties and on Pressing and Firing Behaviour (in Spanish)*. Valencia: University, Faculty of Chemistry, 1987. Doctoral Dissertation.

INFLUENCE OF PRESSING VARIABLES ON AIR PERMEABILITY OF FIRED FLOOR TILE BODIES

Agustín Escardino, José L. Amorós, Mª José Orts
Instituto de Tecnología Cerámica, Asociación de Investigación de las Industrias Cerámicas, Universitat Jaume I. Campus Universitario Riu Sec. 12006. Castellón, Spain.

Vicente Beltrán
Departamento de Ingeniería Química, Universitat Jaume I. 12006. Castellón, Spain.

ABSTRACT

The influence of the pressing variables (pressing pressure and moisture content) and firing temperature on the air permeability of fired materials made from a low porosity industrial tile body composition (water absorption <3%) was studied. Air permeability was shown to rise with firing temperature up to a peak value. The variation in permeability with temperature as well as peak permeability depended on green porosity. The porous texture of the bodies was characterized by mercury porosimetry and scanning electron microscopy, comparing the mean pore radii found from air permeability and pore-size distribution.

INTRODUCTION

On firing ceramic materials, numerous reactions occur involving gases (oxidation, decomposition, etc.). The rate at which these reactions develop largely depends on the gas permeability of the porous solid.[1] Moreover, on heating the tile body in the kiln, trapped air inside the initially open pores expands, producing a gas flow from inside the material outward into the kiln atmosphere. The permeability of a porous solid thus affects the tile-kiln atmosphere gas exchange rate, making this parameter a key tile body characteristic.

Various authors have used this property to characterize body porous structure, given the speed and simplicity of the determination together with the low cost of the measuring equipment involved.[2,3] Drugoveiko and co-workers[3] characterized and compared the porous texture of porcelain bodies formed by isostatic and plastic-pressing on the basis of permeability, hydraulic radius, and pore-size distribution found by mercury porosimetry. Permeability has also been shown to largely determine a ceramic material's frost resistance[4] and given the simplicity of the measurement, several authors recommend it as a measure of the degree of firing of ceramic materials made from natural clays.[5]

Darcy's law introduced the concept of permeability, according to which the flow rate of water per surface unit through a porous bed is proportional to the pressure gradient between the bed's two sides.[6] In the case of gases circulating through a cylindrical porous material of length L and cross section S, as a result of a pressure difference (P_1-P_2), integrating Darcy's law yields:

$$K_p = \frac{2\mu QL}{S}\frac{P_2}{P_1^2 - P_2^2} \tag{1}$$

where Q is the gas flow rate (m^3/s) and K_p is the material's gas permeability coefficient (m^2).

Assuming that the porous solid contains a system of cylindrical, non-tortuous pores of the same length and radius, a mean pore radius can be found from Darcy's and Poiseuille's laws. This radius is called hydraulic or capillary radiu[7] and can be described by:

$$r_h = \sqrt{8K_p / \varepsilon} \tag{2}$$

where ε is the material's porosity.

In this work, the effect of process variables (pressing pressure and moisture content, as well as peak firing temperature) on the air permeability coefficient and hydraulic radius of the fired materials was studied. The microstructure of these pieces was examined by scanning electron microscopy (SEM) and the hydraulic radii were compared with the mean pore radii found by mercury porosimetry.

MATERIALS AND EXPERIMENTAL PROCEDURE

The tests were run on a composition used in industry to manufacture low porosity floor tile. The composition comprised red illitic-kaolinitic clays, with a prevailing illitic structure and abundant quartz (≈36%). The composition's chemical analysis is reported in Table I. The composition was prepared on an industrial scale by wet milling the raw materials mixture and spray drying the resulting suspension. Five series of cylindrical test specimens were formed by pressing at P=15, 25 and 45MPa, using spray-dried powder moisture contents of Xp=0.033, 0.055 and 0.070 kg water/kg dry solid.

The test specimens were oven dried at 110°C and fired in an electric laboratory kiln at a constant heating rate of 25°C/min. to different peak temperatures in the range 850-1100°C. Porosity was determined in the fired test specimens from true density and bulk density, while the air permeability coefficient was determined on a constant load permeability tester schematically depicted in Figure 1. Apparent pore-size distribution (PSD) was found by mercury porosimetry (Micromeritics poresizer 9310), recording the mercury intrusion-extrusion curves. Polished cross-sectional slices were prepared with some test specimens, which were examined by SEM.

Table I. Chemical analysis of the composition used.

Oxide	SiO_2	Al_2O_3	Fe_2O_3	CaO	MgO	Na_2O	K_2O	TiO_2	MnO	P_2O_5	L.O.I.
wt(%)	60.2	17.3	6.34	1.86	0.96	0.66	3.27	0.94	0.44	0.32	7.34

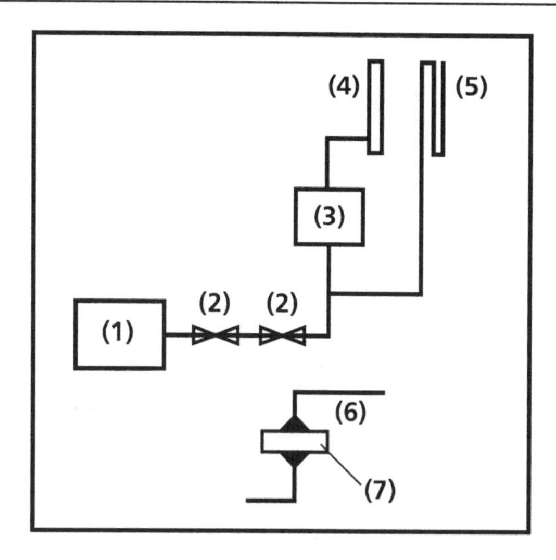

Figure 1. Constant load permeability tester: 1 Compressor. 2 Pressure adjustment valves. 3 Permeability cell. 4 Floweter. 5 Differential gauge. 6 Detail of 3. 7 Test specimen.

RESULTS AND DISCUSSION

Evolution of the air permeability coefficient with peak firing temperature

Table II details the green porosity values (ε_0) of the five series of test specimens. Figures 2 and 3 plot specimen porosity and the air permeability coefficient (K_p) versus peak firing temperature. The evolution of the hydraulic radius (r_h) with peak firing temperature is identical to that of K_p, which is why it has not been included in the plot.

In the experimental determination it was observed that slight variations in green porosity (ε_0), caused by minor changes in pressing pressure and/or moisture content considerably modified the value of K_p. For this reason five specimens were tested in each case, averaging the findings.

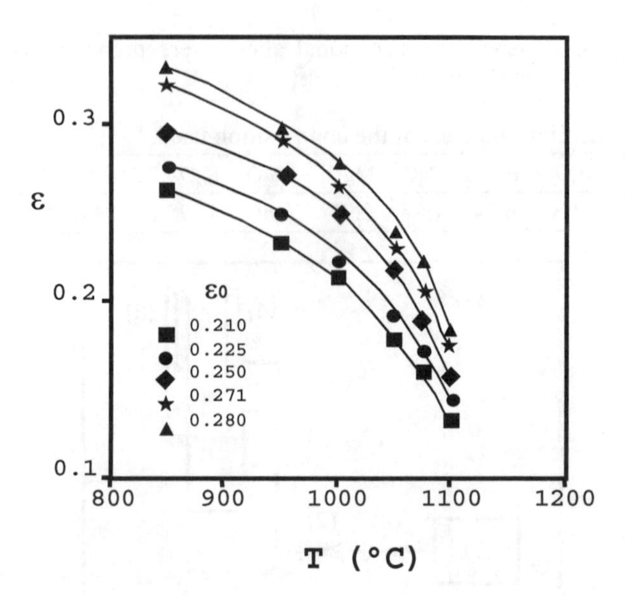

Figure 2. Variation of porosity (ε) with firing temperature for five series with different green porosity (ε_0).

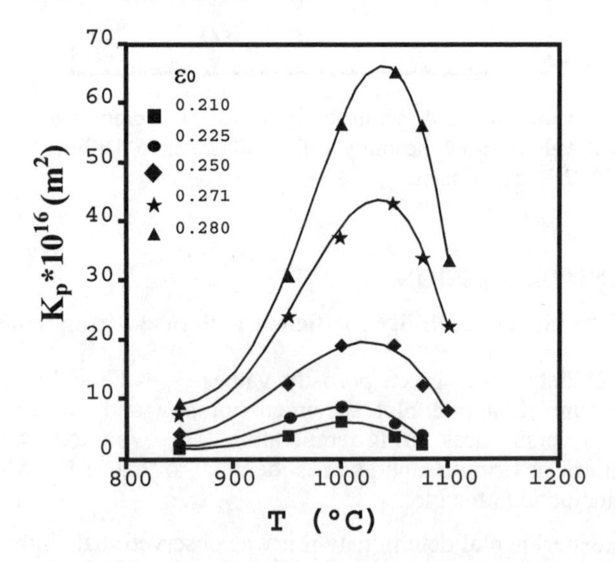

Figure 3. Variation of the air permeability coefficient (K_p) with firing temperature for the five series of test specimens.

Table II. Green porosity (ε_0) of the five series of test specimens.

Series	P(MPa)	X_p (kg water/kg dry solid)	ε_0
1	15	0.055	0.2804
2	25	0.033	0.2712
3	25	0.055	0.2490
4	25	0.070	0.2249
5	45	0.055	0.2101

Figure 2 shows that at the tested temperature range, porosity decreased as firing temperature rose. At the same firing temperature, the specimens formed at higher pressing pressures and/or moisture contents (lower green porosity) were less porous. However (Figure 3), at the same green porosity, the air permeability coefficient (and r_h) rose with temperature up to a peak value. The lower the compactness of the green body (higher starting porosity ε_0), the more pronounced were these effects. The peak air permeability coefficient (and peak r_h), and the temperature at which this value was found, fell as green porosity decreased.

The variation of ε, K_p and hydraulic radius with firing temperature was due to the densification of the compositions by liquid phase sintering.[8,9] Thus as temperature rose, a liquid phase formed while viscosity decreased, reducing porosity. As the green bodies involved consisted of variously sized particles of differing nature, they exhibited a heterogeneous microstructure, and on increasing temperature small pores progressively disappeared. This led to differential shrinkage in microdomains of the piece, enlarging the bigger pores and raising mean pore size. Moreover, on raising the proportion of liquid phase and reducing its viscosity, besides lowering total porosity, the initially interconnected porous system was partly blocked, further lowering permeable porosity. These joint effects yielded a peak value for the variation of air permeability and hydraulic radius with temperature.

Figure 4 depicts the plots of the hydraulic radii of the fired specimens versus their degree of densification. This last variable is expressed as the difference between fired porosity and fired porosity at 850°C (ε_{850}-ε), the temperature at which the breakdown reactions have ended.

At low degrees of sintering, at which most porosity may be assumed to be permeable (interconnected capillary system), the results were found to satisfactorily fit a straight line of positive slope for each green porosity value. However, at advanced sintering stages, the decrease in permeable porosity (rise in tortuosity) lowered the air permeability coefficient, so that the hydraulic radius was unlikely to be a representative value for mean pore size.

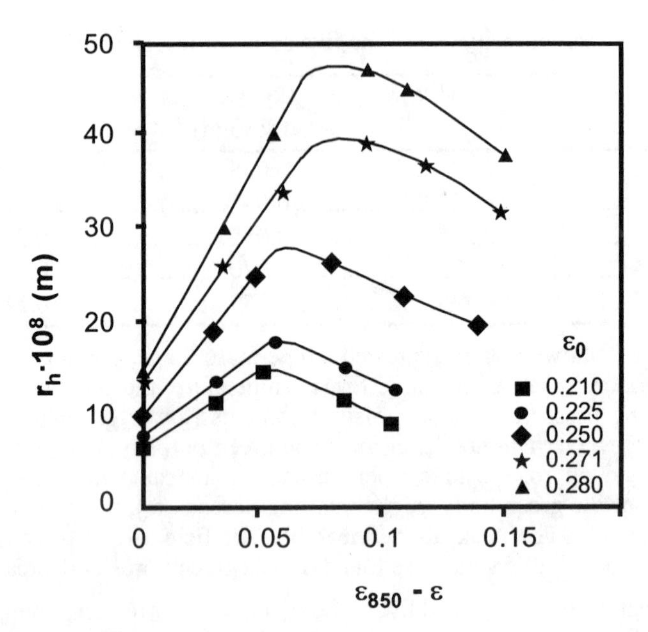

Figure 4. Variation of hydraulic radius (r_h) with the degree of densification (ε_{850}-ε).

Relationship between permeability and porous structure

Figure 5 plots apparent pore-size distribution in differential form, which corresponds to the mercury intrusion curve for series 1, 3 and 5 specimens fired at 850°C. The pressing variables were found to modify the percentage and size of the largest pores, though hardly altering the number and size of the finest ones. The same results were found at all the tested temperatures, except 1100°C. At this temperature the connectivity of the capillary system had considerably decreased and the closed porosity fraction started to become important.

Figure 6 plots the PSD of series 3 specimens fired at different temperatures to visualize the effect of firing temperature on porosity. It can be observed that at the same green porosity, raising firing temperature in the tested temperature range lowered porosity as well as narrowing PSD. The mean pore radius (r_G) found on fitting PSD to a log-normal distribution, and which practically coincided with the maximum value of the curves in Figure 6, rose with temperature up to a peak value. The temperature at which mean pore radius maximized depended on green porosity. This pore growth, due to the progressive elimination of the finest pores and growth of the largest ones increased with starting porosity, i.e., with greater heterogeneity of the body's green microstructure (greater scatter in PSD).

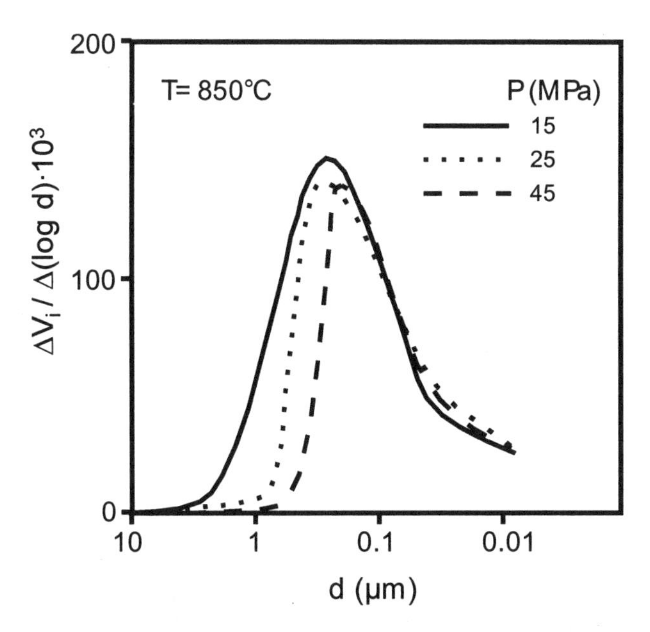

Figure 5. Variation of pore-size distribution with pressing pressure. Test specimens fired at 850°C.

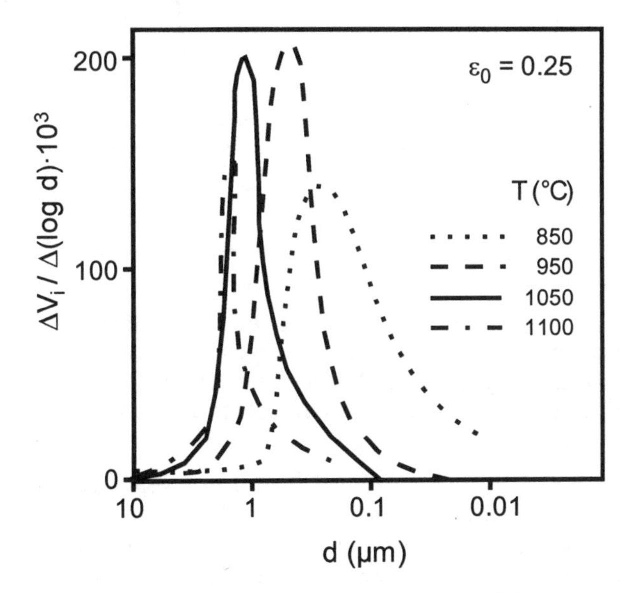

Figure 6. Variation of pore-size distribution with firing temperature for series 3 test specimens (ε_0=0.2490).

Figure 7 plots the mercury intrusion-extrusion curves for series 3 test specimens fired at 850°C, 1000°C and 1100°C. The hysteresis between the mercury intrusion and extrusion curves stemmed from the fact that the pores were not cylindrical and because during sample depressurization in the mercury poresizer, breaks occur in the outward flow of mercury, which lead to mercury being retained in certain areas of the capillary system.[10]

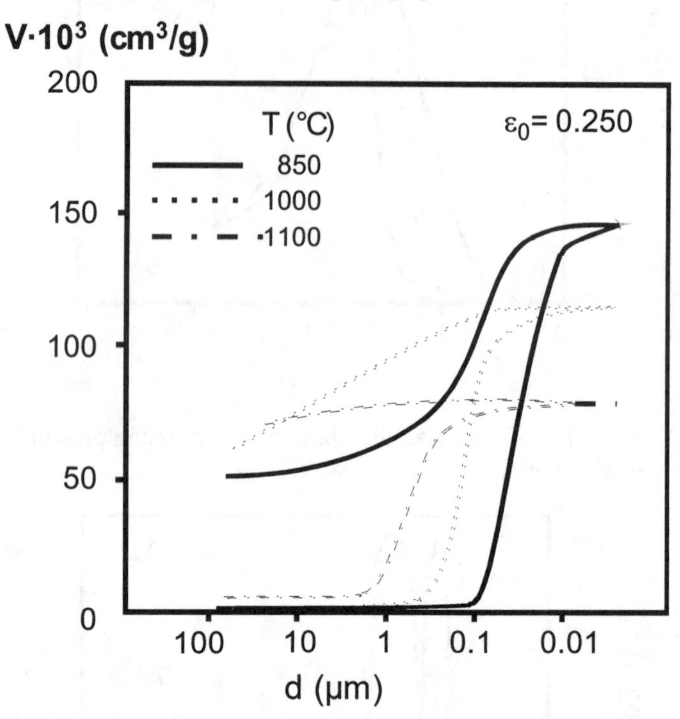

Figure 7. Mercury intrusion-extrusion curves for series 3 test specimens fired at different temperatures.

The hysteresis of these curves was quantified as the amount of mercury retained inside the solid. This was found to depend very little on green porosity, but to vary considerably with firing temperature. These findings confirm that as the specimen sintered, the capillary system lost its connectivity owing to partial pore necking, hence raising tortuosity.

SEM examination of the polished specimen cross sections revealed that except for the specimens fired at 1100°C, the capillary system was interconnected, exhibiting large pores together with much smaller ones. However, the specimens fired at 1100°C exhibited isolated pores, with many tending to sphere, indicating that abundant liquid phase had formed.

Figure 8 plots the hydraulic radii (r_h) found from the air permeability coefficient (K_p) versus mean pore radii (r_G) determined from the fit of the mercury intrusion curves to a log-normal distribution. Note that there is no single relation between the pore sizes obtained by either procedure. At low firing temperatures (850°C, 950°C) the values of r_h and r_G practically coincided, as at these temperatures the specimens largely exhibited apparent porosity with an interconnected capillary system. However at higher firing temperatures, r_h was lower than r_G and the difference became more marked as firing temperature rose. These findings match the variation with temperature of mercury intrusion-extrusion curve hysteresis, stemming as previously mentioned from the drop in interconnected porosity with rising firing temperature and increasing tortuosity of the capillary system.

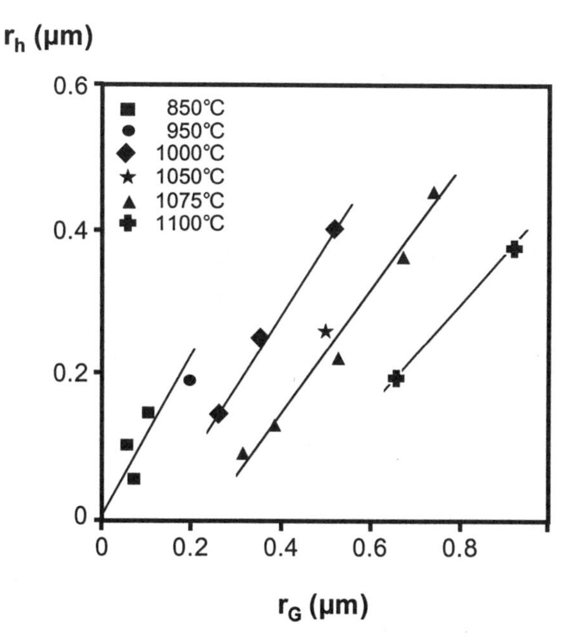

Figure 8. Comparison between the mean pore radii found from the permeability measurements (r_h) and from the mercury intrusion curves (r_G).

CONCLUSIONS

The air permeability coefficient of fired floor tile bodies was shown to be very sensitive to variations in porous texture, stemming from changes in pressing variables (pressing pressure and moisture content) and/or firing temperature. This parameter can therefore be used satisfactorily in manufacturing process control.

The body's air permeability and hydraulic radius increased with firing temperature up to a peak value. This was because densification occurred by liquid-phase sintering in the presence of heterogeneous starting particles and pore sizes, eliminating small pores and enlarging the biggest ones in sintering. The porous system's subsequent loss of connectivity and arising sealed porosity then led to a decrease in air permeability and hydraulic radius.

The variation in permeability and hydraulic radius with firing temperature and their maximum values largely depended on green porosity. Raising pressing pressure and/or moisture content lowered the body's permeability and hydraulic radius.

The hydraulic radii were generally lower than the pore radii found from the mercury intrusion curve. The mean pore sizes determined by both procedures were only virtually identical at low firing temperatures. The existing proportionality between these two parameters appears to be related to the capillary system's connectivity.

REFERENCES

1. A. Barba, *Black core oxidation in firing ceramic tile. Influence of body structure, organic matter and iron oxide contents on process kinetics* (in Spanish), Doctoral Dissertation, Valencia University, Faculty of Chemistry, 1989.
2. B. R. Steele, J. O. Ware and B. W. Olfirld, "Permeability measurements as a means of characterizing powder compacts," *Trans. Brit. Ceram. Soc.,* **65** 17-32 (1966).
3. O. P. Drugoveiko and A. M. Kolpishon, "Porous structure of semifinished electrical porcelain," *Glass Ceram.*, **9** 21-22 (1982).
4. G. C. Robinson, "The relationship between pore structure and durability of brick," *Am. Ceram. Soc. Bull.*, **63** 295-300 (1984).
5. M. R. Arnott and G. G. Litvan, "Quality control test for clay based on air permeability," *Am. Ceram. Soc. Bull.*, **67** 1412-1417 (1988).
6. F. A. L. Dullien, *Porous media. Fluid transport and pore structure*, pp.78-82. Academic Press, New York, 1979.
7. P. C. Carman, "Fluid flow through granular beds," *Trans. Inst. Chem. Eng.*, **15** 150-166 (1937).
8. J. L. Amorós; *Single-fire floor tile bodies. Effect of pressing variables on green tile properties and on pressing and firing behaviour* (in Spanish), Doctoral Dissertation, Valencia University, Faculty of Chemistry, 1987.
9. M. J. Orts; *Sintering of floor tile bodies* (in Spanish), Doctoral Dissertation, Valencia University, Faculty of Chemistry, 1991.
10. S. J. Gregg and K. S. W. Sing, *Adsorption, surface area and porosity*, Academic Press, New York, 1982.

DIE DESIGN, COMPUTED TOMOGRAPHY, AND MODELING

John Lannutti
Ohio State University

Dale Fronk
The Edward Orton, Jr., Ceramic Foundation

ABSTRACT

Commercial dry-pressing operations attempt to minimize the effects of density gradients in the production of shaped ceramics. The associated downstream effects of the many factors that determine the quality of these parts are only just beginning to be understood. We discuss fill, friction during compaction and compaction diagrams in industrial practice and give examples of the often overlooked consequences of density gradients.

INTRODUCTION

Compaction is a simple process made complex by the lack of control of variables at the microscopic level. In the following paper we will attempt to describe many of these variables in light of new discoveries revealed using a unique combination of X-ray Computed Tomography (CT)[1-5] and discrete element modeling*.[6]

Fill

A poorly studied but critical aspect of the dry-pressing process is the method by which powder is added into a die cavity. Commercial processes involve the motion of not one but of thousands of agglomerates at a time out of a "fill shoe" into the die cavity. The head pressure from the supply bin, hose configuration/composition, shoe geometry, shoe size, shoe motion, and press vibration all control the resulting uniformity.

An example of this effect on product uniformity and how it was countered is given as follows. Tempcheks, a pyrometric device providing accurate measurements of the thermal exposure for industrial products, were examined using X-ray Computed Tomography (XRCT) and were found to have the density distributions shown in Figure 1_C†. Besides an overall lack of uniformity it was noted that the 'foot' (lower edge) of the component displayed consistently higher densities. This was traced to the method by which powder flowed from the fill shoe into the die cavity and the subsequent dragging motion of this shoe back over the cavity. The shoe successfully filled the cavity but its base and contents were dragged across the bed of powder during the return stroke. Horizontal forces that lead to enhanced densification of one end of the fill were generated. The

* Particle Flow Code, Itasca Consulting Group, Minneapolis, MN.
† The "C" subscript indicates that the figure is in the color plate section at the end of the book.

volume change associated with this density increase was immediately compensated for by the powder within the shoe and was not visible in the fill.

This problem was solved by instituting the following changes 1) the feed to the shoe was split into two more horizontal streams to slow refilling and decrease head pressure on the 'foot' and 2) via modeling exercises the die was slightly redesigned. The result is shown in Figure 2_C; the density distribution is considerably improved. The resulting sintering shrinkage is twice as uniform and distortion is reduced.

Contact Order

Closely associated with the fill process are the characteristics of the resulting powder surface. In a typical compaction process, an advancing ram contacts a powder surface consisting of many agglomerates. The initial distribution of force in this packed bed is partially determined by the order in which agglomerates first make contact with the ram. Those agglomerates undergoing initial contact will transmit significantly greater levels of force into the bed than their neighbors.[6] Uniformity of fill obviously plays a role in this process.

These early events bias density distributions and govern the continued application of force into the powder assembly. Transmission of force into a given area *predisposes that area to resist additional force*. If large enough, these areas will begin to limit compaction. The most commonly encountered example of this is found between the wall and the advancing ram. The uppermost agglomerates in contact with the wall and the ram are more inhibited in their motion. Even in the early stages, they preferentially absorb applied compaction force. This effect intensifies as these areas become denser and larger and can better resist the advancing ram. The pressures applied during the early stages condition the compact to resist generalized deformation and enhance subsequent non-uniformity.

FRICTION AND COMPACTION

The most important factor in compaction is the friction that develops between an agglomerate and the die wall; this presents the largest barrier to the uniform transmission of applied pressure and results, instead, in its concentration into high-density zones. A primary contributor to this is the thin film of organic additive that typically exists on the surface of synthetic ceramic agglomerates. Under the appropriate conditions this film has the attributes of an adhesive and, not surprisingly, strongly influences friction and compaction efficiency. Modeling shows that inter-agglomerate friction is greatly increased (Figure 3_C) and this prevents both inter-agglomerate motion and sliding relative to the die wall. In addition the pronounced moisture sensitivity of many of these additives provides a source of "seasonal variations" that can plague pressing operations that do not control ambient humidity levels during powder storage and handling.

The effects of inter-agglomerate friction, or the friction *between* agglomerates, are also illustrated in Figure 3_C. The 'resistance' of a collection of identical agglomerates to compaction as their frictional parameters are increased. As inter-particle friction increases, more compaction force is transmitted to the die wall rather than into the bulk.

Aggravating this basic behavior is the existence of fines often not detectable by agglomerate size analysis due to their close association with larger agglomerates. These fines are generated by the agglomerate production process (e.g., as satellites during spray drying) or spontaneously as surface debris resulting from handling. Whatever the source, such fines have a *dominant* effect on compaction efficiency. Figure 4_C shows the changes in compaction as the percentage of fines increases. Under the same levels of applied force, the amount of force directed to the wall (and *not* into the bulk of the powder) increases dramatically with increased fine content. Figure 5_C shows the same behavior in terms of the *distribution* of fines; if fines are segregated (due to vibration, for example) they can greatly change the overall distribution of force, enhancing part-to-part variation.

Friction during compaction must be influenced by the nature of the organic additive(s) present. Organic additives presently marketed often have opposing effects on compaction processes. Softer binders (e.g., PVA-PEG) may allow for greater levels of density but can intensify friction and the formation of HD zones. Harder binders (e.g., acrylates) promote more uniform densification during compaction at the expense of lower overall densities. Figure 6_C compares the compaction behavior of 'hard' (higher modulus) agglomerates versus 'soft' (lower modulus) agglomerates having identical size distributions and frictional characteristics. Clearly, the macroscopic behavior of these powders during production translates into defects and internal density gradients.

Environmental effects upon these additives control many of these variables. As was illustrated previously (Figure 3_C) fluctuating levels of moisture and/or temperature associated with seasonal changes will vary the resulting frictional characteristics altering inter-agglomerate friction, rearrangement, flow, and transference of stress to the die wall. For example, ram sticking and air entrapment typically become more noticeable as humidity increases. However, agglomerates simultaneously become more deformable and overall density may increase. Obviously, moisture content must be carefully monitored to exert appropriate control over both reproducibility and profitability.

COMPACTION DIAGRAMS

An indirect method of measuring the compaction efficiency of a given powder involves compaction diagrams. These serve to gauge the resistance of a mass of agglomerates in a *specific* die to compaction. They provide useful data concerning the combined influence of many factors on compaction. They are *not*, however, able to yield specific information regarding the characteristics of individual agglomerates. The average response of a collection of agglomerates is controlled by many different variables. The following sections describe the different stages of compaction and how these can be related to the density gradients that control this process.

Stage I

The earliest stage of compaction involving the initial stages of agglomerate deformation. Figure 7_C shows that as force is transmitted from the advancing ram into the powder, a region of higher density develops at the very top of the loose powder. The corresponding pressures are very small and involve only very limited amounts of net motion. For a brief period, the powder accommodates the ram and allows motion highly coordinated with its advance.

Obviously, this step is sensitive to any environmental conditions that influence the physical characteristics of the agglomerates that make up such a thin layer. Chief among these is humidity. Variations in humidity (or the binder) can alter the deformation characteristics of a binder-agglomerate system to promote brittle agglomerate fracture or gentle, gradual deformation.[5]

Stage II

The Stage I-II boundary is partially governed by L/D and that point at which wall friction begins to bias further propagation of force into the powder. Instead of a uniform front a densification 'wave' pinned by HD zones moves into the powder (Figure 7_C).

The well-known sensitivity of the Stage I-II transition to relative humidity is, in reality, an outgrowth of the changing characteristics of the agglomerates in the topmost layer.[8] At high levels of humidity, the easy deformation of these soft agglomerates combines with greater wall friction to result in the relatively early transmission of force into the compact. Thus the Stage I-II boundary occurs at lower values of force. In contrast, low humidities will result in harder agglomerates that better resist the advancing ram but do not generate as much wall friction. This can result in a more generalized mode of compaction. Unfortunately, however, higher levels of force are required to reach equivalent levels of density.[5]

Returning to a compact within a simple cylindrical die, the densifying compact is now partially pinned by a ring of high-density compacted agglomerates circling the outer circumference of the ram face. As the compaction front moves into the body of the powder (Figure 7_C), even more of the ram compaction force is transmitted to the wall. This constraint is the origin of double-ended compaction. These allow one ram to propagate density gradients in opposition to another. The result is also non-uniform but the range of densities produced is smaller as the LD zone is sandwiched between two HD zones.

Stage III

The most controversial of the compaction stages is the third and final stage. The uncertainty surrounding it is partially due to the high pressures involved which make detailed experiments of the process difficult. However, both modeling and experimentation suggest that this stage occurs when the advancing gradients contact the opposing surfaces (Figure 7_C).

Die Complexity and Compaction Diagrams

Unfortunately, the relationship between complex dies and compaction diagrams is poorly developed. Even simple deviations beyond a perfectly cylindrical ram advancing into a cylindrical die produces complex density distributions. Figure 8_C shows the density gradients that develop above and around a simple 1" x 1" button sitting on the bottom of a 2" die. An arrow-head shaped HD zone develops above the button and serves to limit the progress of the upper ram into the die by preferentially transmitting force to the button. Thus, Stages I, II, and III will all occur relatively early.

Figure 9_C shows a simulation depicting a) the overall compaction diagram and b) a localized compaction diagram comprising only the upper portion of such a compact. Not surprisingly, the overall compaction diagram conveys little information about the non-uniformities developing within the component.

THE PRODUCT

The product of compaction is by no means a simple, homogeneous solid but is a complex structure containing agglomerates in a variety of stages of deformation, residual stresses, and varying solids (both inorganic and organic) contents. The higher temperature behavior of this mass must reflect these initial inhomogeneities.

Dimensional Tolerance

High density zones in any as-pressed component have both the largest residual stresses and the highest solids concentrations. Any behavior controlled by these variables will be magnified relative to the remainder of the compact.

Heating triggers dimensional change before sintering even begins. This results in dimensional changes sometimes visible as gross defects – warping, cracking – but more often influencing only dimensional tolerance. The seriousness of these deviations greatly depends on product specifications. Proper control of dimensional tolerance is often complicated not only by density gradients but also by the part-to-part variation endemic to dry pressing.[6]

Unfortunately, such changes have historically been difficult to monitor. Standard dilatometry both biases and oversimplifies dimensional change versus temperature by applying a force to monitor the gross shrinkage in one dimension for only very simple geometries. We have sidestepped this constraint by building a non-contact laser dilatometer that allows us to monitor two-dimensional changes in the shape of complex objects. Figure 10_C shows data resulting from the low-temperature firing of a simple cylinder containing HD and LD zones.

At low temperatures (50°C) the compact is fairly uniform although the HD end shows some expansion relative to its room temperature state. At 150°C, the sample has expanded. The portion in contact with the 'setter' has expanded more than the rest of the compact due to the combined effects of restraint and gravity. At the completion of burnout, what was once a net shape compact now has regions both larger and smaller than the desired geometry.

Another illustration of this effect is seen in a much more complex CIPped part; this data is shown below in Figure 11c. Under fairly rapid heating, one end of the part contracts by up to 7% relative to the remainder of the compact. No sintering has taken place. This dimensional change is due solely to the relaxation of internal residual stresses caused by the presence of a liquid phase (the binder).

Sintering Effects

Once beyond the influence of low-temperature changes in dimension, the barriers to maintaining dimensional tolerance become simpler to comprehend but just as difficult to overcome. Existing mass imbalances can only be completely removed by long-term (uneconomical) thermally-driven diffusion. The key questions that must be answered are a) how severe is the problem and b) can the remaining porosity be tolerated in the application? In most cases the answer to b) has been 'yes.' This has, in large part, determined our low regard for the mechanical reliability of commercially produced ceramics.

Such hidden problems are again caused by the presence of density gradients.[6] Even neglecting the influence of setter drag, the relative placement of high-density zones during firing can have an effect on the dimensional response of the compact during heating. For example, Table I contains data traditionally reported for two such specimens before and after sintering. These macroscopic data appear to be fairly similar suggesting that both samples are at the same stage of densification. This could lead us to conclude that the samples behave in similar ways during sintering.

Table I. Comparison of gross data from two samples having different HD vs. LD orientations within the furnace during sintering.

	Green Density (%)	Fired Density (%)	Diametric Shrinkage (%)	Longitudinal Shrinkage (%)
LD zone as free end	49.18	73.26	13.20	13.42
HD zone as free end	49.57	73.31	13.03	13.20

However, dramatic differences (Figure 12c) are apparent in deviation in the shrinkage curves. Dimensional change is clearly sensitive to the relative orientation of the HD and LD zones. When the LD zone is the free end consistent changes are observed (Figure 12A). When the HD zone is the free end, however, substantial part-to-part variation occurs. When the HD zone is the free end the samples display less predictability, likely a measure of the randomness of the zones themselves.

The Consequences of Agglomeration

Agglomeration simplifies and standardizes processes upstream of compaction at the expense of downstream processes and properties. Although the evidence supplied by the naked eye would suggest that an as-compacted body is homogeneous, microscopy tells us otherwise (Figure 13). Agglomerates are forced together but certainly do not disappear. Agglomerate "triple boundaries" are plainly evident and indicate that small-scale variations across these boundaries must still exist. Added to this is the common occurrence of internal defects within the agglomerates themselves. These are not eliminated by compaction but are often protected and persist to limit green and fired (Figure 14) strength.

These defects often survive the initial stages of sintering (and may even grow as agglomerates shrink).[6] This is, in large part, responsible for the lack of utility of much of the early sintering work.[9] These efforts focussed on applying the shrinkage behavior of a few particles to describe macroscopic shrinkage.[10] Unfortunately, large-scale defects exist outside of this approach but provide 'hard' limits to both final densification and ultimate properties.

REFERENCES

1. D. H. Phillips and J. J. Lannutti, "Characterization of Density Gradients During Dry Pressing of Ceramics," *Visions*, **15** [Spring/Summer], 2-5 (1993).
2. D. H. Phillips, "Quantitative Characterization of Density Gradients in Ceramic Processing by X-Ray Computed Tomography," Ph.D. thesis, Ohio State University, (1994).
3. D. H. Phillips and J. J. Lannutti, "Measuring Physical Density with X-ray Computed Tomography," *NDT&E International*, **30**, 339-350 (1997).
4. J. J. Lannutti, T. A. Deis, C. M. Kong, and D. H. Phillips, "Density Gradient Evolution during Dry Pressing," *Amer. Cer. Soc. Bull.*, **76**, 53-58 (1997).
5. T. A. Deis and J. J. Lannutti, "X-ray Computed Tomography for Evaluation of Density Gradient Formation during the Compaction of Spray-Dried Granules," *J. Amer. Cer. Soc.*, **81**, 1237-1247 (1998).
6. J. J. Lannutti, "Characterization and Control of Compact Microstructure," *MRS Bull.*, **22**, 38-44 (1997).
7. D. H. Phillips and J. J. Lannutti, "X-ray Computed Tomography for the Testing and Evaluation of Ceramic Processes," *Amer. Cer. Soc. Bull.*, **72**, 69-75 (1993).
8. C. W. Nies and G. L. Messing, "Effect of Glass-Transition Temperature of Polyethylene Glycol-Plasticized Polyvinyl Alcohol on Granule Compaction," *J. Amer. Cer. Soc.*, **67**, 301-304 (1984).
9. G. Petzow and E. Exner, "Particle Rearrangement in Solid State Sintering," *Zeischrift fur Metallkunde*, 67, 611-618 (1976).
10. W. D. Kingery, H. K. Bowen, and D. R. Uhlmann, *Introduction to Ceramics*, 2nd ed., John Wiley, New York, 368-374, 1976.

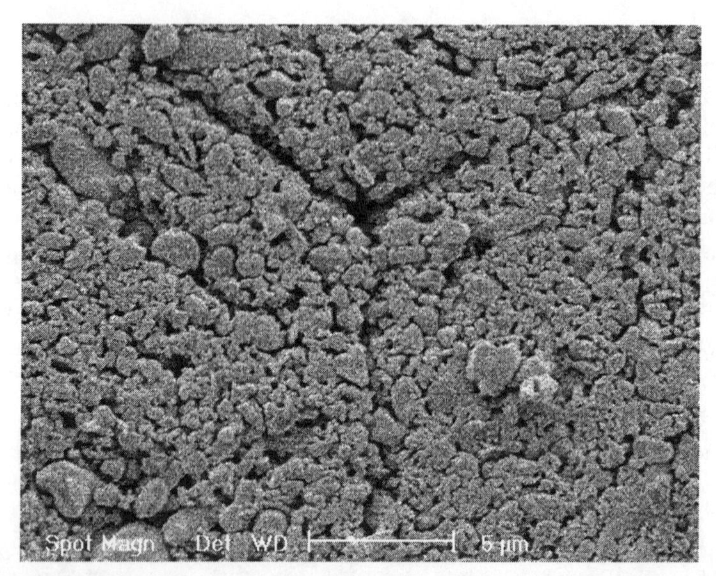

Figure 13. "Triple point" boundary between agglomerates in an industrially-produced compact plainly visible even after compaction to 55% density.

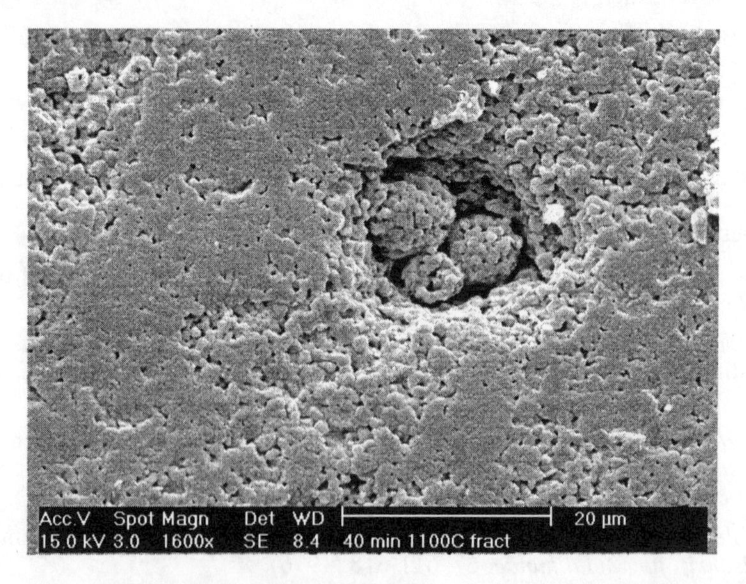

Figure 14. Internal agglomerate pore preserved after firing a compact to 75% theoretical density. The sintered versions of several 'satellites' trapped within the pore are also visible.

DETERMINATION OF THE INFLUENCE OF PRESSING VARIABLES ON TILE COMPACTION BY EQUIPPING AN INDUSTRIAL PRESS WITH SENSORS

Gustavo Mallol, Arnaldo Moreno, Domingo Llorens
Instituto de Tecnología Cerámica. Asociación de Investigación de las Industrias Cerámicas, Universitat Jaume I. Campus Universitario Riu Sec. 12006, Castellón, Spain

Paco Negre
Departamento de Ingeniería Química, Campus Universitario Riu Sec. Universitat Jaume I. 12006, Castellón, Spain

ABSTRACT

A study was undertaken on the effect of the pressing cycle on the green bulk density of tile bodies produced by uniaxial pressing. The following operating variables were modified, monitoring the major process variables on-line:

- Feeding system (position and speed)
- Lower punch (moment of descent)
- Hydraulic circuit (peak pressure)

A non-destructive control system was developed which allows monitoring industrial pressing parameters, determining their influence on-line on the overall and localized bulk density (dimensions and wedging respectively) of green tile bodies.

INTRODUCTION

The most widely implemented forming method in tile manufacture is uniaxial dry pressing of suitably conditioned spray-dried pressing powder on hydraulic presses.[1] Higher quality demands at the end of the 1970s led to the introduction of hydraulic presses. These presses were much steadier in service than the existing friction presses. At the time, little attention was paid to hydraulic press measuring and control systems, possibly because the innovation itself substantially raised tile quality and productivity. However, current trends towards larger sizes, more complex models, and quality requirements call for greater production process control, particularly in the forming stage. The state of the art in industrial control and instrumentation[2] allows on-line monitoring and control of key press feed variables[3-5] and of the press itself[6].

OBJECTIVES

The current relationship between tile green bulk density and pressing variables in hydraulic presses is wholly empirical. Operating variables are manually controlled and adjustments are based on subjective assessment.

The present study was undertaken to measure the major pressing variables using a suitable system, determining the relation between operating variables and resulting bulk density.

SELECTION OF THE VARIABLES TO BE MEASURED AND MEASURING SYSTEM

The following variables were measured: hydraulic pressure (P_h), feeding system position (L_c), feeding system speed (V_c), travelling frame position (L_t), lower punch position (L_p), die specific pressure (P_e) and spray-dried powder moisture content (X).

Transducer electric signals were fed into digital control, data acquisition system amplifiers. Given the high compression cycle speed, data were captured at 4800 measurements/s.

The sensors used provided data on the various pressing movements in the pressing cycle. Figure 1 presents the measured variables.

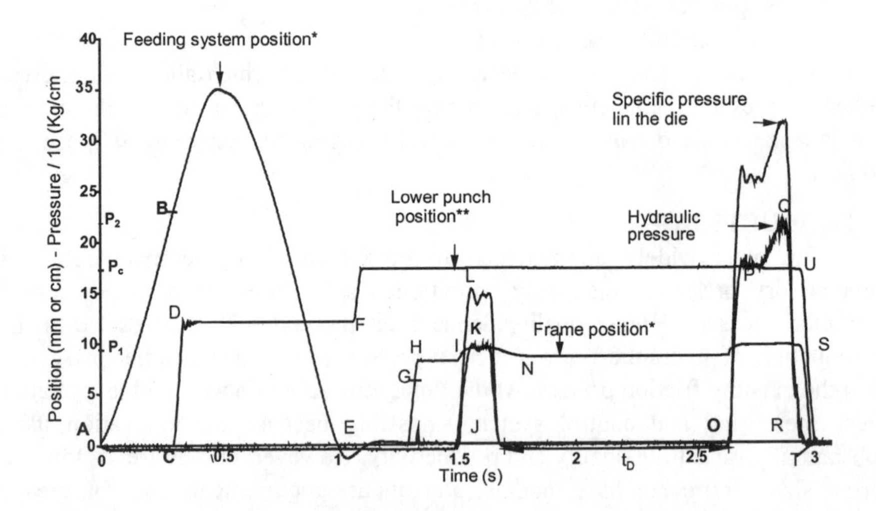

Figure 1. Schematic of the full pressing cycle; all displacements are considered positive. *in cm, **in mm.

The cycle starts with filler advance (A): frame and lower punch are at their highest position, with zero pressure in the main hydraulic circuit and die. When the feeding system reaches a certain point above the die (B), the lower punch (C) drops for the first time and the die fills. The lower punch drops very fast (slope of CD) to the end point (D), determining powder bed height. On ending the filler run (E), the lower punch drops again (F), preparing the powder for compaction, and the frame descends. When the first pressing commences, hydraulic pressure rises in the circuit (JK) and in the die (JL), and the frame drops (IN).

After the first pressing stroke (P_1), pressure in the hydraulic circuit and die drops, while the frame springs back slightly owing to the shock absorbers (N). De-aerating time (t_D) starts, allowing trapped air to escape from the die. After de-aerating, the second pressing stroke starts (O). The rising plot shows two clearly separate regions: a direct pressing domain with pressure supplied by the central hydraulic system (OP), and a second region starting at peak central hydraulic system pressure, in which pressure is provided by a multiplier manifold (PQ), until completing the second pressing stroke (P_2). A plot of the pressure recorded by the force sensors in the die exhibited the same form.

At peak second pressing pressure, pressure drops in the main hydraulic circuit (QR) and the frame rises (ST). Ejection of the compact starts with the rise of the lower punch (UV). The compact is thus ready for removal by the travelling system prior to die filling. This kind of diagram allows determining the synchrony and timing of the different pressing steps for pressing cycle optimization.

EXPERIMENTAL

Modified variables

A series of experiments was designed to study how different operating variables affected the bulk density of green tile compacts, by modifying one variable in each case and determining resulting bulk density.

Four experiments were performed for each modified variable, covering the range of possible press settings, or, when this was not feasible, covering the usual working range. Peak hydraulic pressure was held at 200 kg/cm^2 (according to the fitted Bourdon tube gauge) in the second press stroke, except when the effect of modifying this value was being tested.

The following variables were modified: feeding system speed, moment of first lower punch drop and peak hydraulic pressure.

Sample preparation and data processing

Ten compacts were made for each modified variable, determining bulk density by mercury intrusion in five domains in these specimens, corresponding to the force sensor locations in the upper punch.[8] Peak hydraulic pressure and mean powder moisture content were recorded during sampling for each specimen.

RESULTS AND DISCUSSION

Experiments run on the feeding system

Spray-dried powder feeding system position and speed were determined as shown in Figure 2. The filler accelerates (stretch A), holds an approximately constant speed (stretch B), and slows down (stretch C) before reaching peak displacement (D).

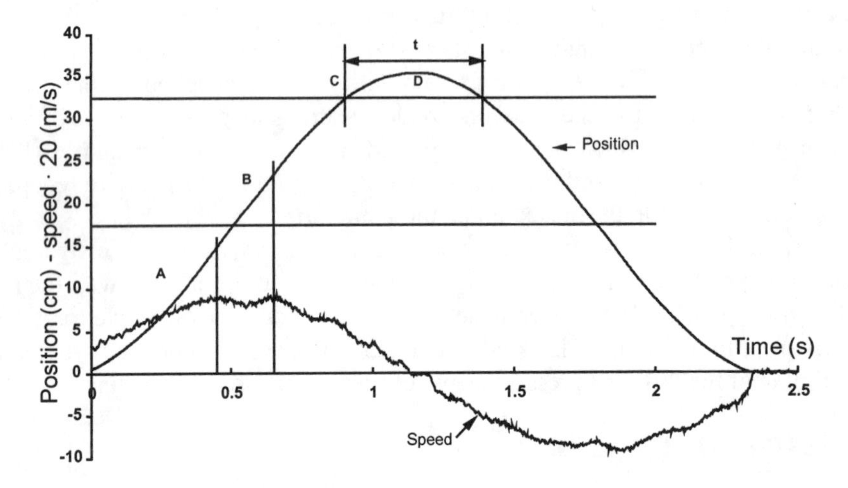

Figure 2. Filler position and speed in die filling.

Filler return mirrors filler advance. Filler speed is therefore only steady over a small part of the run, depending on the programming. The graphs accurately depict filling timing and synchrony, and allow determining filler dwell time over the die, which at a constant powder flow rate in turn determines the quantity of powder that is poured into the die and hence bulk density.

The press used in this study had two combinable feeding system speed settings, a fast and a slow one. Four experiments were designed, dividing the whole filler run into four parts, programming a speed for each. Figure 3 shows filler position in the four settings.

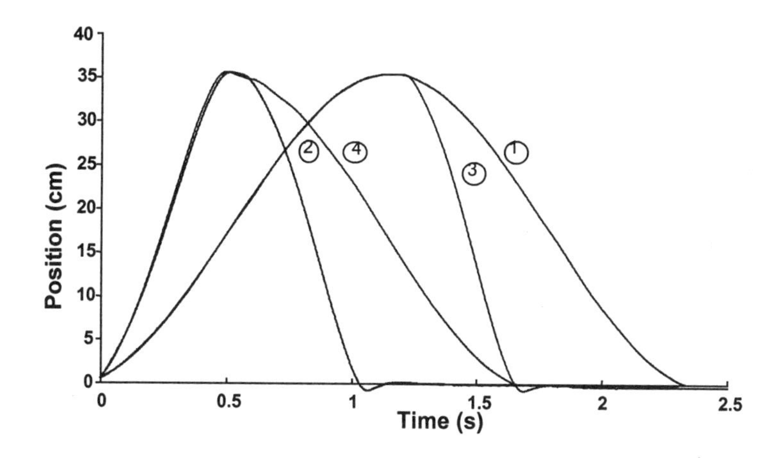

Figure 3. Feeding system position in the four settings.

Filler dwell time over the die, as well as specimen mean and overall bulk densities, were determined for each setting. The values are listed in Table I. It can be seen that compaction tended to increase on raising filling time.

Increasing filler dwell time over the die meant that a greater amount of spray-dried powder was poured into the die, thus slightly raising overall specimen bulk density. Analysis of setting 3 and 4 compaction data (Table I) reveals that although compaction distribution differed, the overall bulk densities were quite alike. Filling time was similar in both cases, although the combination of speeds differed. Filling time (for a spray-dried powder with constant technological characteristics and pressing pressure) affected overall bulk density, and therefore final tile dimensions.

Table I. Specimen localized and overall bulk density (g/cm^3) and filling time (s)

Domain	**Setting**			
	1	**2**	**3**	**4**
Front	2.077	2.039	2.052	2.057
Centre	2.091	2.069	2.072	2.080
Rear	2.065	2.062	2.071	2.058
Overall	2.075	2.054	2.063	2.062
Filling time (s)	1.28	0.58	0.95	0.89

Experiments run on the lower punch

The lower punch serves to limit the height of the powder fed into the die, settle the bed, limit the height of the body together with the frame, and eject the compact. Figure 4 depicts lower punch travel during the pressing cycle, indicating feeding system and lower punch positions versus time during die filling.

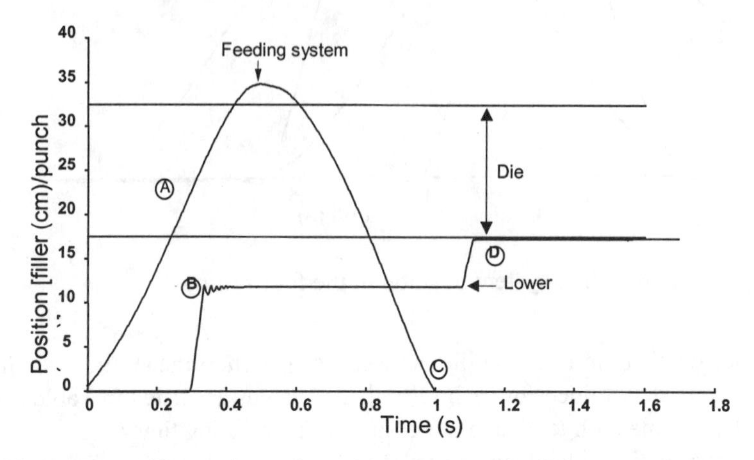

Figure 4. Filler and punch positions during die filling.

When the filler reaches position (A), the lower punch drops to (B). When the filler ends its run (C), the lower punch drops again (D). This drop keeps a partial vacuum from arising in the die when the upper punch first impacts the powder bed. The moment at which the lower punch dropped was modified, covering the full range of press settings, while keeping a constant dropping rate.

Table II reports filler positions over the die when the lower punch dropped, as well as mean density values, for each specimen domain and setting.

Table II. Mean compaction values (g/cm^3) per domain and distance (cm) travelled by the filler across the die when the lower punch dropped.

Domain	Setting			
	1	2	3	4
Rear	2.067	2.068	2.060	2.044
Centre	2.054	2.081	2.069	2.099
Front	2.041	2.042	2.059	2.090
Distance (cm)	4.0	5.7	10.3	12.6

The equations are derived below, which allow filler dwell time over each point of the die to be computed. To calculate this time, the following assumptions were made: punch descent is very fast, filler speed across the die is constant and, in powder fall, transverse movement far exceeds horizontal movement, the latter being virtually negligible.

A situation is assumed as shown in Figure 5, in which the feeding system has travelled a distance (P) from the beginning of the die, when the lower punch drops. In die filling, there are then two clearly differentiated regions:

- The points lying between die beginning and the point at which the lower punch drops (X_1), when filling starts.
- The domains located beyond the point at which the punch drops (X_2), at which no filling occurs until they are reached by the filler.

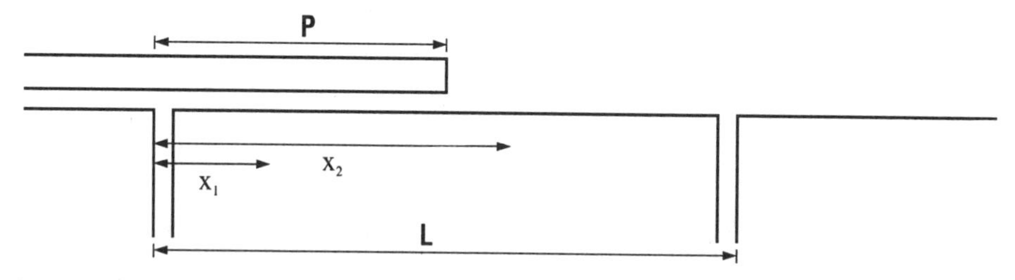

Figure 5. Schematic of die filling.

For the domain between die beginning and the lower punch dropping point, filling time starts when the punch drops, and ends when the filler crosses back over this domain again on its return. Filling time (t_1) for any point in this domain (X_1) will thus be:

$$t_1 = \frac{2(L-P)}{v} + \frac{(P-X_1)}{v} \tag{1}$$

where (v) is filler velocity.

Filling time (t_2) for any point (X_2) beyond the feeding system position, on filling the lower punch will be:

$$t_2 = \frac{2(L-X_2)}{v} \tag{2}$$

Figure 6 plots filler dwell times over each domain versus its position for the different settings. Bulk density in each domain of the compact has been plotted in Figure 7 versus the lower punch dropping point.

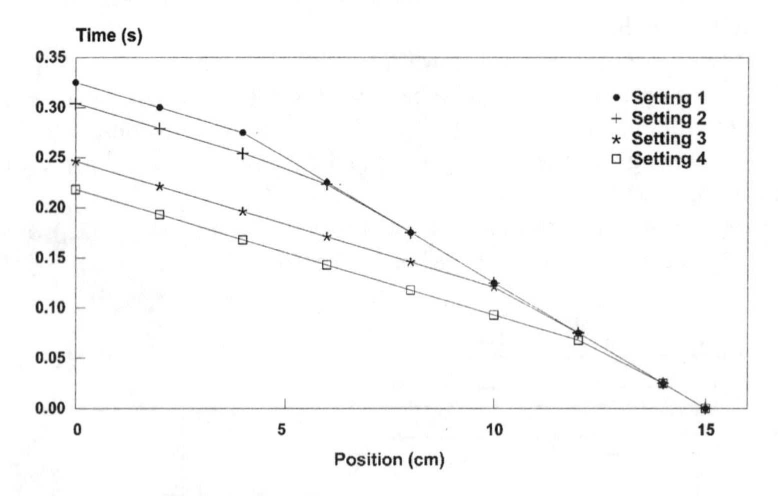

Figure 6. Feeding system dwell times over the die versus system position for each setting.

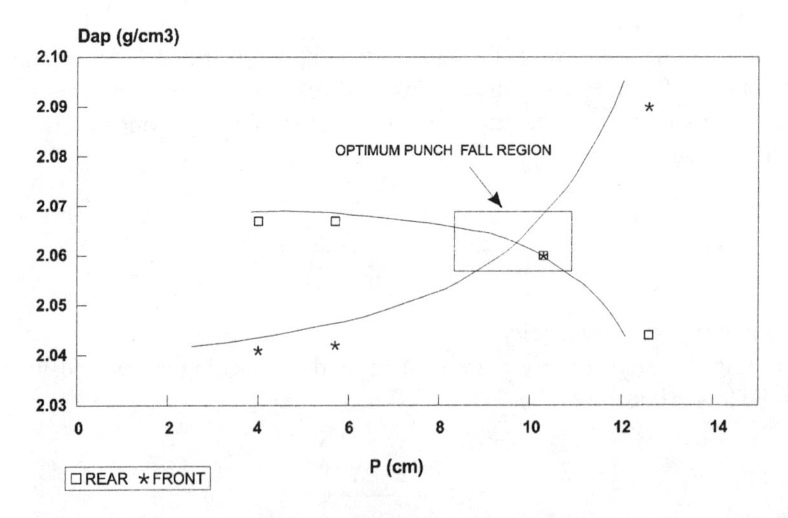

Figure 7. Bulk density of front and rear domains versus lower punch drop.

On delaying lower punch drop (increase of P) relative to feeding system position, filler dwell time over the rear decreased (Figure 6). As was to be expected, bulk density dropped (Figure 7).

Filling time for the region preceding the punch dropping point was independent of this moment (Figure 6), while the parameter only assumed constant compaction in this area. However, when the punch drops, the filler proceeds to fill the die, and smaller amounts of powder would be expected to fall progressively as the filler advances. Hence, the closer to punch dropping point, the higher will bulk density be. Thus, in the front domain, delaying the punch drop raised bulk density (Figure 7).

Specimen centre bulk density depended on the foregoing effects, as the filling system had in some settings already crossed this area, while in others, the filler had not yet reached this domain when the lower punch dropped.

Experiments run on the hydraulic circuit

Peak hydraulic pressure in the compression cycle was related semi-logarithmically to the compact's overall bulk density, making it a key factor in the forming process.

Different compression cycles were designed, modifying peak hydraulic pressure, measured with a strain sensor and the typical Bourdon tube gauge fitted in most hydraulic presses. Table III details mean peak pressure gauge readings (P_M) and corresponding mean peak sensor pressures (P_S).

Table III. Gauge and sensor pressure readings

Gauge pressure P_M (kg/cm^2)		150	170	200	230	260
Sensor pressure P_S (kg/cm^2)	182	188	201	246	277	305

Table III data reveal that the recorded mean sensor pressures were proportionally greater than the gauge readings at every tested set pressure. This could be because the gauge response rate might not have been sufficiently high to reflect the actual variation in hydraulic pressure. The pressure sensor used had a faster response rate and was more suitable for processes with high pressure variation rates, as in a tile body compression cycle (1000 (kg/cm^2)/s). There was a difference between measured and set pressure (Table III). On ramping at a mean pressure of 1000 (kg/cm^2)/s, set pressure was reached very fast (in 40 ms), requiring measuring systems with a high response rate. It can also be observed that there is some uncertainty regarding actual applied pressure. This variation in hydraulic pressure could alter bulk density, and would appear to take place under steady press operating conditions: filling, powder moisture content, peak pressure, etc.

GENERAL CONCLUSIONS

1) The measuring system used allowed monitoring the evolution of the major pressing variables, studying their synchrony and establishing their relation with the green bulk density of tile bodies.
2) The response rate of the hydraulic pressure measuring system currently being used is not fast enough to accurately detect actual peak pressing pressure during compression.
3) The main changes in bulk density distribution occurred when filling system and lower punch operating conditions were changed.
4) The designed control system (incorporating sensors) allows optimizing pressing cycle synchrony and timing, maximizing productivity and tile quality.

REFERENCES

1. J. Reed, *Introduction to the Principles of Ceramic Processing*, Wiley. Interscience Publication, 1988.
2. A. Creus, *Industrial instrumentation* (In Spanish). Marcombo, S.A. 1979.
3. J.L. Amorós; A. Blasco, J.E. Enrique, and A. Escardino, "Study of Tile Compaction. I. Influence of Spray-Dried Granule Size" (In Spanish). *Bol. Soc. Esp. Ceram. Vidrio*, **21** [4-5], 245-250 (1982).
4. J.L. Amorós; V. Beltrán; P. Negre, and A. Escardino, "Study of Tile Compaction. II. Influence of Pressing Pressure and Moisture Content" (In Spanish). *Bol. Soc. Esp. Cera. Vidrio* **22** [1], 9-18 (1983).
5. F. Negre; J.C. Jarque; G. Mallol; and M. Sáez, "On-line Determination in Real Time of Spray-Dried Powder Moisture Content" (In Spanish). *Técnica Cerámica*, Nº 200 (1992).
6. A. Leucona and A. Calvo, Computerizing Testing and Experimental Data Analysis and Measurement".(In Spanish), Department of engine propulsion and thermal fluid dynamics. Madrid Polytechnic University. (1990).
7. A. Blasco; D. Llorens, G. Mallol, and J.C. Jarque, "Experimental Study for the Determination of Dry Compaction of Ware Shaped by Unidirectional Pressing, in Continuous Operation and in True Time", *Tile and Brick Int.*; **8**, [6] (1992).
8. J.L. Amorós, V. Beltrán; A. Blasco; C. Felíu, and R. Sancho-Tello. "Tile Compaction Experimental Control Techniques" (In Spanish). *Técnica Cerámica*, **116**, 1234-1246 (1983).

Banquet Guest Lecture

APPROACHING THE VOID

Anne Currier
School of Art and Design
New York State College of Ceramics
Alfred University
Alfred, NY 14802

ARTIST'S STATEMENT

The ceramic sculptures express my curiosity and need to experience the physical and visual exchange of masses and voids in space. Projection and recession, hard and soft, light and shadow, substance and impression are aspects of the works' subject matter. Allusions to an interaction between human bodies suggest some of the work's narrative content. Positions of the human figure found in Greek and Buddhist temple pediments and friezes are part of an exploration that generates an intriguing sense of scale, intimacy and narrative. It's an obsessive compulsion to experience inside and outside - moving into and through a space and time - using the simplicity of cylinders, cones, planes and edges. Light and shadow are crucial to the visual illusion and exchange of recession and projection. The process of working with the clay shapes provides a clarity and occasions to physically being there at that moment in time when the material shapes intersect, extend, collide or pass through / over / under one another to create a space. Colors for the glaze are about the richness and contrasts of winter colors that I see here in Allegany County: charcoal blacks, slate blue/grays, deep rusts and warm tans. The texture of the glaze was chosen to enhance the ambiguity between the visual perception of a surface and the reality of touch. A glazed surface that looks soft and absorbs light yet actually feels like a 220-grit sand paper is part of the interplay.

TECHNICAL STATEMENT

I use a white stoneware that will be twice-fired with glaze. All pieces are hollow and slab-constructed, using the basic procedures of scoring and reinforcing seams. Slabs are made on a Brent slab roller. Cardboard tubes of various diameters are used for forming the clay cylinders. I also make cone shapes from the flat slabs.

Sculptures are bisque fired to cone 03. The first glaze, usually black, is sponged on and fired to cone 03. The fired surface of this first glaze is a soft, satin finish. The second glaze is sprayed on and while it is still wet, it is scraped off the edges of the piece to expose the first glaze. The piece is then refired to cone 08. The fired surface of the second glaze is comparable to 220 grit sandpaper. The edges of the piece reveal the soft, satin finish of the first glaze, suggesting a drawing within the sculpture.

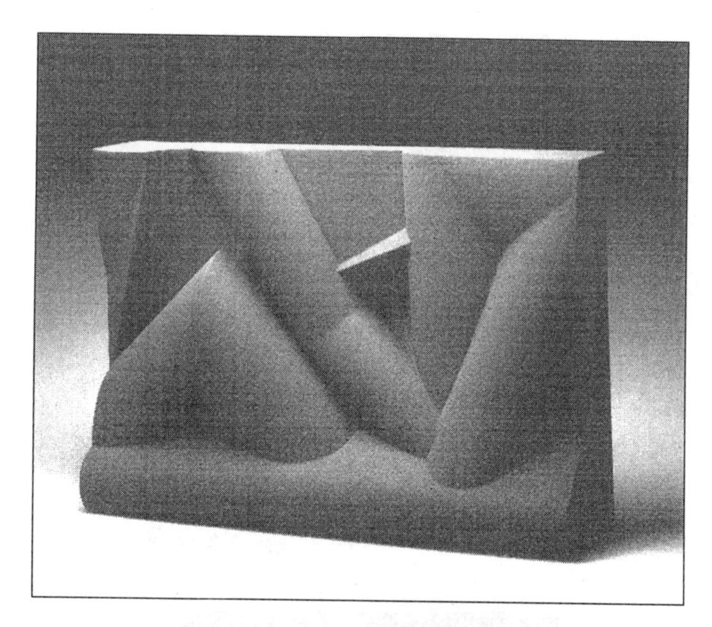

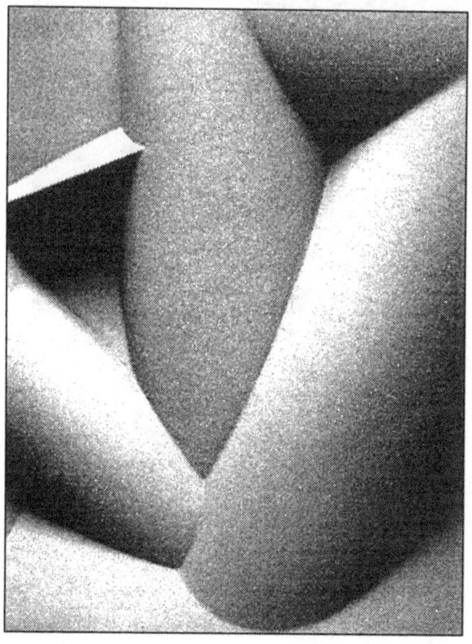

Panel series #2: *In the shallows*. Glazed clay 20 in. height, 30 in. length, 8 in. depth. 1998.

Panel series #2: *Caress*. Glazed clay 20 in. height, 30 in. length, 8 in. depth. 1998.

Color Plates

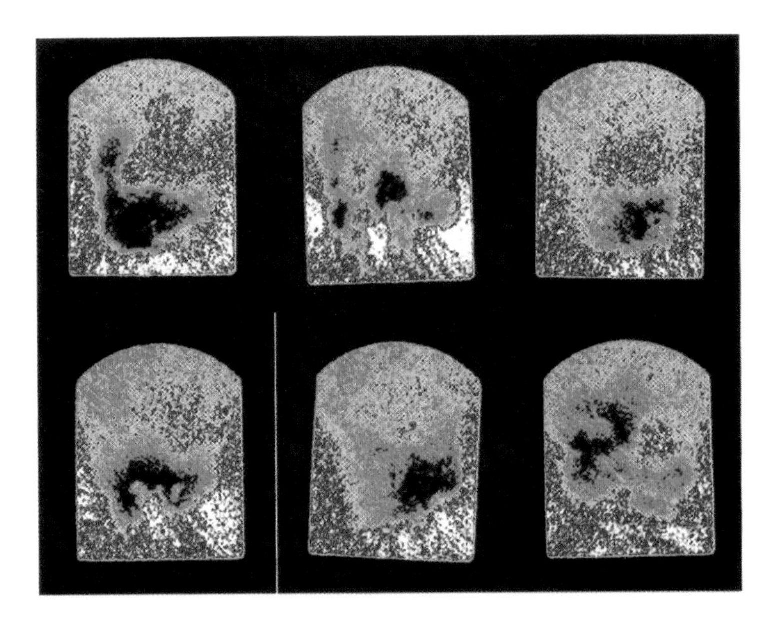

Plate 1. Density distributions in an early version of Tempcheks, a pyrometric product manufactured by Orton. These images were produced utilizing computed tomography (CT).[4,7] The density scheme employed is as follows: the highest densities = white; the lowest = blue. None of the parts are identical, in spite of the fact that they were produced consecutively. In addition, the bottom of each part shows the same line of high density. (Figure 1, Chapter 29)

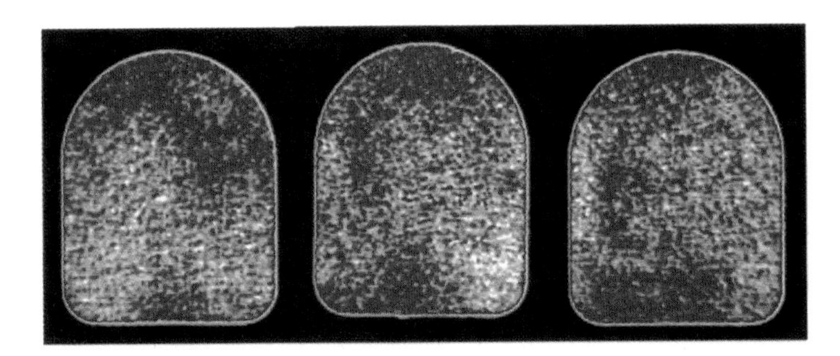

Plate 2. Improved Tempchek density distributions. The net variation in density within a part is considerably reduced compared to Figure 1. These components shrink more uniformly and experience considerably less distortion during firing. (Figure 2, Chapter 29)

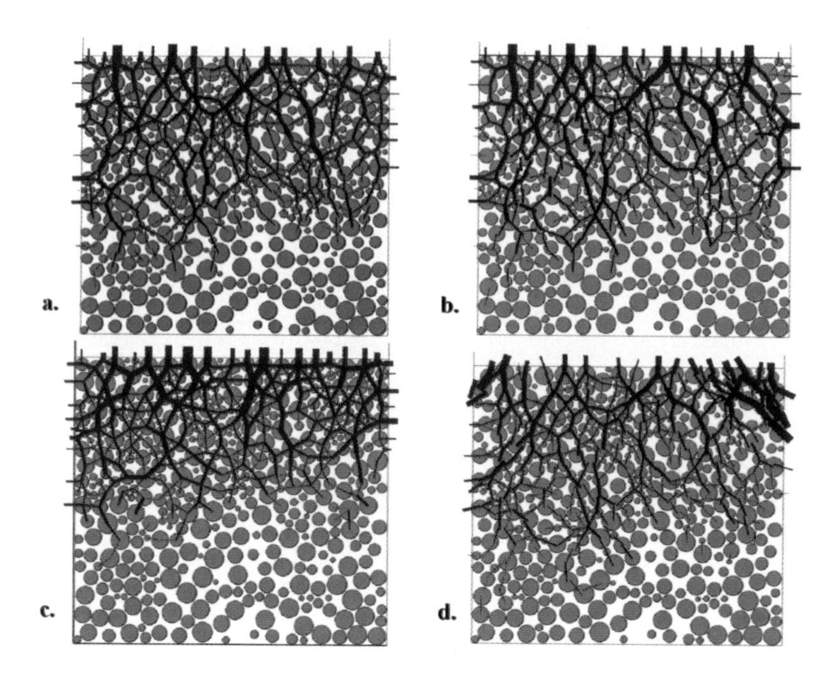

Plate 3. Compaction simulations with different inter-agglomerate friction (F_{IA}) and wall friction (F_W) coefficients. The particle size distribution and placement in the initial fill are exactly the same in each case. (Figure 3, Chapter 29)

a) $F_{IA} = 0.0$; $F_W = 0.0$. Both F_{IA} and F_W are assumed to be negligible. Even with $F_W = 0.0$, normal contacts can still transmit force to the wall. Somewhat generalized (but not uniform) compaction occurs in the bulk of the particles. Inter-agglomerate contact force is uniformly observed in many of the agglomerates although wide force variations develop. Even under these ideal conditions, the lower half of the fill experiences negligible compaction.

b) $F_{IA} = 1.0$; $F_W = 0.0$. Increase F_{IA} causes only a slight diversion of contact force to the die wall.

c) $F_{IA} = 0.0$; $F_W = 1.0$. Increasing F_W has an immediate and serious affect on uniformity. The applied compaction force become much more concentrated in the agglomerates nearest the advancing ram (top).

d) $F_{IA} = 1.0$; $F_W = 1.0$. The presence of high F_{IA} and F_W cause pronounced 'arcing' of the applied force to the wall. Force transmission to the wall becomes especially serious at the corners of the ram-wall interface, a region typically observed to develop high density (HD) zones in real compacts.

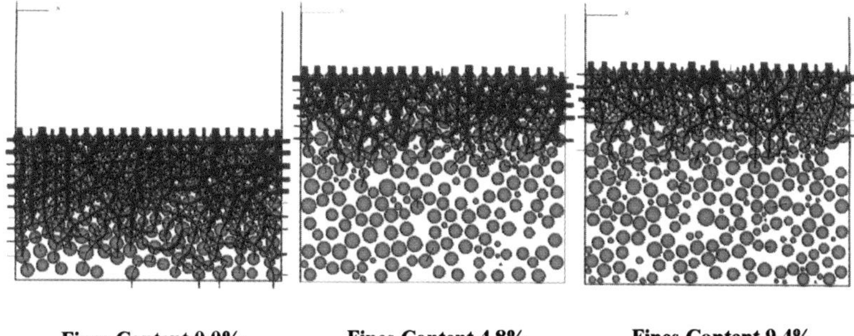

Fines Content 0.0%   Fines Content 4.8%   Fines Content 9.4%

Plate 4. Variation in compaction efficiency with fines content (on an area basis). Here, fines are set at 50% of the coarse particle diameter. Both the fine and the coarse agglomerates occupy a small range of sizes. At 0% fines, compaction is fairly uniform and the compaction pressure reaches the static end of the die. At 4.8% fines a large drop in compaction efficiency occurs due to inter-agglomerate bridging by the fines. Much of the applied compaction pressure is now directed toward the wall; relatively little of it reaches the static end of the die. At 9.4% fines the situation is only slightly worse. (Figure 4, Chapter 29)

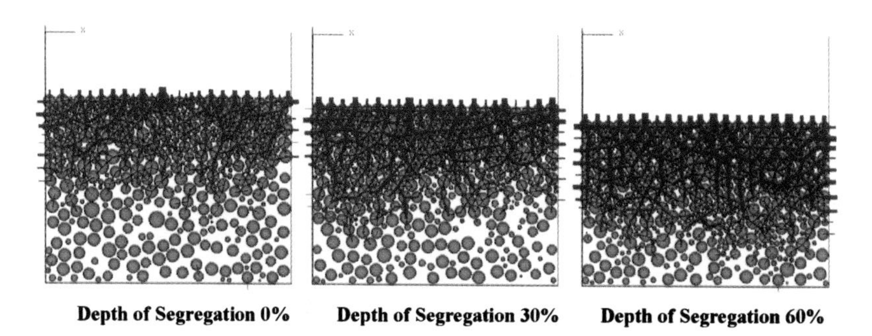

Depth of Segregation 0%   Depth of Segregation 30%   Depth of Segregation 60%

Plate 5. Variations in the efficiency of compaction with fine segregation. Again, fines are approximately 50% of the coarse particle diameter. When the fines are uniformly distributed (at 10% loading) compaction resistance is at a maximum. When the fines are removed from the top 30% of the fill, compaction resistance decreases. Once removed from 60% of the fill, compaction is easier and more uniform in the fine-free area. However, contact force transmission effectively ceases at the interface between the two size distributions. (Figure 5, Chapter 29)

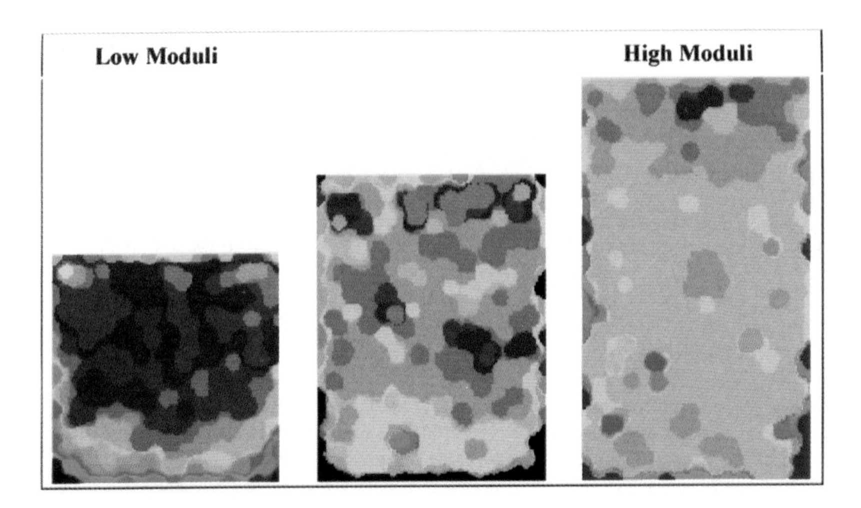

Plate 6. The compaction of 'hard' (high modulus) particles compared to the compaction of softer particles. Under the same levels of final pressure the lower modulus particles compact to a greater value of density. However, the density gradients in this case are the most severe. (Figure 6, Chapter 29)

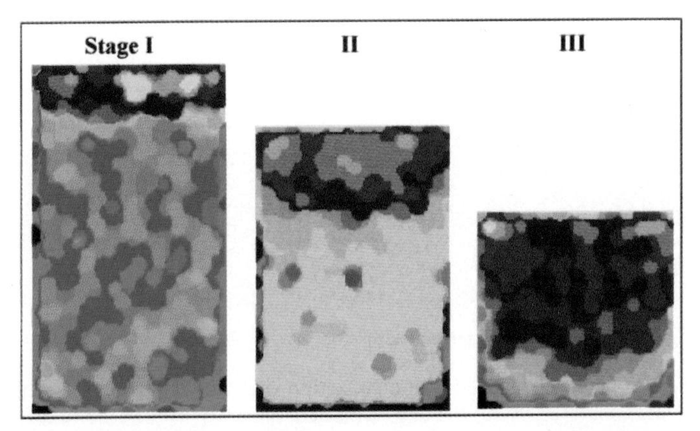

Plate 7. The three different stages of compaction. Stage I involves only the topmost layer of agglomerates. Their collapse triggers the advance of higher density regions into the bulk of the compact, or Stage II compaction. Stage III may be caused by the contact of these higher density regions with opposing surfaces. (Figure 7, Chapter 29)

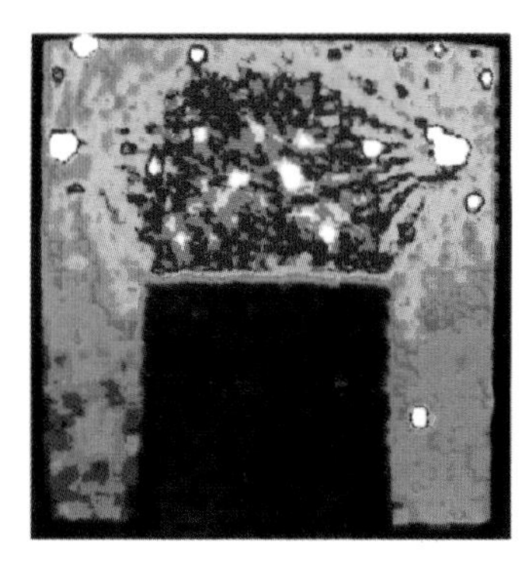

Plate 8. Density gradients that develop within a cylindrical die containing a 1x1" button on the static side of the die. An arrow-headed region of high density develops and limits the progress of the upper ram (Figure 8, Chapter 29).

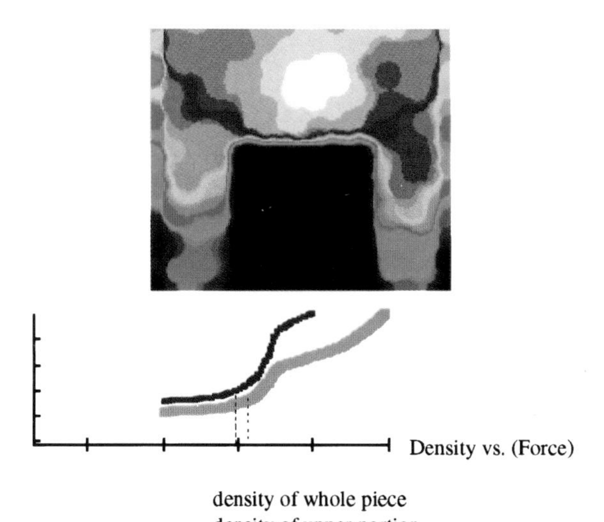

Plate 9. Simulation of overall compaction and a 'split' version of the compaction diagram showing how density in the upper portion of the fill develops differently from that in the lower portion delimited by the 1x1" button. The overall diagram tells us little about the density gradients that develop within the component (Figure 9, Chapter 29).

Index

KEYWORD AND AUTHOR INDEX